VIBRATIONAL SPECTRA AND STRUCTURE

A SERIES OF ADVANCES

VOLUME 13

EDITED BY

JAMES R. DURIG

Department of Chemistry
University of South Carolina
Columbia, South Carolina

ELSEVIER — AMSTERDAM — OXFORD — NEW YORK — TOKYO   1984

ELSEVIER SCIENCE PUBLISHERS B.V.
Molenwerf 1
P.O. Box 211, 1000 AE Amsterdam, The Netherlands

Distributors for the United States and Canada:

ELSEVIER SCIENCE PUBLISHING COMPANY INC.
52, Vanderbilt Avenue
New York, NY 10017

The Library of Congress Cataloged This Serial as
 Follows:
  Vibrational spectra and structure. v. 1–
    New York, M. Dekker.

  V. 4—
  Amsterdam, Elsevier Science Publishers B.V.

      ⌡ v. 23 cm.

    Editor : 1972–₎)J. R. Durig.

    1. Vibrational spectra—Collected works. 2. Chemistry, Physical
  and theoretical—Collected works.   I. Durig, James R., ed.
  QC454.V5V53   537.53'52   72–87850
  ISSN 0090–1911   MARC-S
  Library of Congress   73 ₍2₎

ISBN 0-444-42394-X (Vol. 13)
ISBN 0-444-41708-7 (Series)

© Elsevier Science Publishers B.V., 1984

Printed in The Netherlands

VIBRATIONAL SPECTRA AND STRUCTURE

VOLUME 13

PREFACE TO THE SERIES

It appears that one of the greatest needs of science today is for competent people to critically review the recent literature in conveniently small areas and to evaluate the real progress that has been made, as well as to suggest fruitful avenues for future work. It is even more important that such reviewers clearly indicate the areas where little progress is being made and where the changes of a significant contribution are minuscule either because of faulty theory, inadequate experimentation, or just because the area is steeped in unprovable yet irrefutable hypotheses. Thus, it is hoped that these volumes will contain critical summaries of recent work, as well as review the fields of current interest.

Vibrational spectroscopy has been used to make significant contributions in many areas of chemistry and physics as well as in other areas of science. However, the main applications can be characterized as the study of intramolecular forces acting between the atoms of a molecule; the intermolecular forces or degree of association in condensed phases; the determination of molecular symmetries; molecular dynamics; the identification of functional groups, or compound identification; the nature of the chemical bond; and the calculation of thermodynamic properties. Current plans are for the reviews to vary, from the application of vibrational spectroscopy to a specific set of compounds, to more general topics, such as force-constant calculations. It is hoped that many of the articles will be sufficiently general to be of interest to other scientists as well as to the vibrational spectroscopist.

Most of the recent reviews in the area of vibrational spectroscopy have appeared in other progress series and it was felt that a progress series in vibrational spectroscopy was needed. A flexible attitude will be maintained and the course of the series will be dictated by the workers in the field. The editor not only welcomes suggestions from the readers but eagerly solicits your advice and contributions.

James R. Durig
Columbia, South Carolina

v

PREFACE TO VOLUME 13

It is the Editor's belief that the current volume fills the goals of providing critical summaries of recent work and reviewing some fields of current interest as stated in the preface to the series. In Chapter 1, M. B. Mitchell and W. A. Guillory review recent efforts that have been successful in directly obtaining vibrational frequencies of electronically excited states and species. This Chapter deals with both continuous wave and pulsed infrared techniques, as well as time resolved resonance Raman spectra of transient and specific electronically excited states to obtain the vibrational spectra of these excited states and species. Additionally, the authors have discussed the use of time independent resonance Raman spectroscopy for obtaining changes in excited states and normal coordinates, along with the role that vibronic coupling plays in the systems with its effect on force constants and infrared absorption intensities.

In Chapter 2, Roger Frech discusses in detail the optical constants, internal fields, and molecular parameters obtained from the spectra of crystals. Most of the spectroscopic studies in this area have either been carried out by infrared reflectivity measurements or Raman scattering experiments. The deduction of molecular parameters from such studies is considerably more difficult than gas phase studies due to a combination of factors such as internal field effects, collective effects leading to a complex multiplet structure, and molecular perturbation effects. The author has reviewed in detail spectroscopic studies in this area.

P. L. Polavarapu in Chapter 3 has reviewed in detail the recent advances in model calculations of vibrational optical activity. It is hoped that vibrational optical activity studies will be more informative than electronic optical activity studies since the former should be able to provide stereochemical information on various segments of a chiral molecule. This type of selective information offers the hope that vibrational optical activity experimental data could be used to derive a three-dimensional structure of an entire chiral molecule. Theoretical models are currently being improved in order to better understand the origin of vibrational optical activity in a given molecular structural fragment and to apply that knowledge for deriving stereochemical information in molecular systems where configurational or conformational detail is not available or uncertain at this time. The author has reviewed in detail the theoretical models for vibrational optical activity in terms of recent advances, their interrelations and their ability to explain the existing experimental vibrational optical activity data.

In Chapter 4, L. Nemes has reviewed the theoretical and methodical advances in the derivation and interpretation of empirical molecular structures, their reduction to the equilibrium representation, the various ways of extracting structural information from rotation, and rotation-vibrational spectra induced by various intramolecular effects. In the calculation of microwave substitution structures, the problem of small substitution coordinates is frequently encountered, and these usually carry relatively large vibrational contributions. The author has reviewed in some detail the developments in the assessment of the small r_s structural coordinates.

In Chapter 5, G. N. Zhizhin and A. F. Goncharov have reviewed the applications of Davydov splitting for studies of crystal properties. The theoretical analysis of vibrational exciton states in molecular crystals, as well as in crystals with complex ions, is very similar to that of electronic exciton states. To explain the splitting of electronic exciton states, it is usually sufficient to take into consideration only the dipole-dipole interactions. However, in the vibrational spectra of molecular crystals, it is frequently necessary to take into account multipoles of higher order because the oscillator strength of the electronic transitions is about five to six orders higher than that of the vibrational transitions. The authors have reviewed in detail the utility of Davydov splittings for obtaining intermolecular interaction potentials, the study of phase transitions, and structural analysis of molecular crystals.

In the final chapter, H. J. Jodl has reviewed the experimental work carried out in the field of Raman spectroscopy on matrix isolated species. The author has restricted his review to mainly diatomic molecules isolated in rare gas matrices in order to simplify the interpretation of the experimental data and the theoretical analysis. The technical and physical parameters and the problems encountered in practicing matrix isolation Raman spectroscopy are extensively described. All techniques that have so far been used to produce matrix isolated species or to perform Raman scattering experiments on them are compared, and the new approaches are mentioned. The present status, remaining problems, and future trends in matrix isolation spectroscopy are discussed.

The Editor would like to thank the Editorial Board for suggestions of possible topics for this volume, and the authors for their contributions. The Editor would also like to thank his secretary, Mrs. Janice Long, for superbly typing all of the articles in camera-ready copy, Dr. J. F. Sullivan for her editorial assistance, and his wife, Marlene, for copy editing the manuscripts and for the preparation of both the author and subject indices.

James R. Durig
Columbia, South Carolina

CONTRIBUTORS TO VOLUME 13

ROGER FRECH, Department of Chemistry, The University of Oklahoma, Norman, Oklahoma

A. F. GONCHAROV, Institute for Spectroscopy, USSR Academy of Science, Troitsk, Moscow, USSR

WILLIAM A. GUILLORY, Department of Chemistry, University of Utah, Salt Lake City, Utah

H. J. JODL, Fachbereich Physik, Universität Kaiserslautern, 6750 Kaiserslautern, W. Germany

MARK B. MITCHELL, Department of Chemistry, University of Utah, Salt Lake City, Utah

L. NEMES, Research Laboratory for Inorganic Chemistry, Hungarian Academy of Sciences, Budapest, Hungary

P. L. POLAVARAPU, Department of Chemistry, Vanderbilt University, Nashville, Tennessee

G. N. ZHIZHIN, Institute for Spectroscopy, USSR Academy of Science, Troitsk, Moscow, USSR

CONTENTS

CHAPTER 1.   VIBRATIONAL SPECTRA OF ELECTRONICALLY
             EXCITED STATES   1

             Mark B. Mitchell and William A. Guillory

CHAPTER 2.   OPTICAL CONSTANTS, INTERNAL FIELDS, AND
             MOLECULAR PARAMETERS IN CRYSTALS 47

             Roger Frech

CHAPTER 6.   RAMAN SPECTROSCOPY ON MATRIX ISOLATED
              SPECIES 285

              H. J. Jodl

CONTENTS OF OTHER VOLUMES

VOLUME 12

Chapter 1

VIBRATIONAL SPECTRA OF ELECTRONICALLY EXCITED STATES

Mark B. Mitchell and William A. Guillory
Department of Chemistry
University of Utah
Salt Lake City, Utah

I. INTRODUCTION

During the last two decades, considerable effort has been devoted to the characterization of the potential surfaces of electronically excited states. The generation of accurate molecular potential functions requires the knowledge of vibrational frequencies. Until recently, the principal source of this information for excited state species has been electronic absorption and emission spectroscopy [1]. Excited electronic states differ from the ground state because of changes in the electronic distribution within the molecular framework. Such changes are in turn reflected by vibrational spectra and are accompanied by changes in bond order, vibrational force constants, and dissociation energies. The experimental characterization of excited state potential surfaces is of pivotal importance in understanding the dynamics of radiationless decay processes as well as establishing reaction mechanisms in photochemistry and photobiology.

Various infrared and Raman techniques have been developed to obtain vibrational spectra of both short (<1 nsec) and long ($1\text{-}10^{-3}$ sec) lived electronically excited states. Direct infrared spectra have been obtained with both conventional and Fourier transform infrared (FTIR) spectrophotometers. The principal source of Raman spectra has been by the use of time-resolved resonance Raman (TR^3) spectroscopy. This review will primarily focus on recent efforts during the last five years that have been successful in obtaining direct vibrational spectra of electronically excited states and species. The first section will involve the discussion of both continuous wave (CW) and pulsed infrared techniques that have been used. The second section involves the discussion of time-resolved resonance Raman spectra of transient and specific electronically excited states.

The third section will discuss the use of time independent resonance Raman spectroscopy to obtain changes in excited state normal coordinates. Finally, because of the generally greater excited state density found in the neighborhood of the originating states, vibronic coupling plays a significant role in these systems either directly, as in the case of resonance Raman spectroscopy, or more subtly, as in the effect on force constants and infrared absorption intensities. These points will be discussed in the last section.

II. EXCITED STATE INFRARED SPECTRA

A. Acetone and Naphthalene

One of the earliest accounts of the direct recording of the vibrational spectra of molecules in the triplet state is that of R. M. Hexter [2]. The essential idea is illustrated in Fig. 1, which summarizes the various processes under consideration and involves excitation-modulation spectroscopy. A low-temperature matrix-isolated sample of the species to be studied is illuminated with uninterrupted infrared radiation from a Perkin-Elmer 13 spectrometer. The sample is also illuminated with chopped visible (or ultraviolet) light, using either normal or oblique incidence, and the chopping frequency is that normally utilized in the Perkin-Elmer 13 instrument. The visible or UV source is used to pump $S_1 \leftarrow S_0$ transitions, which in turn undergo inter-system crossing (ISC) into the T_1 state which, for the species reported, have phosphorescence lifetimes, $\tau_p > 1$ sec. When the chopped pumping source is "on", the infrared beam is attenuated as a result of $v_f' \leftarrow v_i'$ transitions within the triplet (T_1) manifold; where $v_i' = 0$ as a result of rapid relaxation in the condensed medium. Only the differential changes in the infrared spectrum of the S_0 and T_1 states are recorded. This technique is similar to the "chemical modulation" technique of Bair, Lund, and Cross [3] and to the "pressure modulation" technique of Gilfert and Williams [4].

Using this excitation-modulation technique, it is important to note that the differential change of the T_1 state is ~99% and that of the S_0 state only 1% from the dark to light period. Tentative assignments for the T_1 vibrational spectra of acetone and naphthalene were reported. Given the triplet radiative lifetime of acetone (~0.014 sec), the corresponding concentration buildup possible, and the reported presence of impurity in the naphthalene study, it is rather doubtful that the numerous infrared absorptions reported are actually due to the triplet states of these species.

B. Naphthalene

In a subsequent study involving CW illumination with a 200 watt Hg high pressure ultraviolet lamp, the infrared spectrum of $[^2H_8]$-naphthalene in a Nujol solution at ca. 80K was recorded [5]. The experiment was performed by use of a Perkin-Elmer 180 spectrophotometer in conjunction with simultaneous focused UV pumping to produce the long lived triplet state ($\tau_p \sim 20$ sec). The infrared spectrum revealed a new peak at 535 cm^{-1} which decayed

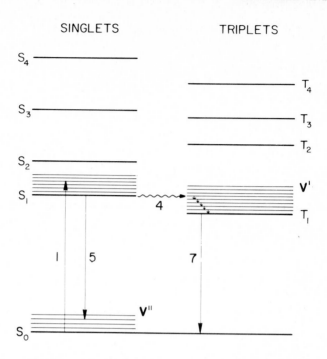

FIG. 1. Singlet and triplet energy manifolds; vibrational levels are indicated only for states S_0, S_1, and T_1. See text for explanation.

with a lifetime of 20.6 ± 1.2 sec when the UV lamp was shuttered. In addition, the e.s.r. signal of $[^2H_8]$-naphthalene in Nujol at 77K was observed to decay with a lifetime of 19.8 ± 0.7 sec and the measured phosphorescent decay of the same solution gave a lifetime of 20.4 ± 2.4 sec. The sum total of these lifetime measurements would seem to confirm that the UV light-dependent absorption observed at 535 cm^{-1} originates in the lowest triplet state of $[^2H_8]$-naphthalene.

The authors further suggest that the observed vibrational band might be the corresponding b_{3u} out-of-plane ring mode occurring at 630 cm^{-1} in the ground electronic state. This assignment is further rationalized by the expected reduced π bonding character in the triplet state, thus a lowering of the observed triplet state mode. In addition, the intensity of this mode is expected to be conserved barring gross geometry changes, and the reduction of the vibrational frequency by 95 cm^{-1} is consistent with that observed in the singlet-triplet absorption spectrum of the naphthalene crystal [6].

C. Acridine and Phenazine

1. Triplet State IR Spectra

In more recent investigations, M. B. Mitchell et al. [7,8] have reported the infrared spectra of the lowest triplet states of acridine and phenazine in matrices of argon. In these experiments, the concentration of the lowest triplet state sufficient to perform infrared spectroscopy is generated by indirect excitation with a chopped Ar^+ laser source.

The expression characterizing the steady-state concentration of a metastable triplet of a typical three-level system is obtained by considering the following condensed phase processes (Fig. 1).

Process	Rate	Description	
$S_0^0 + h\upsilon \to S_1^V$	$I_a = k_r S_0^0$	Excitation	(1)
$S_1^V \to S_1^0$	$k_r^{S_1} (S_1^V)$	Relaxation (S_1^V)	(2)
$S_1^0 \to S_0^V$	$k_{ic} (S_1^V)$	Internal Conversion	(3)
$S_1^0 \to T_1^V$	$k_{isc} (S_1^V)$	Intersystem Crossing	(4)
$S_1^0 \to S_0^V + h\upsilon_f$	$k_f (S_1^V)$	Fluorescence	(5)
$T_1^V \to T_1^0$	$k_r^{T_1} (T_1^V)$	Relaxation (T_1^V)	(6)
$T_1^0 \to S_0^V + h\upsilon_p$	$k_p (T_1^V)$	Phosphorescence	(7)
$T_1^V + M \to S_0^V + M^*$	$k_q (T_1^V)(M)$	Quenching (T_1^V)	(8)

where I_a is the average number of quanta absorbed per unit volume per second and $k_r = 2.3 I_0 \varepsilon$ (with I_0 in quanta/cm^2-sec and ε, the molecular extinction coefficient, in cm^3/molecule-cm) [9]. The superscripts refer to the vibronic level(s) of a given electronic state and the ISC process $S_0 \leftarrow T_1^0$ is assumed to be unimportant. Reaction (8) is only considered to be important when a dopant is added to purposely quench T_1. Assuming vibrational relaxation within S_1 and T_1 (processes (2) and (6)) to be very rapid and not rate determining, we obtain for the steady-state ratio,

$$\frac{(T_1^0)_{ss}}{(S_0^0)_{ss}} = 2.3 \ I_0 \varepsilon \phi_{isc} \tau_p \qquad (9)$$

Thus, this ratio is maximized by a relatively intense radiative pumping source, efficient intersystem crossing, and a relatively long phosphorescent lifetime ($\tau_p > 10^{-3}$ sec). Using the reported parameters of acridine and phenazine in Eq. (9) [10,11] and a laser intensity of 120 mW, their calculated steady-state ratios are 0.058 and 0.116, respectively. Although the excitation source is chopped, the chopping rate is sufficiently slow that the steady-state concentration is achieved to the extent of 75%.

The advantages of carrying out these experiments in the condensed state are the following: (1) experimental spectroscopy can be performed with extremely small quantities of sample ($\sim 10^{14} - 10^{17}$ molecules); (2) in the cases where the metastable and ground electronic states have vibrational absorption frequencies which involve small relative shifts, these absorptions are more easily resolvable in matrix-isolation experiments; (3) condensed-phase experiments can be designed to favor various processes (such as ISC for maximum triplet production), thus simplifying the steady-state rate equation which characterizes the metastable concentration; (4) excitation and luminescence spectra (fluorescence and phosphorescence) involve vibronic transitions where rotational motion is quenched, which considerably simplifies the observed spectra.

In these experiments, the displex sample compartment was placed in the sample chamber of a Perkin-Elmer 180 (PE 180) double beam infrared spectrophotometer. The UV lines (3638 Å, 3514 Å, 3511 Å, 3336 Å) of a Coherent Radiation CR-18 argon-ion laser were collectively used as the excitation source. The laser beam was chopped and expanded in order to match the infrared beam diameter. The source, gratings, filters, and detector of the spectrophotometer were used in the usual manner, but the choppers of the instrument were disengaged and the infrared source beam directed continuously to the sample. After the detector signal had been through a transformer and ~140 dB of amplification in the PE 180, it was further amplified with an Ithaco 397EO lock-in amplifier (or a model 393 for phase sensitive measurements) and displayed on an X-Y recorder. The laser beam chopper provided a reference signal for the lock-in amplifier. The detector amplification circuits in the PE 180 are tuned to 15 Hz so that one is limited to this chopping frequency in these experiments.

Figures 2 and 3 are laser-induced infrared absorption spectra of matrix-isolated acridine and phenazine, respectively. Since both the ground and triplet states undergo laser-induced concentration changes, the peaks in the laser-induced spectra correspond to both ground and triplet state vibrational frequencies. The excited singlet state is not considered since its population at any given time is so small, due to rapid fluorescence and ISC, as to make it negligible compared to either the ground or triplet state populations.

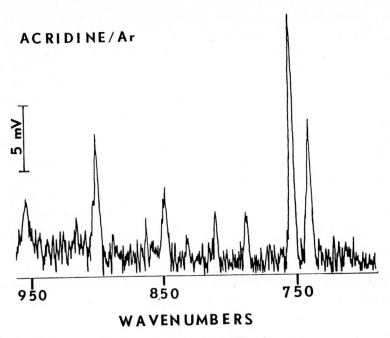

FIG. 2. Laser-induced spectrum of acridine in argon.

The lock-in amplifier measures changes in detector output which can be related to changes in transmittance.

$$\frac{I}{I_0} - \frac{I'}{I_0} = 10^{-\varepsilon bc} - 10^{-\varepsilon bc(1 - \frac{\Delta c}{c})} \tag{10}$$

where I is the transmitted IR intensity when the laser beam is blocked by the chopper, I' is the transmitted IR intensity when the laser beam is passed by the chopper, and $\Delta c/c$ is the fractional change in ground state concentration. By rearranging Eq. (10), one obtains

$$I - I' = I_0(T - T^{(1 - \frac{\Delta c}{c})}) \tag{11}$$

where $T = 10^{-\varepsilon bc}$ is the IR transmittance.

I_0 was determined by setting up the spectrophotometer as it would be for a laser-induced spectrum, chopping the sample beam, and monitoring the detector output with the lock-in. This procedure generated an I_0 spectrum that had all the source, grating, filter, and detector characteristics folded into it.

For completeness, a relation corresponding to Eq. (11) for a triplet state peak is

$$I - I' = I_0(1 - 10^{-\varepsilon_S bcr(\frac{\Delta c}{c})})$$

$$= I_0(1 - T^r{}^{\frac{\Delta c}{c}}) \tag{12}$$

where $r = \varepsilon_T/\varepsilon_S$ is the ratio between triplet and ground state IR extinction coefficients, c is the ground state concentration of the sample, and $T = 10^{-\varepsilon_S bc}$ is the transmittance of the ground state peak.

It should be noted that no relationship is assumed to exist between ground and triplet state vibrational motions in Eq. (12). Figure 4 shows the relative intensity of ground and triplet state features with the same IR extinction coefficient (r = 1) plotted against the transmittance of the ground state feature. Qualitatively, as the sample becomes more concentrated, triplet peaks increase in intensity while, due to the finite width of the absorptions, the more intense ground state features show "dips" at their respective absorption maxima.

If completely overlapping ground and triplet state absorptions are to be considered, a similar derivation for the resulting peaks yields

$$I - I' = I_0 T(1 - T^{(r-1)(\frac{\Delta c}{c})}) \tag{13}$$

where the variables have the same meaning as in Eq. (12). If $r > 1$, then $I - I' > 0$, and the resulting peak will have the same phase as a single triplet peak, and if $r < 1$, then $I - I' < 0$, and the resulting peak will have the same phase as a single ground state peak. Examples of these effects are the 1518 cm^{-1}, 1369 cm^{-1}, 1138 cm^{-1}, and 1112 cm^{-1} bands of phenazine and the 1518 cm^{-1} band of acridine. With higher resolution spectra, several cases of what appear to be exact coincidence between ground and triplet absorptions would be resolved to two peaks. At the present time, however, S/N limitations prevent the use of spectral slit widths smaller than the 3 cm^{-1} resolution typically used in the laser-induced spectra.

The ground and triplet state IR spectra are 180° out-of-phase, providing unambiguous identification of the excited metastable state absorptions. A portion of the IR spectrum (730-765 cm^{-1}) of matrix-isolated acridine is shown in Fig. 5. Spectrum (a) is with the lock-in-amplifier in the auto-phasing mode and shows the ground (737 cm^{-1}) and triplet (751 cm^{-1}) absorptions. Spectrum (b) is a phase sensitive scan with the phase adjusted to maximize the peak at 751 cm^{-1}. By monitoring the amplitude of each of the peaks

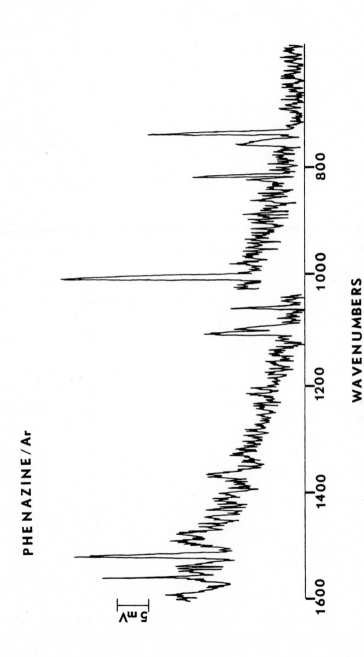

FIG. 3. Laser-induced spectrum of phenazine in argon.

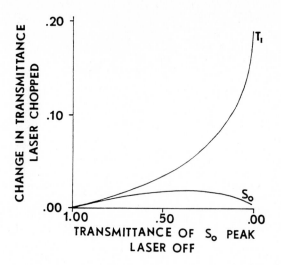

FIG. 4. Change in transmittance of ground and triplet state peaks ($\varepsilon_T/\varepsilon_S = 1$) plotted against the transmittance of the ground state peak. The curves were calculated using Eqs. (11) and (12).

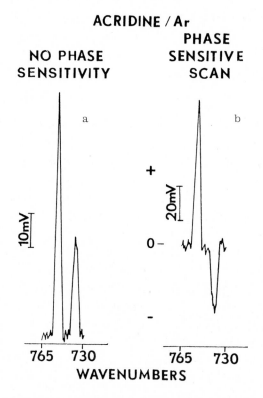

FIG. 5. Comparison of laser-induced spectra of acridine in argon measured with the vector-summing lock-in (spectrum a) and with the phase-sensitive lock-in (spectrum b).

TABLE 1

Triplet State (T_1) IR Active Vibrational Frequencies (cm^{-1}) of Acridine and Phenazine

Approximate Group Mode	Acridine	Phenazine
Ring stretch	1571	1570
	1518	1518
	1470	1479
	1435	1435
		1369
C-H in-plane bend		1138
		1112
C-H out-of-plane bend	1104	1095
C-H in-plane bend	1048	1048
	1004	1001
Skeletal deformation	920	
	903	905
		820
C-H out-of-plane bend	751	734
Skeletal deformation		583

while changing phase angle, it was found that the 751 cm^{-1} absorption was 180° out-of-phase with the S_0 737 cm^{-1} absorption. In each of the three situations, i.e., non-overlapped, partially overlapped, and totally overlapped IR absorptions of the ground and triplet states, unambiguous assignment of all of the triplet state absorptions was possible. In addition to the phase information, the decay rate of the infrared absorption was found to be 41 ± 16 sec^{-1} for the 751 cm^{-1} peak, which is in good agreement with the decay rate of the triplet state calculated from phosphorescence measurements to be 39.9 ± 0.2 sec^{-1}.

All of the observed excited state spectral features are summarized in Table 1. There are more excited state features listed for phenazine than for acridine. A Hückel calculation indicated that the $^1\pi\pi^*$ excited state symmetries of acridine and phenazine are the same as those of their respective ground states, and the same is probably true for their corresponding triplet states. Thus, more IR active features would be expected in the laser-induced spectrum of acridine (of lower symmetry) than that of phenazine. However, many of the ground state modes of acridine are much less absorbing than those of phenazine. Assuming an analogous situation occurs for the triplet states, such an explanation would account for the greater number of phenazine triplet state absorptions.

2.   Stretched Polymer Linear Dichroic IR Spectra

The usual approach taken in the analysis of excited state vibrations is to relate the observed absorptions to ground state vibrations for which the normal coordinates are known. In the work discussed in the previous section, the only evidence available for relating a triplet state vibration to one of the ground state was the proximity of their frequencies. Using the technique of absorbing solute molecules into stretched polymer films, thereby inducing infrared linear dichroism [12], it has been possible to determine the symmetries of several of the normal vibrations of the excited triplet state acridine molecule.

The unstretched polyethylene sheets were prepared as in Ref. [12]. The sheets were soaked in a saturated solution of acridine in chloroform for two days or longer, then dried on a vacuum line and stretched 600%. The stretched polymer sample was mounted between two CsI windows on the cold tip of a closed-cycle refrigerator (Air Products CS202 Displex). The experimental arrangement [8] was modified only to the extent that an aluminum grid IR polarizer was inserted in the IR beam before the sample and a 1 kW Hg-Xe high pressure arc lamp was used as the excitation source.

The Eqs. (11-13) are useful for comparing the intensities of peaks due to ground and triplet state species. However, it was necessary to determine the relationship of the induced signals to the absorbances of the species in order to obtain the dichroic ratios needed for the symmetry assignments. A detailed mathematical model of the experiment, including a Fourier expansion of the input signal to the lock-in amplifier, showed that for triplet state absorption peaks in the induced spectra the signal intensity is

$$S_{T_1}(\omega) = k'I_0(\omega)A_{T_1}(\omega) \qquad\qquad (14)$$

and the intensity for ground state absorption peaks in the induced spectra is

$$S_{S_0}(\omega) = -kI_0(\omega)T_{S_0}(\omega)A_{S_0}(\omega). \qquad\qquad (15)$$

In the above expressions, S is the signal intensity, k is a proportionality constant that only depends upon the molecule and the rate of UV excitation, I_0 is the infrared beam intensity incident on the detector with no sample present, T is transmittance, and A is absorbance. A_{T_1} is the absorbance of the steady-state triplet concentration that would be obtained if the excitation source were not chopped, and A_{S_0} is the absorbance of the ground state population with no excitation. Equations (14) and (15) apply to separated or non-overlapping peaks. Since the corresponding signals are 180° out of phase, peaks due to a combination of ground and triplet state absorptions have an intensity proportional to the difference in the respective signals over the frequency range of the overlap.

The dichroic ratio, $d(\omega)$, is the ratio of the absorbance of a band measured parallel to the polymer stretching direction to that measured perpendicular to the polymer stretching direction. For separated triplet state peaks, the ratio of the signal measured parallel to the polymer stretching direction to that measured perpendicular is related in a very simple way to the dichroic ratio:

$$d(\omega) = \frac{A_{T_1}^{\|}(\omega)}{A_{T_1}^{\perp}(\omega)} = \frac{S_{T_1}^{\|}(\omega)}{S_1^{\perp}(\omega)} \cdot \frac{I_0^{\perp}(\omega)}{I_0^{\|}(\omega)} \tag{16}$$

or, more accurately, for the ith transition:

$$d_i = \frac{\int\limits_{i \text{ band}} [S_{T_1}^{\|}(\omega)/I_0^{\|}(\omega)]d\omega}{\int\limits_{i \text{ band}} [S_{T_1}^{\perp}(\omega)/I_0^{\perp}(\omega)]d\omega} = \frac{A_{T_1}^{\|}(i)}{A_{T_1}^{\perp}(i)} . \tag{17}$$

The corresponding equations for ground state and overlapping ground and triplet state bands do not yield formulas with such simple relationships to the dichroic ratio and are most successfully modeled numerically.

Dichroic ratios and orientation factors were determined for ground state absorptions in regular double beam spectra and in photo-induced spectra. The two sets of orientation factors were the same within experimental error, indicating that photoselection by the unpolarized UV light had a negligible effect on the orientation of the excited triplet state acridine molecules. This result is reasonable since two overlapping absorption bands of the acridine molecule of opposite in-plane polarization are being pumped [13].

Polarization of the infrared beam causes two problems, both of which originate with the direct dependence of the signal strength on the infrared beam intensity. Insertion of the polarizer into the infrared beam decreases the beam intensity which reduces the S/N ratio. In addition, the measured infrared linear dichroism is the product of the linear dichroism of the sample multiplied by the wavelength dependent polarization bias of the gratings in

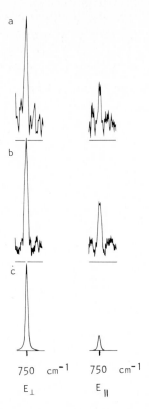

750 cm^{-1} 750 cm^{-1}

$E_\perp$ $E_{||}$

FIG. 6. Comparison of the spectra of acridine in stretched polyethylene before and after data reduction. Part (a) is the raw data, part (b) is four averaged scans, and part (c) is after the I_0 dependence has been accounted for.

the infrared monochromator. These problems were handled by digitizing the analog output of the lock-in amplifier and storing the data in the dynamic RAM of a microcomputer. In this way, it was possible to add any number of scans together to overcome S/N problems. It was also possible to digitize the perpendicular and parallel spectra of pure stretched polyethylene and divide the polarization effects of the gratings and the polyethylene from the spectra.

Part (a) of Fig. 6 shows the raw data corresponding to the 751 cm^{-1} triplet state absorption of acridine with the IR radiation polarized parallel and perpendicular to the stretching direction. Part (b) shows the result of four averaged scans under the same conditions. Part (c) is the corresponding Lorentzian peak after deconvolution of the slit function and accounting for the I_0 dependence. The integrated absorbance ratio, $A_{T_1}^{||}/A_{T_1}^{\perp}$, is 0.2.

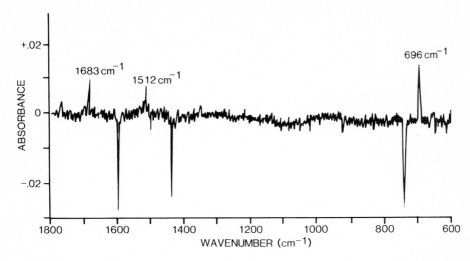

FIG. 7. Computer-subtracted infrared difference spectrum of triphenyl-ene in an N_2 matrix at 15K: pump source on minus pump source off. Excited triplet state modes have positive peaks while ground state vibrations display negative peaks. (Reproduced from Ref. [14], used by permission)

D. Fourier Transform Infrared Spectroscopy of Triphenylene

Triplet state infrared absorption spectroscopy has also been successfully performed using the combination of Fourier transform infrared spectroscopy (FTIR) as the probe and a CW xenon lamp source as the pump [14]. The FTIR computer-subtracted infrared difference spectrum of N_2 matrix-isolated triphenylene, between 600 and 1800 cm^{-1}, is shown in Fig. 7. The positions of the triplet state absorptions with the excitation source on are designated at 696, 1512, and 1683 cm^{-1}. In this spectrum, excited triplet state vibrational modes show upward peaks while ground state modes, whose intensities are not the same in the source-on and source-off runs, display downward peaks. Preliminary assignments of the observed triplet state absorptions based on ground state correlations are the following: the 696 cm^{-1} mode to the CH out-of-plane wag and the 1512 and 1683 cm^{-1} absorptions to the C-C stretching motions. The authors rationalize the observed downward frequency shift of the former and upward frequency shift of the latter as resulting from a decrease in CH bond electron density and an increase in the triplet state's C-C bond electron density. Therefore, it is suggested that there is a flow of electron density from the CH bonds to the CC bonds in the excited triplet state.

TABLE 2

Matrix-Isolated Anthracene Triplet State ($^3B_{1u}$) Vibrational Absorptions
and Preliminary Mode Assignments [20-22]

Observed Frequency (cm^{-1})	Preliminary Mode Assignment
1450	C-C ring stretch
1434	C-C ring stretch
1283	CH in-plane bend
899	skeletal deformation
886	CH out-of-plane bend
778	CH out-of-plane bend
719	CH out-of-plane bend

*Correlated with the IR crystal spectrum [22]

E.   Anthracene

Using the technique previously described [7,8], Hoesterey et al. [15]
have obtained the lowest triplet state infrared spectrum of matrix-isolated
anthracene, [structure], in N_2 at 20K. In this case, the anthracene
$S_1 \leftarrow S_0$ UV absorption was pumped by a 1000 watt Hg-Xe lamp with a wide-
band monochromator dispersed output. The UV absorption spectrum showed
strong absorptions at 3710, 3650, 3520, 3360 Å, and weaker absorptions at
shorter wavelengths. These can be compared with previously reported values
for $S_1 \leftarrow S_0$ of 3790 Å (n-heptane solution) [16] and 3590 Å (vapor phase)
[17], and is assigned as the $^1B_{1u} \leftarrow {}^1A_{1g}$ transition. Intersystem crossing to
$T_1(^3B_{1u})$ subsequently occurs which is the assumed metastable intermediate
which gives rise to the new absorption features in the modulated IR spectrum.
The previously reported triplet lifetimes for phosphorescence are 90 msec [18]
and 40 msec [19]. When the eight triplet absorptions are correlated with the
assigned ground state IR active vibrational frequencies of anthracene [20-22],
the preliminary assignments shown in Table 2 are made. The intense absorptions at 899 and 778 cm^{-1}, which are either weak or absent in the ground
state matrix spectrum, suggest the possibility of a significant geometry change
in the excited triplet state.

F. Benzophenone

Preliminary time-resolved triplet state IR spectra of N_2 matrix-isolated benzophenone at 20K have been obtained [23]. Confirmatory [23] phosphorescence spectra of the matrix-isolated sample was performed using the excitation of an Ar^+ laser. The IR absorptions found in the photo-induced spectra were 1675, 1595, 1318, 1308, and 1280 cm^{-1}. Further investigation and assignment of these absorptions are in progress.

III. EXCITED STATE RAMAN SPECTROSCOPY

A. Time-Resolved Raman Spectroscopy

In addition to the pulse-modulation infrared technique described above, excited state vibrational spectroscopy is also being performed by the technique of time-resolved Raman spectroscopy (TR^2S) [24,25] which is rapidly gaining popularity. This technique is particularly powerful when the excitation source is resonant with a strong absorption of a given species (resonance Raman) since the scattered Raman signals are increased by several orders of magnitude as compared to ordinary Raman scattering. Additionally, resonance Raman is frequency selective as compared to regular Raman spectroscopy and retains the sharpness and structural detail of the vibrational band.

1. p-Terphenyl

One of the earliest reported uses of this technique as a means of recording the time-resolved resonance Raman spectrum (TR^3S) of an excited triplet state is of p-terphenyl, [26]. The experimental setup consisted of a 30 nsec pump pulse of 2 MeV electrons from an electron accelerator followed by a probe flashlamp-pumped dye laser as the excitation source for resonance Raman scattering. The electron irradiated sample of a 0.01 molar solution of p-terphenyl in liquid benzene is known to produce triplet states of organic molecules [27]. The mechanism proposed in the literature is the formation of a radical cation-electron pair of a solvent molecule followed by fast recombination into a low energy excited state and subsequent energy transfer to solute molecules leaving these in the lowest excited triplet state. After establishing triplet-triplet absorption spectra, the lifetime of the triplet state (4.4 μsec), and the approximate triplet concentration, the recorded TR^3S was analyzed by means of an image intensifier

TABLE 3

Vibrational Wavenumbers (cm^{-1}) of p-Terphenyl in Its

Ground State and Lowest Excited Triplet State

Symmetry Species in D_{2h}	Ground State	Triplet State
A_g	1613	
		1540
	1599	
A_g	1505	
	1494	
A_g	1284	1350
A_g	1195	1227
A_g	1038	
A_g	1008	
		921
	993	
A_g	761	
	593	587

coupled to a TV-camera gated synchronously with the dye laser pulse. Spectra were subsequently stored on a magnetic-tape for further data handling.

The resonance Raman spectrum of a solution of p-terphenyl in benzene, with and without the pump irradiation source, is shown in Fig. 8 and the corresponding triplet state absorptions are listed in Table 3.

The authors assigned all the bands of the transient triplet state to the totally symmetric species with the exception of the 587 cm^{-1} absorption, suggesting that A-term scattering predominates. Based on normal coordinate calculations of biphenyl by Zerbi and Sandroni [28], the triplet absorption at 1350 cm^{-1} is assigned to the ring C-C stretching vibration, which is correlated to the parent absorption at 1284 cm^{-1}. The 66 cm^{-1} increase in frequency is believed to be due to an increased bond order and higher degree of planarity in the triplet state as compared to the ground state. This explanation is apparently supported by the work of Wagner [29] which suggests that, on the basis of spectroscopic and quenching results, the ground state of biphenyl is twisted with an interplanar angle of 25°, whereas the triplet state seems to be planar.

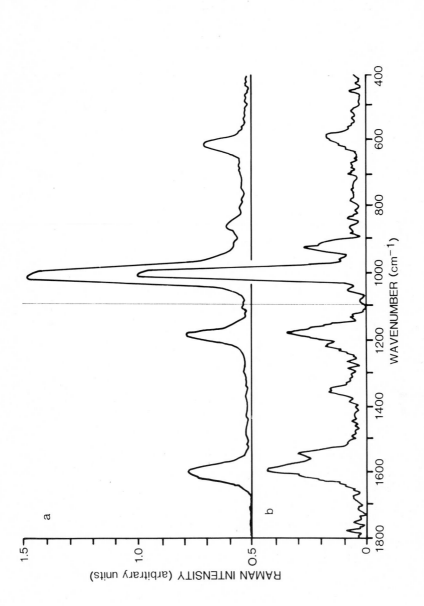

FIG. 8. (a) Raman spectrum of a 0.01 mol dm^{-3} solution of p-terphenyl in benzene; (b) resonance Raman spectrum of the same solution 600 ns after irradiation with a 30 ns pulse of 2 MeV electrons. Each spectrum recorded with two laser pulses (as indicated by the vertical line), excitation wavelength 459.0 nm. (Reproduced from Ref. [26], used by permission)

TABLE 4

Comparison of CARS with and without SH Excitation and

of Spontaneous Raman Scattering of Chrysene (Slit-width ~5 cm^{-1})[a]

CARS, chrysene 5 x 10^{-3} M		Spontaneous Raman Scattering, chrysene-powder	
without SH (cm^{-1})	with SH (cm^{-1})	cm^{-1}	rel. int.
		1431	27
1378 m	1378 m	1382	100
	1367 m		
1360 w		1364	23
	1336 m		
		1330	4
	1237 w	1228	5
	1164 w	1160	5
		1136	1
		1045	5
		1019	31
		892	5

[a] m-medium, w-weak.

2.   Chrysene

Coherent anti-Stokes Raman scattering (CARS) and TR^3S have been respectively used to study the vibrational modes of the singlet [30] and triplet [31] states of the chrysene molecule. In the CARS study, chrysene enclosed in polymethylmethacrylate was excited to the lowest singlet state with the second harmonic (SH) of a ruby laser (347 nm, 5 MW, 15 nsec) and simultaneously with a tunable dye laser source in the spectral range 735-780 nm. In a previous study [32], it was demonstrated that transient visible absorptions corresponding to the lowest excited singlet and triplet states of chrysene were time-resolvable; the triplet appeared ~325 nsec after SH ruby pumping the ground to singlet. The scattered radiation was detected with an optical multichannel analyzer (OMA) after spectral resolution by a polychromator. Using simultaneous pump and probe, it was possible to obtain a multiplex CARS spectrum in an interval of 200 cm^{-1} by one laser pulse [33], which is attributed solely to the singlet state. This assumption is additionally confirmed by $S_n \leftarrow S_1$ (20 nsec delay) and $T_n \leftarrow T_1$ (325 nsec delay) transient absorption spectra. The triplet absorptions in the frequency range 800 to 1600 cm^{-1} are summarized in Table 4 and compared with a chrysene powder spectrum obtained by spontaneous Raman scattering. Based on the observed differences in the CARS frequencies with (triplet state) and without (ground state) SH excitation, the authors suggest a possible difference in structure between the two states, but conclude that the lack of vibrational assignments makes this suggestion tenuous at present.

Atkinson and Dosser [34] subsequently reported the TR^3S spectrum of the excited $^3B_u^+$ state of chrysene-h_{12} and chrysene-d_{12} dissolved in oxygen-free n-hexane. Experimentally, the frequency-tripled output of a Nd:YAG laser at 354.7 nm (50 nsec pulsewidth FWHM) was used to populate the first excited singlet $^1B_{3u}^-$ state over the nsec time regime as discussed above. Resonance Raman scattering was then generated by tuning the output (0.01 nm bandwidth, 1 μsec pulsewidth FWHM) of a pulsed dye laser into resonance with the transient triplet-triplet ($^3A_g^-$ - $^3B_u^+$) absorption transition between 470 and 640 nm. The resultant resonance Raman spectrum monitored the vibrational degrees of freedom of chrysene in the excited $^3B_u^+$ state. By adjusting the time delay to 1 μsec, the observed chrysene-h_{12} triplet state Raman bands were 982, 1095, and 1342 cm^{-1}. When the same experiment was performed on the identical 10^{-4} M solution of chrysene-d_{12} in n-hexane, the observed bands were 952 and 1012 cm^{-1}. Although this publication reported excellent experimental and supporting data suggesting that these time-dependent Raman bands arise from the enhancement effects of the $^3A_g^-$ - $^3B_u^+$ transition, no vibrational assignments were made.

3.   Rhodamine 6G and Rhodamine B

The CARS spectra of the S_1 states of the dyes rhodamine 6G (R6G) [35,36] and rhodamine B (RB) [36] have recently been reported. In the original study [35], dilute solutions of R6G in methanol, ethanol, and acetonitrile were pumped to S_1 by a preresonant wavelength of 4880 Å. The observed spectrum, over the range of 1500-1650 cm^{-1}, gave rise to an intense absorption at 1605 cm^{-1} which was assigned to an S_1 state vibrational mode.

The subsequent study [36] of both 10^{-3}-10^{-4} M R6G and RB in ethanol reported a variety of vibrational frequencies for both S_0 and S_1 over the range 1200 to 1700 cm^{-1}. In these experiments, the two dye lasers (10 KW, 4 nsec, 20 pm bandwidth) were simultaneously pumped by a nitrogen laser (500 KW, 5.5 nsec) and directed into the sample at angles of 2-3°. The pumping wavelengths were held constant at 588 nm (preresonant) and 541 nm (resonant) and preceded the Stokes pulse by ~1 nsec. Thus, the 588 nm pulse was used to generate the CARS spectrum of the ground state and the 541 nm radiation specifically generated vibrational bands of both S_0 and S_1. These authors also point out that the vibrational mode cited above in the original CARS study at 1605 cm^{-1} is equally explainable as a changed CARS line shape without any frequency variation. From the list of frequencies observed and compared for R6G and RB in terms of inverse Raman scattering (IRS), infrared (IR), and CARS, the authors conclude that the xanthene skeleton does not undergo any significant structural change in the S_1 state and that the conjugation is not extended to the benzene ring by levelling of the molecule.

4.   Tris (2,2'-Bipyridine) Ruthenium (II) and Related Complexes

The resonance Raman vibrational spectrum of the lowest triplet state of tris(2,2'-bipyridine) ruthenium (II), *$Ru(bpy)_3^{2+}$ was performed in pulsed and CW modes and reported in a series of publications [37-39]. The *$Ru(bpy)_3^{2+}$ triplet state is generated by direct laser pumping of the S_0 ground state with wavelengths between 265 and 550 nm to the first excited singlet, S_1. This is a metal to ligand charge-transfer (MLCT) state with lifetime $\tau_{A_2} \leq 10^{-10}$ sec [40] and undergoes intersystem crossing in less than a nanosecond to the lowest triplet MLCT state, T_1. The triplet has a lifetime of ~665 nsec and a strong ligand-centered $\pi^* \leftarrow \pi$ absorption maximum at 360 nm [41]. Therefore, in all three studies cited above, the excitation source was used both for optional pumping and as the Raman scattering probe. In the original study of this system [37], the third harmonic of a Nd:YAG laser (354.5 nm, 5 nsec pulse width, ~5 mJ/pulse) was used as the pump and probe sources. Experiments were also performed with 457.9 nm CW excitation, which corresponded exclusively to the ground state spectrum.

The pulsed spectrum revealed seven new peaks which were attributed to ligand vibrations of the *$Ru(bpy)_3^{2+}$ triplet. The authors further assumed that the new peaks were probably due to the seven C-C and C-N stretching vibrations which exist for chelated bipyridine in either C_{2v} or D_3 symmetry. Based on the large frequency shifts of the correlated ligand vibrations to those of the bipyridine ligand, the authors tentatively suggested that the MLCT electron density in the excited state was predominantly localized on one of the bipyridine ligands, rather than delocalized over all three, on the vibrational time scale.

The original report of the TR^3 spectrum of this species was complemented by a continuous-wave resonance Raman (CWRR) study [38]. Part of this publication discusses the general use of CW laser sources for the study of excited states using the RR technique. Formulas are derived for an idealized four-level system, and subsequently applied to the $Ru(bpy)_3^{2+}$ - $*Ru(bpy)_3^{2+}$ system. All spectra reported in this publication were performed with a 65 mW Spectra Physics Model 170 Kr^+ (350.6 nm) laser source in focused and defocused modes. The comparison of the calculated and experimental concentrations of the steady-state triplet was 9.2 x 10^{-4}M and 2.2 x 10^{-4}M, respectively, and the difference attributed to the possible reabsorption of the laser light by the triplet. In addition to the bands reported in the original study [37], new singlet $Ru(bpy)_3^{2+}$ bands were observed at 663, 766, 1041, 1109, and 1276 cm^{-1}. New triplet $*Ru(bpy)_3^{2+}$ bands were observed at 341, 375, 741, 1098, and 1477 cm^{-1}. Table 5 lists the characterized RR bands of $Ru(bpy)_3^{2+}$ and $*Ru(bpy)_3^{2+}$ obtained from polarized spectra with focused and defocused laser beams and appropriate subtraction. For comparison, the data of $Fe(bpy)_3^{2+}$ and $Na^+bpy^{\bar{\cdot}}$ are also included. This table represents the most complete tabulation of the vibrational bands of the triplet species. The excited triplet frequencies were generally lower than the ground state frequencies by ~60 cm^{-1}, confirming the reduced character of the ligands, in agreement with reference [37]. Therefore, only one bpy in $*Ru(bpy)_3^{2+}$ is reduced, and it carries a full unit charge, whereas the other two bpy ligands are formally uncharged. The authors also cite supporting evidence from electronic absorption spectra of electrochemically reduced $Ru(bpy)_3^{2+}$ [42,43], thus $*Ru(bpy)_3^{2+}$ is more precisely written as $[Ru^{IV}(bpy)_2(bpy^{\bar{\cdot}})]^{2+}$.

In the most recent and comprehensive study of the $Ru(bpy)_3^{2+}$ system and the related complexes, $Os(bpy)_3^{2+}$, $[Ru(bpy)_2(NH_3)_2]^{2+}$, and $[Ru(bpy)_2(en)]^{2+}$, both CWRR and TR^3S are used to confirm the previous findings and provide the basis for the more general application of these techniques for the structural characterization of electronically excited states in solution [39]. Both visible and UV excitation sources were used in obtaining the CW Raman spectra of these complexes. The TR^3 instrumental setup used the harmonics of a pulsed Nd:YAG laser and a Nd:YAG pumped dye laser as sources, both to create the excited states and to affect the Raman scattering. The TR^3 spectra of all three complexes were obtained with the third harmonic of the Nd:YAG laser at 354.5 nm and produced good yields of the MLCT excited states. The laser is also sufficiently energetic and resonant with the excited state for enhancement of Raman scattering of all three complexes, that the same laser pulse is used as the probe source. The TR^3 spectrum of $Ru(bpy)_3^{2+}$, which is approximately 90% saturated, from 900 to 1700 cm^{-1} is shown in Fig. 9. Table 6 lists the ground state and TR^3 vibrational bands of cis-$[Ru(bpy)_2(en)]^{2+}$ and the shifted bands of the $(bpy^{\bar{\cdot}})$ ligand in $*Ru(bpy)_3^{2+}$. The frequency shifts of the observed bipyridine modes in the MLCT state of this complex are qualitatively the

TABLE 5

Resonance Raman Band Frequencies $\nu(cm^{-1})$ of $Fe(bpy)_3^{2+}$, $Ru(bpy)_3^{2+}$, $*Ru(bpy)_3^{2+}$ and bpy^{-}, and Depolarization Ratios ρ

$Fe(bpy)_3^{2+a}$	$Ru(bpy)_3^{2+b}$		$*Ru(bpy)_3^{2+b}$		Na^+bpy^{-c}
ν	ν	ρ	ν	ρ	ν
					1598 w
1607 ms,p[d]	1605 s	0.2	1545 s	0.3	1558 s
1563 s,p	1560 vs	0.2	1499 m	0.3	1497 vs
1490 vs,p	1489 s	0.1	1477 w	p	1478 vs
					1450 w
	1448 w	0.3	1425 s	0.3	1429 w
1321 s,p	1317 s	0.2	1363 w	p	1357 w
1277 m,p	1276 w[e]		1283 vs	0.3	1273 m
	1254 w[b,e]	p	1211 s	0.3	1205 m
	1173 m	0.2			1151 w
					1116 w
1109 mw,p	1109 vw	---	1098 w	p	1090 w
1067 w,p	1067 w	p			
1035 mw,p	1041 m	0.3	1030 w	p	1033 w
1025 m,p	1027 w	p	1012 w	0.1	982 m
					952 s
768 w,p	766 w	p	741 m	0.2	746 m
662 m,p	663 w	0.3			
642 vw,dp					
370 w,p			375 w		
			341 w	0.3	
					189 w
146 w,p					

[a] λ_{exc} = 514.5 and 647.1 nm
[b] λ_{exc} = 350.6 nm
[c] λ_{exc} = 406.7 nm, ~10^{-3}M in THF
[d] p = polarized, s = strong, m = medium, w = weak, v = very
[e] λ_{exc} = 406.7 nm

same as those observed in $*Ru(bpy)_3^{2+}$. Although the TR^3 spectra of $*Os(bpy)_3^{2+}$ and $[Ru(bpy)_2(NH_3)_2]^{2+}$ are not as definitive in matching the $*Ru(bpy)_3^{2+}$ spectra, the observed features strongly support a localized structure for the MLCT states of the other complexes as well.

TABLE 6

Ground State (Continuous-Wave-Excited) and Pulse-Excited Resonance Raman Frequencies (cm^{-1}) of cis-$[Ru(bpy)_2(en)]^{2+}$, Compared to Shifted Frequencies in the MLCT State of $Ru(bpy)_3^{2+}$

Ground State λ_0 = 356.4 nm	Pulse Excited λ_0 = 354.5 nm	Shifted Peaks[a] of *$Ru(bpy)_3^{2+}$
	1015[b]	1016
	1031[b]	1035
1037	1041	
1174	1176	
	1217[b]	1214
1275	1272	
	1288[b]	1288
1321	1318	
	1430[b]	1429
1450	1448	
1490	1506/1498[b]	1500/1482
	1554	1550
1562	1564	
1606	1607	

[a] Peaks due to "bpy$\overline{\cdot}$" in the MLCT state
[b] Excited state peaks of $[Ru(bpy)_2(en)]^{2+}$

5.   The Carotenoids

The TR^3 technique has also been applied, with great success, to the study of excited states of the conjugated polyenes such as the carotens [44-46]. In this section, we describe the study of all-trans-β-carotene and eight additional carotenoids. The first report of the resonance Raman spectrum of the lowest triplet state of all-trans-β-carotene was that of Dallinger et al. [44]. A 10^{-4} M solution of β-carotene containing 10^{-2} M of naphthalene was irradiated with 4-MeV electron beam pulses of 800 nsec duration, which in turn produced the β-carotene triplets by sensitized energy transfer via naphthalene. Transient $T_n \leftarrow T_1$ absorption spectra showed that the β-carotene triplet concentration was maximized at ~1 μsec after completion of the radiolysis pulse. The Raman scattering probe, the second harmonic of

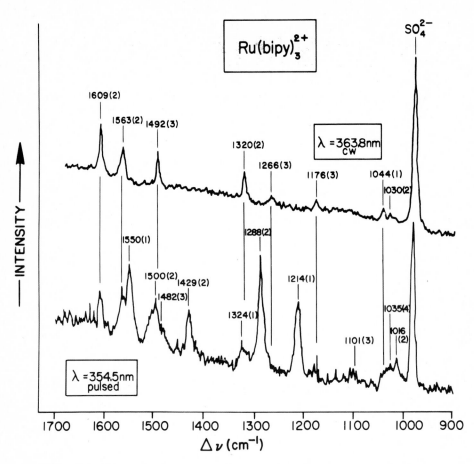

FIG. 9.   Upper trace: ground state (continuous wave excited) resonance Raman spectrum of $Ru(bpy)_3^{2+}$.   Lower trace: TR^3 spectrum of the
MLCT state, $*Ru(bpy)_3^{2+}$.   The peaks at 984 cm-1 are due to 0.50 M SO_4^{2-}
added as an internal intensity reference.   The numbers in parentheses are
standard deviations.   (Reproduced from Ref. [39], used by permission)

a Nd:YAG laser (532 nm, 7 nsec), was synchronized with the electron beam
to pump the sample at maximum triplet concentration.   The difference spectrum with and without electron beam irradiation revealed a new set of bands
at 1495, 1126, and 1014 cm^{-1} attributed to the triplet state.   The formation of
the β-carotene triplet involves a $\pi^* \leftarrow \pi$ excitation, resulting in the production
of greater conjugation in the excited triplet state.   Such a change should also
involve a decrease in the frequency of the C=C double bond and a corresponding increase in the C-C single bond frequency, simply on the basis of
electron density redistribution.   In this study, the 1495 cm^{-1} band, which is
assigned to the C=C stretching motion, does in fact decrease by 26 cm^{-1}
relative to the ground state (1521 cm^{-1}).   The other two bands appear to

TABLE 7

Vibrational Frequencies (cm^{-1}) of all-<u>trans</u>-β-Carotene

Ground State				Triplet Excited State	
Rimai et al.[48]	Jensen et al.[45]	Dallinger et al.[44]	Jensen et al.[45]	Dallinger et al.[44]	Jensen et al.[45]
CW-excit n-hexane 488 nm	CW-excit benzene 514.5 nm	pulsed benzene 531.8 nm	pulsed benzene 531.1 nm	pulsed benzene 531.8 nm	pulsed benzene 531.1 nm
961 vw					965 ± 13 w
1006 w	1005 w	1003		1014	1009 ± 3 m
1158 vs	1157 vs	1157	1150	1126	1125 ± 2 vs
1193 w	1190 w				1188 ± 3 m
1215 vw	1213 vw				1236 ± 3 s
1527 vs	1520 vs	1521	1521	1495	1496 ± 2 vs

have correlated shifts of -31 and +11 cm^{-1} relative to ground state absorptions, and were not clearly assignable to either the C-C stretch or possibly the C-C-H in-plane deformation, which also occurs in this spectral region.

The follow-up study of this system [45], wherein a greater concentration of the triplet was generated, confirmed the results of the previous study [44] in addition to the identification of three additional bands at 965, 1188, and 1235 cm^{-1}. The major difference in this work and the previous one are (1) variable higher irradiation doses were used resulting in greater concentration of the excited triplet state, (2) a longer laser pulse of 600 nsec, and (3) the spectra were recorded with single pulses, with a change of solution after each pulse. It was estimated that ~42% of the triplet naphthalene molecules were quenched by β-carotene, resulting in an estimated 50% β-carotene molecules excited to the triplet at the maximum dose used for resonance Raman experiments. The transient triplet bands obtained from the difference spectrum are tabulated in Table 7, along with data on ground state β-carotene and the results of Dallinger et al. [44]. Both the dose and time-dependence of the transient Raman absorptions were observed in order to confirm that the difference bands were in fact due to the lowest excited triplet of β-carotene. Based on the vibrational analyses of the polyenes, including β-carotene by Rimai et al. [47], vibrational mode assignments were discussed for the observed triplet features listed in Table 7. The triplet bands at 965 and 1009 cm^{-1} were assumed to reasonably correlate with the ground state C-C-C and C-C-H angular deformation modes at 961 and 1005 cm^{-1} on the basis of their frequencies and relative intensities. The bands at 1125, 1188, and 1236 cm^{-1} showed an intensity pattern which could not be correlated with the ground state spectrum, and was therefore assumed to reflect a conformational change of all-trans-β-carotene in the triplet state compared to the ground state. It was further suggested that there is substantial C=C bond twisting in the equilibrium conformation in the triplet state, localized at the 15,15' double bond. The remaining band at 1496 cm^{-1} was reasonably correlated with the C=C in-phase stretching vibration which occurs at 1520 cm^{-1} in the parent; the -24 cm^{-1} shift is reasonably rationalized by the decreased π-bond order in the excited state.

The most recent publication in this series is the extensive study of nine carotenoids including all-trans-β-carotene by Dallinger et al. [46]. The nine carotenoids shown in Fig. 10 involve systematic structural variations from β-carotene and thus provide a more extensive view of the excited-state vibrational spectroscopy and corresponding molecular structure of these polyenes. The carotenoid triplets were generated by the previously described technique above and, additionally, by optical excitation of a sensitizer (anthracene) and subsequent energy transfer. The experimental apparatus used to perform the optical studies is shown in Fig. 11. This set-up allowed the performance of experiments having time-resolution of μsecs to psecs. The μsec delays required the use of two separate lasers, while the psec experiments were performed with a single source Quantel NG50 34 psec laser having split beams, with one delayed in time by an optical delay line. All of the carotenoids studied showed similar vibrational shifts between the S_0 and T_1 states which could be reasonably correlated. Of the six T_1 bands observed, the ones thought to be most reasonably assigned were the C=C and C-C stretches, where the two retinal homologues (β-apo-8'-carotenal and ethyl β-apo-8'-carotenoate) exhibited anomalously higher frequencies for the C=C stretch, and the two carotenoids with longer all-trans C=C conjugated chains (dihydroxylycopene and decapreno-β-carotene) have anomalously low C=C stretch frequencies. When the C-C bond stretching frequency versus the C=C bond stretching frequency of the remaining five carotenoids are plotted (those having all-trans C=C bands and terminal cyclohexane rings) a linear relationship with a negative slope is obtained. In addition, both of these bands decrease by ~20 to 30 cm^{-1}, with the shift in the C-C band being as large as, or sometimes larger than, the C=C band. Again, such a shift is expected for the C=C band as a result of the increased conjugation in the π^*-π excited state, however, the effect is opposite to that expected for the C-C bond. The authors rationalize this observation by suggesting that the interaction (off diagonal) force constant connecting the primary force constants for the

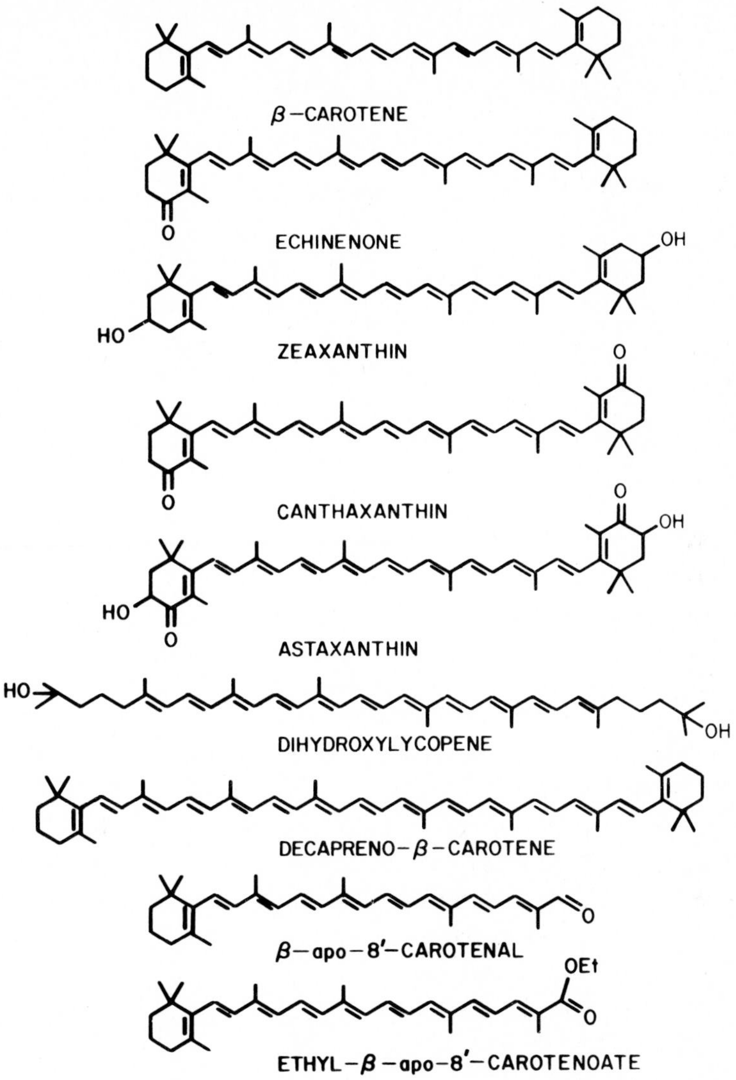

FIG. 10.   Structures of the carotenoids examined in Ref. [46].   (Used by permission)

C-C and C=C stretches, and perhaps other interaction constants, may change sign between S_0 and T_1 in polyenes.   Such a change in interaction force constants could explain the apparently anomalous frequency shifts for the C-C stretch without having to invoke large structural changes.   At the present time, resolution of the problem requires unambiguous clarification of the C-C stretch assignment and further model calculations on the four atom C-CH=C system as is presently being performed by Woodruff and co-workers.

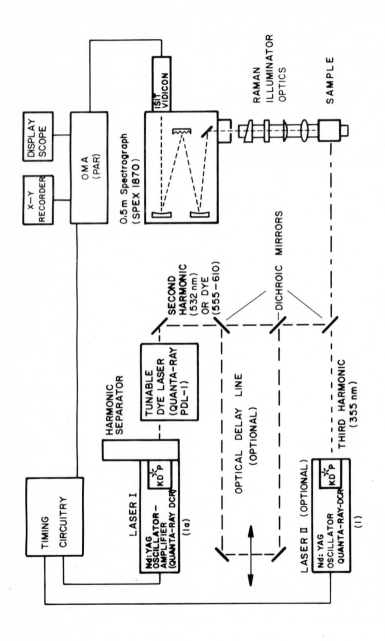

FIG. 11. Simplified diagram of the experimental apparatus for the optically sensitized triplet generation/ TR³ studies. Requisite time delays were obtained either by optical delay or by the two-laser experimental configuration. The optical delay line is not to scale; its actual length was ca. 120 ft. (Reproduced from Ref. [46], used by permission)

B. Time-Independent Raman Spectroscopy - Chromium(III) Complexes

The optically pumped excited states of ruby ($Al_2O_3:Cr^{3+}$) and aqueous solutions of $[Cr(bpy)_3][ClO_4]_3$ have been studied with electronic [49,50] and vibrational [51] time-independent Raman spectroscopy and UV, IR [52,53], and FIR [54] electronic absorption spectroscopy. The transition pumped is the visible $^4A_2 \rightarrow {}^4T_2, {}^4T_1$ transition of the Cr^{3+} ion. The excited states generated relax to the ^2E state which is split into the $\bar{E}$ and $\bar{A}_1 + \bar{A}_2$ states separated by ~29 cm^{-1} in Al_2O_3. By using an Ar^+ laser as the excitation source and the Raman probe, the separation between these two states was studied. Shifts in the vibrational frequencies of bipyridine ligands of the $[Cr(bpy)_3]^{3+}$ complex in solution were studied in the same manner. The observed shift in the C-C vibration at ~1043 cm^{-1} was 1.0 ± 0.5 cm^{-1} and that for the C-N vibration at ~1614 cm^{-1} was 2.0 ± 0.5 cm^{-1}.

In the visible/near-IR electronic absorption spectra of the doublet manifold, Xe flashlamps were used as the pump and probe sources. In the FIR absorption spectra, a Hg-Xe lamp was used as the pump source, and 29 cm^{-1} absorptions (HCN laser, 337μ) $\bar{E} \rightarrow 2\bar{A}$ in the doublet manifold were detected as emission $2\bar{A} \rightarrow {}^4A_2$ at 6922 Å.

IV. RESONANCE RAMAN SPECTROSCOPY

Time-independent resonance Raman spectroscopy can be used to determine shifts of certain normal mode potential energy minima that result from the promotion of a molecule to an excited state. The theory has been known for several years [55,56] and a recent review in this series contained information related to this topic [57]. The basis of the new method is the use of the Kramers-Krönig transform to determine relationships between resonance Raman excitation profiles and absorption spectra [58-60].

Following Hassing and Mortensen [61], the Kramers-Krönig relationships for the polarizability are

$$Re\alpha_{\rho\sigma}(\omega) = \frac{2}{\pi} P\int_0^\infty \frac{\omega' Im\alpha_{\rho\sigma}(\omega')}{\omega'^2 - \omega^2} d\omega' \qquad (18)$$

and

$$Im\alpha_{\rho\sigma}(\omega) = -\frac{2\omega}{\pi} P\int_0^\infty \frac{Re\alpha_{\rho\sigma}(\omega')}{\omega'^2 - \omega^2} d\omega' \qquad (19)$$

where P is the induced dipole moment and $\alpha_{\rho\sigma}$ is the $\rho\sigma^{th}$ polarizability tensor element. For the absorption cross-section, $\sigma_a(\omega)$, it is found that

$$\sigma_a(\omega) = \frac{4\pi}{3}\frac{\omega}{c}\operatorname{Im}(\alpha_{xx} + \alpha_{yy} + \alpha_{zz}). \tag{20}$$

For Raman scattering from totally symmetric vibrations the relevant quantity, $\theta(\omega)$, is

$$\theta(\omega) \equiv \left(\frac{d\sigma}{d\Omega}\right)_{\parallel} - \frac{4}{3}\left(\frac{d\sigma}{d\Omega}\right)_{\perp} = \left(\frac{\omega_s}{c}\right)^4\frac{1}{9}\,|\alpha_{xx} + \alpha_{yy} + \alpha_{zz}|^2 \tag{21}$$

where $\frac{d\sigma}{d\Omega}$ is the differential scattering cross-section with a polarization that is either parallel or perpendicular to that of the incident light, and ω_s is the frequency of the scattered light. Equations (18-21) define the relationship between Raman scattering from a molecule and the absorption spectrum of that molecule. By making the usual assumptions, i.e., the Born-Oppenheimer approximation, the Condon approximation, only one coupling excited state, harmonic vibrational wavefunctions, identical symmetry in the ground and excited states, and no vibrational force constant changes in the excited state, it is possible to arrive at expressions for the resonance Raman excitation profiles of fundamental, overtone, and combination levels that are based on the electronic absorption spectrum. Peticolas et al. [58-60] have used these relationships to determine shifts in excited state potential minima of several totally symmetric vibrational modes of uracil and two other pyrimidines.

A. Uracil and Related Compounds

The $\rho\sigma^{th}$ element of the polarizability tensor for the transition $\psi_i \to \psi_f$ is given by

$$\alpha_{fi}^{\rho\sigma} = \sum_e{}' \left[\frac{\langle\psi_f|\hat{\mu}^\rho|\psi_e\rangle\langle\psi_e|\hat{\mu}^\sigma|\psi_i\rangle}{\omega_e - \omega_i - \omega_\ell + i\Gamma_e} + \frac{\langle\psi_f|\hat{\mu}^\sigma|\psi_e\rangle\langle\psi_e|\hat{\mu}^\rho|\psi_i\rangle}{\omega_e - \omega_f + \omega_\ell + i\Gamma_e}\right]. \tag{22}$$

With the assumptions mentioned above and with the Born-Oppenheimer wave-functions given by

$$\psi_{nu} = \phi_n\chi_{nu} \tag{23}$$

we have for totally symmetric vibrations

$$\alpha_{g1,g0} = |\langle\phi_g|\hat{\mu}|\phi_e\rangle|^2$$

$$\times \sum_{v(\nu_1)} \sum_{v(\nu_2)} \cdots \sum_{v(\nu_{3N-6})} \langle\chi^1_{g,\nu_1}|\chi^v_{e,\nu_1}\rangle \langle\chi^v_{e,\nu_1}|\chi^0_{g,\nu_1}\rangle$$

$$\times \frac{\displaystyle\prod_{j=2}^{3N-6} |\langle\chi^0_{g,\nu_j}|\chi^v_{e,\nu_j}\rangle|^2}{(\omega^0_{eg} + \displaystyle\sum_{j=1}^{3N-6} v_j\Omega_j - \omega_\ell + i\Gamma_{ev})} \tag{24}$$

where the total vibrational wavefunction has been treated as a product of harmonic oscillator wavefunctions and only one mode, ν_1, is assumed to be active.

Using the same approximations, the electronic absorption intensity for the two coupled states is

$$I = |\langle\phi_g|\mu|\phi_e\rangle|^2 \sum_{v(\nu_1)} \sum_{v(\nu_2)} \cdots \sum_{v(\nu_{3N-6})}$$

$$\times \frac{\displaystyle\prod_{j=1}^{3N-6} |\langle\chi^0_{g,\nu_j}|\chi^v_{e,\nu_j}\rangle|^2 \cdot \Gamma_{ev}}{[(\omega^0_{eg} + \displaystyle\sum_{j=1}^{3N-6} v_j\Omega_j - \omega_\ell)^2 + \Gamma^2_{ev}]} \tag{25}$$

where the Franck-Condon overlap integrals are given by [57]

$$|\langle\chi^v_{\nu_1}|\chi^0_{\nu_1}\rangle|^2 = \frac{\Delta^{2v}_{\nu_1}}{2^v v!} e^{(-\Delta^2_{\nu_1}/2)} \tag{26}$$

$$\langle\chi^1_{\nu_1}|\chi^v_{\nu_1}\rangle\langle\chi^v_{\nu_1}|\chi^0_{\nu_1}\rangle = \frac{\Delta^{2v+1}_{\nu_1} - 2v\Delta^{2v-1}_{\nu_1}}{2^v\sqrt{2}\, v!} e^{(-\Delta^2_{\nu_1}/2)} \tag{27}$$

and ω^0_{eg} is the energy difference between ϕ^0_e and ϕ^0_g, Ω_j is the frequency of ν_j, and ω_ℓ is the incident laser frequency. Δ is defined by the following:

$$\chi^{v}_{g,v_j} = N_v H_v(\xi_{v_j}) e^{(-\xi^2_{v_j}/2)}$$

$$\chi^{v}_{e,v_j} = N_v H_v(\xi_{v_j} - \Delta_{v_j}) e^{(-(\xi_{v_j} - \Delta_{v_j})^2/2)}$$

where $\xi_{v_j} = (w_{v_j}/\hbar)^{\frac{1}{2}} Q_{v_j}$ and w is the circular frequency of v_j.

If a separation of the real and imaginary parts of the polarizability is carried out and a function $S(v_1\Omega_1, w_\ell)$ defined such that

$$S(v_1\Omega_1, w_\ell) = |\langle \phi_g|\mu|\phi_e\rangle|^2 \sum_{v(v_2)} \cdots \sum_{v(v_{3N-6})}$$

$$\times \frac{\prod_{j=2} |\langle \chi^{v}_{e,v_j}|\chi^{0}_{g,v_j}\rangle|^2}{[(w^0_{eg} + v_1\Omega_1 + \sum_{j=2} v_j\Omega_j - w_\ell)^2 + \Gamma^2_{ev}]} \tag{28}$$

then

$$\alpha_{10} = \sum_{v(v_1)} \langle \chi^{1}_{g,v_1}|\chi^{v}_{e,v_1}\rangle \langle \chi^{v}_{e,v_1}|\chi^{0}_{g,v_1}\rangle [T(v_1\Omega_1, w_\ell) + iS(v_1\Omega_1, w_\ell)] \tag{29}$$

and

$$I(w_\ell) = \sum_{v(v_1)} |\langle \chi^{v}_{g,v_1}|\chi^{0}_{g,v_1}\rangle|^2 S(v_1\Omega_1, w_\ell) \tag{30}$$

where $T(v_1\Omega_1, w_\ell)$ is the Kramers-Krönig transform of $S(v_1\Omega_1, w_\ell)$.

The computational details may be found elsewhere, but basically the analysis proceeds as follows. An $S(v_1\Omega_1, w_\ell)$ is constructed from the absorption spectrum via Eq. (30) and the Franck-Condon factors, Eq. (26), with an assumed value of Δ. A resonance Raman excitation profile (RREP) is then calculated from these values using Eq. (29) once the Kramers-Krönig transform of $S(v_1\Omega_1, w_\ell)$ has been calculated. It turns out that the shape of the RREP is independent of the values chosen for Δ, and only the relative magnitude of the profile is affected by the values chosen for the various Δ's. By also calculating the RREP for combination and overtone vibrations, using the counterpart to Eq. (29) for combination and overtone lines, it was

TABLE 8

Calculated Displacements, Δ_{e,v_j}, for Several Totally Symmetric Modes

of the Molecules Studied

	$\Omega_j(cm^{-1})$	Δ_{e,v_j}
	784	1.3 ± 0.2
cytidine	1243	1.0 ± 0.2
5'-monophosphate	1294	1.0 ± 0.2
	1529	0.8 ± 0.2
	783	1.0 ± 0.1
uridine	1321	1.3 ± 0.2
5'-monophosphate	1396	0.7 ± 0.1
	1630	0.5 ± 0.1
	1880	0.8 ± 0.1
	785	1.0
	1235	1.3
uracil	1390	0.6
	1630	0.5
	1680	0.7

possible to determine a value for the normal coordinate displacements without resorting to the assumption of a value for Γ and without the use of an internal standard.   Table 8 shows the calculated values of Δ for several vibrational modes of the three molecules that have been studied.

B.   Polydiacetylene

Batchelder and Bloor [62] used time-independent resonance Raman spectroscopy to determine excited state normal coordinate shifts of a polydiacetylene.   The RREPs of four Raman lines and six combinations and overtones of those fundamentals were measured.   Equations equivalent to Eqs. (24), (26), and (27), and the counterpart of Eq. (24) for combination and overtone lines, were used with assumed values of the displacements and the observed excited state vibrational frequencies to calculate the RREPs.   The four excited state normal coordinate displacements were varied to fit the ten

TABLE 9

Calculated Displacements for Four Raman Modes of a Polydiacetylene

$\nu_j(cm^{-1})$	Δ_{e,ν_j}
2086	0.57
1485	0.78
1203	0.30
952	0.44

observed RREPs as well as possible. Table 9 gives the calculated displacements and the associated vibrational frequencies. Using the excited state
vibrational frequencies and displacements, they then attempted to calculate
excited state bond lengths and force constants.

V.   VIBRONIC COUPLING

The interpretation of vibrational spectra typically involves the use of two
standard assumptions of molecular spectroscopy. These are the Born-
Oppenheimer and Condon approximations. Vibronic coupling results from a
breakdown of one or both of these assumptions and is manifested in spectra
as the appearance of symmetry forbidden bands or anomalously large
intensities of symmetry allowed bands. The two approximations can be
developed as follows. The complete molecular Hamiltonian is

$$H = T(Q) + T(q) + U(q,Q) + V(Q) \tag{31}$$

where

$$T(Q) = - \sum_a \frac{\hbar^2}{2} \frac{\partial^2}{\partial Q_a^2},$$

$$T(q) = - \sum_\ell \frac{\hbar^2}{2m_e} \frac{\partial^2}{\partial q_\ell^2},$$

$$U(q,Q) = \sum_s \sum_{t>s} \frac{e^2}{r_s - r_t} - \sum_s \sum_a \frac{Z_a e^2}{R_a - r_s},$$

and

$$V(Q) = \sum_a \sum_{b>a} \frac{Z_a Z_b e^2}{R_a - R_b} .$$

The exact wavefunction can be expanded in the form

$$\Psi_{ex} = \sum_k \phi_k(q,Q)\chi_k(Q) \tag{32}$$

where the exact wavefunction satisfies

$$H\Psi_{ex} = W\Psi_{ex} \tag{33}$$

and the functions $\phi_k(q,Q)$ satisfy

$$[T(q) + U(q,Q)]\phi_k(q,Q) = E_k(Q)\phi_k(q,Q) \tag{34}$$

and

$$\langle\phi_k(q,Q)|\phi_\ell(q,Q)\rangle = \delta_{k\ell} . \tag{35}$$

The vibrational wavefunctions are defined by a set of coupled differential equations:

$$[T(Q) + V(Q) + E_n(Q) + \langle\phi_n(q,Q)|T(Q)|\phi_n(q,Q)\rangle - W]\chi_n(Q)$$

$$+ \sum_{n\neq m} [\langle\phi_n(q,Q)|T(Q)|\phi_m(q,Q)\rangle$$

$$- 2\sum_a \frac{\hbar^2}{2} \langle\phi_n(q,Q)|\frac{\partial}{\partial Q_a}|\phi_m(q,Q)\rangle\frac{\partial}{\partial Q_a}]\chi_m(Q) = 0. \tag{36}$$

The Born-Oppenheimer approximation is made by neglecting the coupling terms in Eq. (36) [63-71]. The wavefunctions can then be expressed as simple product functions

$$\psi_{nv} = \phi_n(q,Q)\chi_{nv}(Q). \tag{37}$$

If the matrix elements of the complete molecular Hamiltonian are calculated using the Born-Oppenheimer wavefunctions, the off-diagonal elements of the Hamiltonian are given by

$$W_{nm,uv} = \langle \chi_{nu}(Q) | [\langle \phi_n(q,Q) | T(Q) | \phi_m(q,Q) \rangle$$

$$- 2 \sum_a \frac{\hbar^2}{2} \langle \phi_n(q,Q) | \frac{\partial}{\partial Q_a} | \phi_m(q,Q) \rangle \frac{\partial}{\partial Q_a}] | \chi_{mv}(Q) \rangle . \tag{38}$$

Using Eq. (34)

$$\langle \phi_n(q,Q) | \frac{\partial}{\partial Q_a} | \phi_m(q,Q) \rangle = \frac{\langle \phi_n(q,Q) | \frac{\partial U(q,Q)}{\partial Q_a} | \phi_m(q,Q) \rangle}{E_m(Q) - E_n(Q)} . \tag{39}$$

Also

$$\langle \phi_n(q,Q) | \frac{\partial^2}{\partial Q_a^2} | \phi_m(q,Q) \rangle = \sum_\ell \langle \phi_n(q,Q) | \frac{\partial}{\partial Q_a} | \phi_\ell(q,Q) \rangle \langle \phi_\ell(q,Q) | \frac{\partial}{\partial Q_a} | \phi_m(q,Q) \rangle$$

$$+ \frac{\partial}{\partial Q_a} \langle \phi_n(q,Q) | \frac{\partial}{\partial Q_a} | \phi_m(q,Q) \rangle . \tag{40}$$

It can be seen from Eqs. (39) and (40) that in some cases the Born-Oppenheimer approximation can fail, especially when the energy difference between the coupled states tends towards zero. When the coupled electronic states are degenerate because of symmetry, Jahn-Teller coupling results, and when the two states are nearly or accidentally degenerate, the result is pseudo Jahn-Teller coupling. Neither of these two situations should be treated by perturbation theory. However, if non-adiabatic or Born-Oppenheimer coupling is small, corrections to the wavefunctions may be found using perturbation theory and the corrected wavefunctions are of the form

$$\phi_n'(q,Q) = \phi_n(q,Q) + \sum_{m \neq n} \sum_a \frac{\langle \phi_m(q,Q) | \frac{-\hbar^2}{2} \frac{\partial^2}{\partial Q_a^2} | \phi_n(q,Q) \rangle}{E_m(Q) - E_n(Q)} \phi_m(q,Q) . \tag{41}$$

The other approximation used in molecular spectroscopy is the Condon approximation, which neglects the effect of nuclear motion on the electronic transition dipole moment. The failure of this approximation is called Herzberg-Teller coupling [72-75]. The Condon approximation amounts to expanding the electronic transition dipole moment integral in a Taylor series about the equilibrium nuclear configuration and neglecting the higher order terms: if $\psi_i(q,Q) = \phi_i(q,Q)\chi_i(Q)$ then

$$\langle\psi_i(q,Q)|\hat{\mu}|\psi_f(q,Q)\rangle = (\langle\phi_i(q,Q)|\hat{\mu}|\phi_f(q,Q)\rangle)_0 \, \langle\chi_i(Q)|\chi_f(Q)\rangle$$

$$+ \sum_a \frac{\partial}{\partial Q_a} (\langle\phi_i(q,Q)|\hat{\mu}|\phi_f(q,Q)\rangle)_0 \langle\chi_i(Q)|Q_a|\chi_f(Q)\rangle + \ldots \qquad (42)$$

and only the first term is retained. The expression

$$(\langle\phi_i(q,Q)|\hat{\mu}|\phi_f(q,Q)\rangle)_0$$

only makes sense if the two electronic wavefunctions have been fixed at the equilibrium nuclear configuration; i.e., the electronic wavefunctions are eigenfunctions of the Hamiltonian

$$H(q,Q_0) = T(q) + U(q,Q_0). \qquad (43)$$

Thus the term $(\langle\phi_i(q,Q)|\hat{\mu}|\phi_f(q,Q)\rangle)_0$ can be written as

$$\langle\phi_i(q,Q_0)|\hat{\mu}_{el}|\phi_f(q,Q_0)\rangle. \qquad (44)$$

The contributions to the transition dipole moment integral from the nuclei, contributions that are not considered in Eq. (42), can be expressed as

$$\langle\chi_i(Q)|\hat{\mu}_{nuc}|\chi_f(Q)\rangle = \langle\chi_i(Q)|e \sum_a Z_a R_a|\chi_f(Q)\rangle$$

$$= \langle\chi_i(Q)|e \sum_a Z_a R_a^0|\chi_f(Q)\rangle + \langle\chi_i(Q)|e \sum_a Z_a \sum_b (\frac{\partial R_a}{\partial Q_b}) Q_a|\chi_f(Q)\rangle. \qquad (45)$$

The first term of Eq. (45) is zero if $\chi_i \neq \chi_f$ and the second term is

$$e \sum_a Z_a \sum_b \frac{\partial R_a}{\partial Q_b} \langle\chi_i(Q)|Q_b|\chi_f(Q)\rangle. \qquad (46)$$

Neither of these terms makes any contribution to the transition dipole moment integral if $\phi_i \neq \phi_f$.

With Eqs. (44-46), the transition dipole moment integral, Eq. (42), can be written

$$\langle\hat{\mu}\rangle_{fi} = \langle\phi_i(q,Q_0)|\hat{\mu}_{el}|\phi_f(q,Q_0)\rangle\langle\chi_i(Q)|\chi_f(Q)\rangle$$

$$+ \sum_a[\langle(\frac{\partial}{\partial Q_a}\phi_i(q,Q_0))_0|\hat{\mu}_{el}|\phi_f(q,Q_0)\rangle\langle\chi_i(Q)|Q_a|\chi_f(Q)\rangle$$

$$+ \langle\phi_i(q,Q_0)|\hat{\mu}_{el}|(\frac{\partial}{\partial Q_a}\phi_f(q,Q_0))_0\rangle\langle\chi_i(Q)|Q_a|\chi_f(Q)\rangle]$$

$$+ e\sum_a Z_a\sum_b\frac{\partial R_a}{\partial Q_b}\langle\chi_i(Q)|Q_b|\chi_f(Q)\rangle. \qquad (47)$$

The derivation of Herzberg-Teller coupling terms generally proceeds through the use of perturbation theory with

$$\sum_a\frac{\partial U(q,Q_0)}{\partial Q_a}$$

as the perturbation operator. Eq. (47) can easily be shown to be equivalent to the perturbation theory approach with the use of Eq. (39) and the projection operator

$$P = \sum_j|\phi_j\rangle\langle\phi_j|.$$

Several authors have compared Herzberg-Teller to Born-Oppenheimer coupling [76-79] and it appears that in some cases the two effects may be of the same magnitude. Generally, however, because of the additional factor of ΔE in the denominator of the Born-Oppenheimer coupling term, Herzberg-Teller coupling is perceived as the larger of the two, especially when the coupled states are well separated.

The effect of vibronic coupling on Raman intensities is well documented in the literature [55,80-82] and no further discussion is necessary here. The effect of vibronic coupling on vibrational absorption intensities can be seen by consideration of Eq. (47) [80,83-87]. The first term will be zero since $\chi_i(Q) \neq \chi_f(Q)$. The second and third terms are the ones that will give rise to increased infrared absorption through vibronic coupling. The second term has the form:

$$\sum_{a}\sum_{j\neq i}\frac{\langle\phi_i(q,Q_0)|\dfrac{\partial U(q,Q_0)}{\partial Q_a}|\phi_j(q,Q_0)\rangle}{E_j(Q_0)-E_i(Q_0)}\langle\phi_j(q,Q_0)|\hat{\mu}_{el}|\phi_f(q,Q_0)\rangle$$

$$\times \langle\chi_i(Q)|Q_a|\chi_j(Q)\rangle \tag{48}$$

and the third term considers states vibronically coupled to ϕ_f. These terms can become large if vibronic coupling is large between two neighboring states coupled by an allowed electronic transition.

The effect of Herzberg-Teller type coupling on vibrational force constants has been noted since 1960 [88] and several authors have discussed it at length [86,88,89]. Generally, the effect is regarded as allowing the electron cloud to follow the nuclei during a vibration rather than keeping the cloud fixed as in the classical approximation. This is done using perturbation theory as mentioned before, and correction of the classical force constant yields

$$\frac{\partial^2 E(Q_0)}{\partial Q_a^2}=\langle\phi_n(q,Q_0)|\frac{\partial^2 H(q,Q_0)}{\partial Q_a^2}|\phi_n(q,Q_0)\rangle$$

$$+\sum_{m\neq n}\frac{\langle\phi_n(q,Q_0)|\dfrac{\partial H(q,Q_0)}{\partial Q_a}|\phi_m(q,Q_0)\rangle}{E_m(Q_0)-E_n(Q_0)}. \tag{49}$$

VI.   CONCLUSIONS

It is clear that the direct recording of excited state vibrational spectroscopy is in its infancy. The pulsed infrared (IR) technique appears to be ideally suited for excited state species with lifetimes over the msec and μsec time scales; in particular, the lowest triplet states of numerous systems satisfy the necessary criteria. The IR technique is also complementary with the various Raman scattering techniques.

Time-resolved Raman spectroscopy (TR^2S) is emerging as one of the most powerful techniques for the direct study of excited state vibrational transitions. Unlike the pulsed IR technique, its only practical limitations in time resolution are the temporal nature of the pump and probe pulses and the sophistication of the sensitive electronic multichannel detection and processing devices. At the present time, these studies are being extended to excited states and transients having psec lifetimes. The anticipation is that TR^2S studies will expand and involve the systematic characterization of excited states and identification of transients in reaction mechanisms, particularly in biological systems, where nsec and psec times are involved.

As for the theoretical interpretation of intensities and the calculation of force constants from excited state spectra, it can be seen from the last section that the complete separation of terms involving electrons and nuclei, which allows relatively straightforward analysis of ground state spectra, is no longer generally valid for excited state species. Any sort of rigorous analysis must proceed from a thorough knowledge of the ground state force constants and intensities. The molecules that have been studied thus far are too large for this type of information to be available at this time, and a systematic analysis of smaller molecules should be undertaken, now that the techniques for obtaining excited state spectra have been demonstrated.

ACKNOWLEDGEMENTS

The authors would like to thank Dr. George R. Smith and Barbara Hoesterey who made valuable contributions to this manuscript. Financial support by the National Science Foundation and the Petroleum Research Fund is gratefully acknowledged for the contributions from this laboratory. Grateful acknowledgement is also extended to Mrs. Linda Black for the typing of the original manuscript.

REFERENCES

1. D. C. Moule, Vibrational Spectra and Structure, Vol. 6, (J. R. Durig, ed.), Elsevier, New York, 1977, Ch. 4.

2. R. M. Hexter, J. Opt. Soc. Am., 53, 703 (1963).

3. E. J. Bair, J. T. Lund, and P. C. Cross, J. Chem. Phys., 24, 961 (1956).

4. J. C. Gilfert and D. Williams, J. Opt. Soc. Am., 48, 765 (1958); 49, 212 (1959).

5. R. H. Clarke, P. A. Kosen, M. A. Lowe, R. H. Mann, and R. Mushlin, J.C.S. Chem. Comm., 528 (1973).

6. G. Castro and G. W. Robinson, J. Chem. Phys., 50, 1159 (1969).

7. M. B. Mitchell, G. R. Smith, K. Jansen, and W. A. Guillory, Chem. Phys. Lett., 63, 475 (1979).

8. M. B. Mitchell, G. R. Smith, and W. A. Guillory, J. Chem. Phys., 75, 44 (1981).

9. W. A. Noyes, Jr. and P. A. Leighton, The Photochemistry of Gases, Dover Publications, New York, pg. 17.

10. R. P. Steiner and J. Michl, J. Am. Chem. Soc., 100, 6861 (1978).

11. A. Kellmann, J. Phys. Chem., 81 1195 (1977).

12. J. Radziszewski and J. Michl, J. Phys. Chem., 85, 2934 (1981).

13. R. P. Steiner and J. Michl, J. Am. Chem. Soc., 100, 6861 (1978).

14. J. Baiardo, R. Mukherjee, and M. Vala, J. Mol. Struct., 80, 109 (1982).

15. B. Hoesterey, M. Mitchell, and W. A. Guillory, "The Triplet State IR Spectrum of Matrix-Isolated Anthracene," submitted for publication.

16. H. B. Klevens and J. R. Platt, J. Chem. Phys., 17, 470 (1949).

17. E. E. Koch, A. Otto, and K. Radler, Chem. Phys. Lett., 21, 501 (1973).

18. S. P. McGlynn, T. Azumi, and M. Kasha, J. Chem. Phys., 40, 507 (1964).

19. G. Heinrich and H. Güsten, Z. Phys. Chem. Neue Folge, 118, 31 (1979).

20. J. Vodehnal and V. Stepán, Coll. Czech. Chem. Comm., 36, 3980 (1971).

21. A. Bree and R. A. Kydd, J. Chem. Phys., 48, 5319 (1968).

22. S. Califano, J. Chem. Phys., 36, 903 (1962).

23. B. Hoesterey, M. Mitchell, and W. A. Guillory, preliminary experimental results.

24. R. E. Hester, "Time-Resolved Raman Spectroscopy," The SPEX Speaker, 27, 2 (1982).

25. R. E. Hester, in Advances in Infrared and Raman Spectroscopy, (R. J. H. Clark and R. E. Hester, eds.), Heyden, 1978, Vol. 4, Chap. 1.

26. R. Wilbrandt, N. H. Jensen, P. Pagsberg, A. H. Sillesen, K. B. Hansen, Nature (London), 276, 167 (1978).

27. R. Cooper and J. K. Thomas, J. Chem. Phys., 48, 5097 (1968).

28. G. Zerbi and S. Sandroni, Spectrochim. Acta, 24A, 483, 511 (1968).

29. P. J. Wagner, J. Am. Chem. Soc., 89, 2820 (1967).

30. W. Werncke, H.-J. Weigmann, J. Patzold, A. Lau, K. Lenz, and M. Pfeiffer, Chem. Phys. Lett., 61, 105 (1979).

31. G. H. Atkinson and L. R. Dosser, J. Chem. Phys., 72, 2195 (1980).

32. K. A. Hodgkinson and I. H. Munro, Chem. Phys. Lett., 12, 281 (1971).

33. A. Lau, W. Werncke, J. Klein, and M. Pfeiffer, Opt. Commun., 21, 399 (1977).

34. G. H. Atkinson and L. R. Dosser, J. Chem. Phys., 72, 2195 (1980).

35. J. P. Devlin and M. G. Rockley, Chem. Phys. Lett., 56, 608 (1978).

36. R. König, A. Lau, and H.-J. Weigmann, Chem. Phys. Lett., 69, 87 (1980).

37. R. F. Dallinger and W. H. Woodruff, J. Am. Chem. Soc., 101, 4391 (1979).

38. M. Forster and R. E. Hester, Chem. Phys. Lett., 81, 42 (1981).

39. P. G. Bradley, N. Kress, B. A. Hornberger, R. F. Dallinger, and W. H. Woodruff, J. Am. Chem. Soc., 103, 7441 (1981).

40. F. Bolletta, M. Maestri, L. Moggi, and V. Balzani, J. Phys. Chem., 78, 1374 (1974).

41. R. Bensasson, C. Salet, and V. Balzani, J. Am. Chem. Soc., 98, 3722 (1976).

42. C. Creutz, M. Chou, T. L. Netzel, M. Okumura, and N. Sutin, J. Am. Chem. Soc., 102, 1309 (1980).

43. G. A. Heath, L. J. Yellowlees, P. S. Braterman, J.C.S. Chem. Commun., 287 (1981).

44. R. F. Dallinger, J. J. Guanci, Jr., W. H. Woodruff, and M. A. J. Rodgers, J. Am. Chem. Soc., 101, 1355 (1979).

45. N. H. Jensen, R. Wilbrandt, P. B. Pagsberg, A. H. Sillesen, and K. B. Hansen, J. Am. Chem. Soc., 102, 7441 (1980).

46. R. F. Dallinger, S. Farquharson, W. H. Woodruff, and M. A. J. Rodgers, J. Am. Chem. Soc., 103, 7433 (1981).

47. L. Rimai, M. E. Heyde, and D. Gill, J. Am. Chem. Soc., 95, 4493 (1973).

48. L. Rimai, R. G. Kilponen, and D. Gill, J. Am. Chem. Soc., 92, 3824 (1970).

49. L. V. Haley, B. Halperin, and J. A. Koningstein, Chem. Phys. Lett., 54, 389 (1978).

50. B. Halperin and J. A. Koningstein, J. Chem. Phys., 69, 3302 (1978).

51. M. Asano, D. Mongeau, D. Nicollin, R. Sasseville, and J. A. Koningstein, Chem. Phys. Lett., 65, 293 (1979).

52. G. K. Klauminzer, P. L. Scott, and H. W. Moos, Phys. Rev., 142, 248 (1966).

53. T. Kushida, J. Phys. Soc. Japan, 21, 1331 (1966).

54. H. Lengfellner, G. Pauli, W. Heisel, and K. F. Renk, Appl. Phys. Lett., 29, 566 (1976).

55. J. Tang and A. C. Albrecht, in Raman Spectroscopy, (H. A. Szymanski, ed.), Plenum, New York, Vol. 2, 1970, p. 33.

56. F. Inagaki, M. Tasumi, and T. Miyazawa, J. Mol. Spect., 50, 286 (1974).

57. P. V. Huong, in Vibrational Spectra and Structure, (J. R. Durig, ed.), Elsevier, Amsterdam, Vol. 9, 1981, p. 143.

58. W. L. Peticolas and D. C. Blazej, Chem. Phys. Lett., 63, 604 (1979).

59. D. C. Blazej and W. L. Peticolas, J. Chem. Phys., 72, 3134 (1980).

60. L. Chinsky, A. Laigle, W. L. Peticolas, and P.-Y. Turpin, J. Chem. Phys., 76, 1 (1982).

61. S. Hassing and O. S. Mortensen, J. Chem. Phys., 73, 1078 (1980).

62. D. N. Batchelder and D. Bloor, J. Phys. C: Solid State Phys., 15, 3005 (1982).

63. I. Ozkan and L. Goodman, J. Chem. Phys., 72, 6777 (1980).

64. I. Ozkan and L. Goodman, Chem. Rev., 79, 275 (1979).

65. R. Englman, The Jahn-Teller Effect in Molecules and Crystals, Wiley-Interscience, London, 1972.

66. H. C. Longuet-Higgins, in Advances in Spectroscopy, (H. W. Thompson, ed.), Interscience, New York, Vol. 2, 1961, p. 429.

67. P. A. Geldof, R. P. H. Rettschnick, and G. J. Hoytink, Chem. Phys. Lett., 10, 549 (1971).

68. J. A. Koningstein, in Vibrational Spectra and Structure, (J. R. Durig, ed.), Elsevier, Amsterdam, Vol. 9, 1981, p. 115.

69. W. D. Hobey, J. Chem. Phys., 43, 2187 (1965).

70. W. Moffitt and A. D. Liehr, Phys. Rev., 106, 1195 (1957).

71. R. L. Fulton and M. Gouterman, J. Chem. Phys., 35, 1059 (1961).

72. H. Sponer and E. Teller, Rev. Mod. Phys., 13, 75 (1941).

73. A. D. Liehr, Z. Naturforsch., 13a, 311 (1958).

74. A. C. Albrecht, J. Chem. Phys., 33, 156 (1960).

75. J. N. Murrell and J. A. Pople, Proc. Phys. Soc. (London), A69, 245 (1956).

76. W. G. Breiland and C. B. Harris, Chem. Phys. Lett., 18, 309 (1973).

77. G. Orlandi and W. Siebrand, Chem. Phys. Lett., 15, 465 (1972).

78. S. H. Lin and H. Eyring, Proc. Nat. Acad. Sci. USA, 71, 3415 (1974).

79. G. Orlandi and W. Siebrand, J. Chem. Phys., 58, 4513 (1973).

80. W. L. Peticolas, L. Nafie, P. Stein, and B. Fanconi, J. Chem. Phys.,
     52, 1576 (1970).

81. A. C. Albrecht, J. Chem. Phys., 34, 1476 (1961).

82. T. G. Spiro and P. Stein in Ann. Rev. Phys. Chem. (B.S. Rabinovitch,
     J. M. Schurr, and H. L. Strauss, eds.), Annual Reviews, Inc., Palo
     Alto, CA, Vol. 28, 1977, p. 501.

83. S. Bratoz, C. R. Acad. Sci., (Paris), 243, 1493 (1956).

84. W. D. Jones and W. T. Simpson, J. Chem. Phys., 32, 1747 (1960).

85. J. Burdett, Chem. Phys. Lett., 5, 10 (1970).

86. J. K. Burdett, Appl. Spec. Rev., 4, 43 (1970).

87. W. B. Person and D. Steele, Specialist Periodical Reports-Molecular
     Spectroscopy, The Chemical Society (R. F. Burrow, D. A. Long, and D.
     J. Miller, eds.), Vol. 2, 1974, p. 357.

88. J. N. Murrell, J. Mol. Spect., 4, 446 (1960).

89. R. F. W. Bader, Mol. Phys., 3, 137 (1960).

Chapter 2
OPTICAL CONSTANTS, INTERNAL FIELDS, AND MOLECULAR
PARAMETERS IN CRYSTALS

Roger Frech

Department of Chemistry
The University of Oklahoma
Norman, Oklahoma

I. INTRODUCTION

One of the most important goals of a molecular spectroscopist is to deduce information about a molecule of interest from its observed spectrum.
For a vibrational spectroscopist this information usually includes the vibrational frequency and some measure of how strongly the molecule interacts with
electromagnetic radiation when vibrating in a particular normal mode. The
vibrational frequency of a normal mode can be viewed as a molecular parameter providing information about the intramolecular potential energy surface.
The molecular parameter which characterizes the intensity of interaction with
electromagnetic radiation is usually either a dipole moment derivative or a
polarizability derivative taken with respect to a normal coordinate.

The measurement of spectra in the gas phase is relatively straightforward. In an absorption experiment the decrease in intensity of a beam of
monochromatic radiation is measured after passing through a sample. The
absorption coefficient $\alpha(\omega)$ is defined by

$$\alpha(\omega) = \frac{1}{d} \, \ell n \, \frac{I_o(\omega)}{I(\omega)} \tag{1}$$

where I_o is the initial intensity of the incident beam and I is the intensity of
the beam after traversing a sample of thickness d. The integrated absorption
coefficient may then be written [1]

$$\int \alpha(\omega) \, d\omega = \frac{N\pi}{3c} \, [(\frac{\partial \mu_x}{\partial q_j})^2 + (\frac{\partial \mu_y}{\partial q_j})^2 + (\frac{\partial \mu_z}{\partial q_j})^2] \tag{2}$$

where the integration is to be taken over a band. The quantities $(\partial \mu_\sigma / \partial q_j)$,
where $\sigma = x, y, z$, are the molecular dipole moment derivatives with respect to
the jth normal coordinate and, along with the frequency of the normal mode,
are the molecular parameters of interest. The molecular dipole moment derivatives may then be related to the change of the dipole moment as the molecule
or part of the molecule undergoes a particular kind of motion such as a bond
stretch or a valence angle bend. The development of this branch of spectroscopy was traced by Crawford in his 1982 Priestly Medal Address [2].

The reported vibrational frequency of a normal mode is obtained from an
appropriate analysis of the rotational-vibrational structure of the band and it
may be the harmonic frequency after correction for anharmonic effects or it
may simply be the observed fundamental transition frequency. In either case
it is obtained simply and directly from an observed spectral band shape.

The deduction of molecular parameters from spectral studies of molecular crystals or crystals containing polyatomic ions is considerably more difficult than gas phase studies due to a combination of factors. (Here polyatomic ions will be viewed as "molecules" or molecular units). Since the molecular concentrations encountered in these crystals precludes the observation of a fundamental transition by a simple transmission experiment, usually either a reflectivity measurement or a Raman scattering experiment is required. In addition there are other solid state effects which may be categorized as internal field effects, collective effects leading to a complex multiplet structure, and molecular perturbation effects.

There is a rich background of measurement techniques and analysis methods which is relevant to the determination of molecular parameters from spectroscopic studies of molecular crystals and crystals containing polyatomic ions. Although much of the early work in the area of optical constant measurement and analysis was concerned with simple ionic crystals composed of monatomic ions, obviously the measurement techniques which were developed were oblivious to the structural nature of the vibrating unit interacting with the electromagnetic radiation. However, most of the dynamic models of a crystal, for which optical constants can be derived specifically, consider a lattice of monatomic ions vibrating about their equilibrium sites. Nevertheless, a large part of the resulting formalism can be used almost directly by the solid state spectroscopist wishing to obtain molecular parameters from a particular molecular crystal.

Undoubtedly the most complete text of interest in this area is the classic work by Decius and Hexter [3]. Other very useful treatments of vibrations and the related optical constants in crystals include the work by Born and Huang [4] and Maradudin et al. [5]. The general topic of optical constants is discussed by Landau and Lifshitz [6], while one of the most complete treatments of the optics of crystals is probably the text by Born and Wolf [7]. In addition there are numerous other works in the solid state literature, too numerous to mention, which are of direct relevance to the topics considered here.

II.   THE OPTICAL CONSTANTS

A.   Basic Equations in Non-Absorbing Media

An optical constant may be broadly defined as a quantity which describes the response of matter to a field. In this sense the optical constants will also include the dielectric constants since the two quantities are simply related. Although it is usually the electric field in an electromagnetic wave interacting with an absorbing sample which is of interest to the spectroscopist, it is useful to begin by considering the electromagnetic field in a non-absorbing, homogeneous, isotropic medium. The electromagnetic field in free space can be characterized by the electric vector E and the magnetic induction B. In a medium, the effect of the fields is described through the electric displacement D, the magnetic vector H, and the electric current density, J. These quantities are related through Maxwell's equations (written here in Gaussian units):

$$\nabla \times E + \frac{1}{c} \dot{B} = 0 \qquad \text{(Faraday's Law)} \qquad (3a)$$

$$\nabla \cdot B = 0 \qquad \text{(absence of free magnetic poles)} \qquad (3b)$$

$$\nabla \times H - \frac{1}{c} \dot{D} = \frac{4\pi}{c} J \qquad \text{(Ampere's Law)} \qquad (3c)$$

$$\nabla \cdot D = 4\pi\rho \qquad \text{(Coulomb's Law)} \qquad (3d)$$

where ρ is the charge density and c is the speed of light in vacuum. To allow a unique determination of the field vectors from a given distribution of fields and charges, Maxwell's equations must be supplemented by relations which describe the behavior of matter in a field. These are the material equations or constitutive relations

$$J = \sigma E \qquad (4a)$$

$$D = \varepsilon E \qquad (4b)$$

$$B = \mu H \qquad (4c)$$

In Eqs. (4), σ is the specific conductivity, ε is the dielectric constant, and μ is the magnetic permeability. If one considers that part of the field which contains no charges or currents (non-absorbing medium), Maxwell's equations suggest the existence of traveling waves which transport energy. If Eq. (4c) is substituted in Eq. (3a) and the curl of both sides of the resulting equation is taken, there is obtained

$$\nabla \times (\nabla \times E) + \frac{\mu}{c} \nabla \times \dot{H} = 0. \qquad (5)$$

Combining Eq. (3c) with $J = 0$ (non-absorbing medium) and Eq. (4b),

$$\nabla \times H = \frac{\varepsilon}{c} \dot{E} . \qquad (6)$$

The derivative with respect to time is taken and the result is substituted into Eq. (5), giving

$$\underset{\sim}{\nabla} \times (\underset{\sim}{\nabla} \times \underset{\sim}{E}) + \frac{\mu\varepsilon}{c^2} \ddot{\underset{\sim}{E}} = 0. \tag{7}$$

Invoking the well-known vector identity,

$$\underset{\sim}{\nabla} \times (\underset{\sim}{\nabla} \times \underset{\sim}{E}) = \underset{\sim}{\nabla}(\underset{\sim}{\nabla} \cdot \underset{\sim}{E}) - \nabla^2 \underset{\sim}{E} \tag{8}$$

Eq. (7) becomes, in the absence of free charges ($\rho = 0$ and hence $\underset{\sim}{\nabla} \cdot \underset{\sim}{E} = 0$ through Eqs. (3d) and (4b)),

$$-\nabla^2 \underset{\sim}{E} + \frac{\mu\varepsilon}{c^2} \ddot{\underset{\sim}{E}} = 0. \tag{9}$$

Finally, if the harmonic solution $\underset{\sim}{E}(\underset{\sim}{r},t) = \underset{\sim}{E}(\underset{\sim}{r}) \exp(-i\omega t)$ is assumed,

$$\nabla^2 \underset{\sim}{E} + \frac{\omega^2 \mu\varepsilon}{c^2} \underset{\sim}{E} = 0 \tag{10}$$

and it is apparent that each Cartesian component of $\underset{\sim}{E}$ obeys a wave equation whose solution is

$$\underset{\sim}{E}(\underset{\sim}{r},t) = \underset{\sim}{E}_o \exp(i\underset{\sim}{k} \cdot \underset{\sim}{r} - i\omega t). \tag{11}$$

The propagation direction of the wave is given by the wave vector $\underset{\sim}{k}$, whose magnitude is easily seen to be

$$|\underset{\sim}{k}| = \frac{\omega}{c} \sqrt{\mu\varepsilon}. \tag{12}$$

The quantity $c/\sqrt{\mu\varepsilon}$ is the phase velocity of the wave which is the velocity of energy flow. It is easy to show that the direction of energy flow is the direction of wave propagation.

B. Absorbing Media-Complex Optical Constants

These results can now be easily generalized to the more important case of an absorbing medium (one with a non-zero conductivity) by substituting Eqs. (4a) and (4b) into Ampere's law (Eq. 3c). Then

$$\underset{\sim}{\nabla} \times \underset{\sim}{H} = \frac{\varepsilon}{c} \dot{\underset{\sim}{E}} + \frac{4\pi\sigma}{c} \underset{\sim}{E}. \tag{13}$$

Upon taking the time derivative of Eq. (13), substituting into Eq. (5), and invoking Eq. (8), one has

$$-\nabla^2 \underset{\sim}{E} + \frac{\mu}{c} [\frac{\varepsilon}{c} \ddot{\underset{\sim}{E}} + \frac{4\pi\sigma}{c} \dot{\underset{\sim}{E}}] = 0. \tag{14}$$

This equation differs from its counterpart (Eq. (9)) for the non-absorbing case only in the additional term involving $\dot{\underset{\sim}{E}}$. Here the assumption of a harmonic solution leads to

$$\nabla^2 \underset{\sim}{E} + \frac{\omega^2\mu}{c^2} [\varepsilon + i \frac{4\pi\sigma}{\omega}] \underset{\sim}{E} = 0. \tag{15}$$

The complex dielectric constant is defined as

$$\hat{\varepsilon} = \varepsilon_1 + i\varepsilon_2 \tag{16}$$

where the real part, ε_1, is the ordinary dielectric constant for a non-absorbing medium ($\varepsilon_1 = \varepsilon$) and the imaginary part of $\hat{\varepsilon}$ is related to the specific conductivity of the medium ($\varepsilon_2 = 4\pi\sigma/\omega$). Eq. (15) may then be written

$$\nabla^2\underset{\sim}{E} + \frac{\omega^2\mu\hat{\varepsilon}}{c^2} \underset{\sim}{E} = 0. \tag{17}$$

The wave equation for an absorbing medium has exactly the same form as that for a non-absorbing medium providing the real dielectric constant ε in the latter case is replaced by a complex dielectric constant $\hat{\varepsilon}$ in the former case. This is the basis for the often-quoted and extremely useful statement that the equations for non-absorbing (non-conducting) media can be used to describe the corresponding phenomena in conducting media provided that the dielectric constants and optical constants are taken to be complex quantities.

In a non-absorbing medium the absolute refractive index is defined as the index for refraction from vacuum into the medium and is the ratio of the phase velocity of the wave in vacuum, c, to the phase velocity in the medium, v,

$$n = \frac{c}{v} . \tag{18}$$

The relationship between the index of refraction and the dielectric constant (Maxwell's formula) follows easily;

$$n = \sqrt{\varepsilon\mu} \tag{19}$$

and for an absorbing, non-magnetic medium,

$$\hat{n} = \sqrt{\hat{\varepsilon}}. \tag{20}$$

Normally the complex refractive index is written

$$\hat{n} = n + ik \tag{21}$$

where n is the usual refractive index and k is the extinction coefficient (not to be confused with the magnitude of the wave vector). Occasionally the refractive index is written

$$\hat{n} = n(1 + i\kappa) \tag{22}$$

where κ is the attenuation index, although Eq. (21) is by far the most common usage. It follows from Eqs. (12) and (19) that the wave vector in an absorbing medium is a complex quantity. From Eqs. (20) and (21) it is easy to see that

$$\varepsilon_1 = n^2 - k^2 \tag{23a}$$

$$\varepsilon_2 = 2nk = \frac{4\pi\sigma}{\omega}. \tag{23b}$$

The nature of k, the extinction coefficient, is most clearly seen by considering the passage of a monochromatic light wave through an absorbing medium. The intensity of the wave at some point $\underset{\sim}{r}$ in the medium is given by the time-averaged normal component of the Poynting vector and may be written

$$I(\underset{\sim}{r}) = \frac{c}{4\pi} \text{Re}[\hat{n} \langle \underset{\sim}{E}(\underset{\sim}{r},t) \cdot \underset{\sim}{E}^*(\underset{\sim}{r},t) \rangle] \tag{24}$$

where $\underset{\sim}{E}(\underset{\sim}{r},t)$ is given by Eq. (11). Then

$$I(\underset{\sim}{r}) = \frac{c}{8\pi} \text{Re}[\hat{n} \, |\underset{\sim}{E}_o|^2 \exp(i\hat{\underset{\sim}{k}} \cdot \underset{\sim}{r} - i\hat{\underset{\sim}{k}}^* \cdot \underset{\sim}{r})] \tag{25}$$

and utilizing Eqs. (12) and (21) for a wave propagating in the x direction

$$I(x) = \frac{nc}{8\pi} |\underset{\sim}{E}_o|^2 \exp(-2 \frac{\omega}{c} kx) \tag{26}$$

where k is the extinction coefficient. If x measures the distance the wave has traveled through the medium, then the initial intensity of the wave is

$$I_o = \frac{nc}{8\pi} |\underset{\sim}{E}_o|^2 \tag{27}$$

and Eq. (26) becomes

$$I(x) = I_o \exp(-2 \frac{\omega}{c} kx). \tag{28}$$

Comparing this with the well-known Lambert's law expression which also describes the intensity of a wave propagating in an absorbing medium

$$I(x) = I_0 \exp(-\alpha x) \tag{29}$$

where α is the Bouger-Lambert absorption coefficient, it is easy to identify

$$\alpha = 2 \frac{\omega}{c} k = 4\pi\nu k. \tag{30}$$

In Eq. (30), ν is the frequency in units of cm^{-1}.

C. Anisotropic Media

One of the most comprehensive treatments of optical constants in aniso-tropic media is found in the classic optics text by Born and Wolf [7], and the following discussion utilizes the notation and arguments in Chapter 14 of their book. In an electrically anisotropic, transparent medium the proportionality between the electric displacement $\underset{\sim}{D}$ and the electric field $\underset{\sim}{E}$ is no longer given by Eq. (4b). Instead the scalar ε must be replaced by a dielectric tensor $\underset{\approx}{\varepsilon}$,

$$\underset{\sim}{D} = \underset{\approx}{\varepsilon} \underset{\sim}{E}, \tag{31}$$

where $\underset{\approx}{\varepsilon}$ is a symmetric, second rank tensor. The dielectric tensor is diagonal after transformation to principal dielectric axes and the relations implied by Eq. (31) simplify.

Crystals can be divided into three classes, dependent on their optical properties which in turn depend on the nature of the principal dielectric axes. Cubic crystals have three mutually orthogonal principal dielectric axes which may be freely chosen and along which the principal dielectric constants are equal. Trigonal, hexagonal, and tetragonal crystals are uniaxial crystals in which one principal dielectric axis coincides with the highest symmetry axis (the optic axis), while the other two mutually orthogonal axes may be chosen anywhere in the plane perpendicular to the high symmetry axis. Biaxial crystals are crystals of orthorhombic, monoclinic, or triclinic symmetry. None of the principal dielectric constants are necessarily equal to each other. The directions of the principal dielectric axes are fixed by symmetry considera-tions only in the case of an orthorhombic crystal, since in monoclinic or triclinic crystals the directions of the principal axes are frequency dependent (dispersion of the axes).

In the case of an absorbing anisotropic crystal, Eq. (4a) needs to be replaced by

$$\underset{\sim}{J} = \underset{\approx}{\sigma} \underset{\sim}{E} \tag{32}$$

where the conductivity tensor $\underset{\approx}{\sigma}$ can also be shown to be a symmetric second-rank tensor. The directions of the principal axes of the conductivity tensor necessarily coincide with those of the dielectric tensor only in crystals of orthorhombic symmetry or higher.

In an isotropic medium the vectors $\underset{\sim}{D}$ and $\underset{\sim}{E}$ are collinear, as are $\underset{\sim}{H}$ and $\underset{\sim}{B}$ which can be seen from Eqs. (4). As a consequence, the direction of the energy flow is coincident with the direction of the phase velocity, or wave normal velocity $\underset{\sim}{k}$. However, in an anisotropic medium, recognizing that the vectors $\underset{\sim}{D}$, $\underset{\sim}{E}$, $\underset{\sim}{B}$ and $\underset{\sim}{H}$ are proportional to exp $(i\underset{\sim}{k} \cdot \underset{\sim}{r} - i\omega t)$, Eqs. (3a) and (3c) can be written for the transparent medium,

$$\underset{\sim}{k} \times \underset{\sim}{E} = \frac{\omega}{c} \mu \underset{\sim}{H} \tag{33}$$

and

$$\underset{\sim}{k} \times \underset{\sim}{H} = -\frac{\omega}{c} \underset{\sim}{D} \tag{34}$$

where Eq. (4c) has been used to write $\underset{\sim}{B}$ in terms of $\underset{\sim}{H}$. Since $\underset{\sim}{H}$ is orthogonal to $\underset{\sim}{k}$, $\underset{\sim}{E}$, and $\underset{\sim}{D}$, the latter three quantities must be co-planar. However, $\underset{\sim}{D}$ is not necessarily collinear with $\underset{\sim}{E}$, as can be seen by taking the curl of Eq. (3a), using the vector identity for the curl of a curl, and substituting into (3c) to yield

$$\underset{\sim}{D} = \frac{c^2}{\omega^2} [k^2 \underset{\sim}{E} - \underset{\sim}{k}(\underset{\sim}{k} \cdot \underset{\sim}{E})]. \tag{35}$$

As a consequence, the direction of the energy flow which is given by the Poynting vector and is proportional to $\underset{\sim}{E} \times \underset{\sim}{H}$ does not necessarily coincide with the direction of the wave normal which is proportional to $\underset{\sim}{D} \times \underset{\sim}{H}$.

If Eq. (35) is simplified by transformation to principal dielectric axes, substitution of the constitutive equation (4b), and use of relations (12) and (19), the σ Cartesian component of $\underset{\sim}{E}$ may be written

$$E_\sigma = \frac{\frac{c^2}{\omega^2} k_\sigma (\underset{\sim}{k} \cdot \underset{\sim}{E})}{n^2 - \varepsilon_\sigma} . \tag{36}$$

Multiplying by k_σ, summing over σ, and rearranging the equation results in

$$\sum_\sigma \frac{(k_\sigma/k)^2}{n^2 - \varepsilon_\sigma} = \frac{1}{n^2} \ . \tag{37}$$

This is a quadratic equation in n which means that to every direction of wave normal propagation $\underset{\sim}{k}$ there corresponds two values of n. Recalling that the phase velocity, which is the velocity of energy transport in the wave, is equal to c/n, leads to the very important conclusion that there must be two phase velocities and therefore two monochromatic waves associated with each direction of wave normal propagation. It can be shown that the directions of $\underset{\sim}{D}$ associated with each of these waves are perpendicular to each other [7].

D. The Kramers-Kronig Relations

There exists a relationship between the dispersive (real) part of an optical constant and its absorptive (imaginary) part which is independent of any model. These relations are called the Kramers-Kronig relations [8,9] and are discussed by a number of authors [6,10,11]. These relationships depend on the linearity of the electric displacement $\underset{\sim}{D}$ and the polarization $\underset{\sim}{P}$ with the electric field $\underset{\sim}{E}$, the boundness of the complex optical constant, and a causal condition that there can be no polarization response of a medium before the electric field is turned on. The dispersion relation between the real and imaginary parts of the dielectric constant is

$$\varepsilon_1(\omega) = \varepsilon_0 + \frac{2}{\pi} P \int_0^\infty \frac{\omega' \varepsilon_2(\omega')}{\omega'^2 - \omega^2} \, d\omega' \tag{38a}$$

$$\varepsilon_2(\omega) = \frac{2\omega}{\pi} P \int_0^\infty \frac{\varepsilon_1(\omega')}{\omega^2 - \omega'^2} \, d\omega' \tag{38b}$$

where ε_0 is the static dielectric constant and P indicates that the principal value of the integral should be taken. The corresponding dispersion relation for the index of refraction is

$$n(\omega) - 1 = \frac{2}{\pi} P \int_0^\infty \frac{\omega' \ k(\omega')}{\omega'^2 - \omega^2} \, d\omega' \tag{39a}$$

$$k(\omega) = \frac{2\omega}{\pi} P \int_0^\infty \frac{n(\omega')}{\omega^2 - \omega'^2} \, d\omega' . \tag{39b}$$

These relations have occasionally been used to check the self-consistency of optical constant data [12]. In addition, a number of useful sum rules can be derived [13].

III. VIBRATIONAL DYNAMICS AND OPTICAL CONSTANTS

A. Vibrations in Crystals - Review

The vibrational dynamics of crystals have been described in depth and detail by a number of authors including Born and Huang [4], Maradudin et al. [5], and Weinreich [14], and will only briefly be summarized here. In the usual treatment, the kinetic energy of the crystal lattice is written in terms of time derivatives of Cartesian displacement coordinates

$$T = \frac{1}{2} \sum_{\tau,\ell,\sigma} M_\ell \, \dot{u}_\sigma(\tau\ell)^2 \tag{40}$$

where $u_\sigma(\tau\ell)$ is the σ Cartesian component of the displacement coordinate describing the motion of mass M_ℓ at site ℓ in the unit cell labeled by τ. In the harmonic approximation the vibrational potential energy expanded as a Taylor's series in the displacement coordinates is

$$V = \frac{1}{2} \sum_{\tau\tau'} \sum_{\ell\ell'} \sum_{\sigma\sigma'} \Phi_{\sigma\sigma'}(\tau\ell,\tau'\ell') \, u_\sigma(\tau\ell) \, u_{\sigma'}(\tau'\ell') \tag{41}$$

where

$$\Phi_{\sigma\sigma'}(\tau\ell,\tau'\ell') = [\partial^2\Phi/\partial u_\sigma(\tau\ell) \, \partial u_{\sigma'}(\tau'\ell')]_0 . \tag{42}$$

The equation of motion for $u_\sigma(\tau\ell)$ is

$$M_\ell \ddot{u}_\sigma(\tau\ell) + \sum_{\tau'\ell'\sigma'} \Phi_{\sigma\sigma'}(\tau\ell,\tau'\ell') \, u_{\sigma'}(\tau'\ell') = 0 \tag{43}$$

and upon substitution of the plane wave solution

$$u_\sigma(\tau\ell) = \frac{1}{\sqrt{M_\ell}} \, u_\sigma(\ell) \, \exp[i\underset{\sim}{k} \cdot \underset{\sim}{r}(\tau) - i\omega t] \tag{44}$$

with wave vector $\underset{\sim}{k}$ there results

$$-\omega^2 u_\sigma(\ell) + \sum_{\ell'\sigma'} u_{\sigma'}(\ell') \left\{ \frac{1}{\sqrt{M_\ell M_{\ell'}}} \sum_{\tau'} \Phi_{\sigma\sigma'}(\tau\ell,\tau'\ell') \right.$$

$$\left. \times \exp(i\underset{\sim}{k} \cdot [\underset{\sim}{r}(\tau') - \underset{\sim}{r}(\tau)]) \right\} = 0. \tag{45}$$

By invoking the translational invariance of the lattice, the quantity in the curly brackets can be seen to depend only on the wave vector and the site indices, but not on the unit cell indices. The quantity is identified as the dynamical matrix of the crystal and the equation of motion becomes

$$- \omega^2 u_\sigma(\ell) + \sum_{\ell'\sigma'} D_{\sigma\sigma'}(\underset{\sim}{k};\ell,\ell') u_{\sigma'}(\ell') = 0. \tag{46}$$

The condition that this set of equations has non-trivial solutions is that the determinant of the associated coefficient matrix vanish, i.e.,

$$|D_{\sigma\sigma'}(\underset{\sim}{k};\ell,\ell') - \omega^2\delta_{\sigma\sigma'} \delta_{\ell\ell'}| = 0. \tag{47}$$

In a crystal with n atoms in a primitive unit cell there will be 3n solutions for ω^2 which depend on the value of $\underset{\sim}{k}$; these are designated $\omega^2(\underset{\sim}{k}j)$ and correspond to the normal mode or phonon frequencies where j is a branch index labeling one of the 3n solutions. A phonon is quantized lattice vibrational energy which propagates as a traveling wave through the crystal. The distribution of normal vibrational modes in a crystal of N unit cells consists of 3n phonon branches j with N values of $\underset{\sim}{k}$ on each branch. The dependence of the normal mode frequency on the wave vector is the dispersion of the mode. A phonon coordinate is

$$Q(\underset{\sim}{k}j) = \frac{1}{\sqrt{N}} \sum_{\tau\ell\sigma} \sqrt{M_\ell}\, u_\sigma(\tau\ell)\, e^*_\sigma(-\underset{\sim}{k}j;\ell) \exp[-i\underset{\sim}{k} \cdot r(\underset{\sim}{\tau})] \tag{48}$$

where $e_\sigma(\underset{\sim}{k}j;\ell)$ is the normalized vibrational amplitude which results from the substitution of the root $\omega^2(\underset{\sim}{k}j)$ into the set of equations (46), i.e., an element of the normalized eigenvector of the dynamical matrix of the crystal.

In crystals containing atoms bound by covalent forces into molecules or polyatomic ions, the branches originating in intramolecular motion are higher in frequency than are branches involving librational or translational motion of the molecules or ions moving as a rigid body. These higher frequency branches are the internal optic mode branches in which the molecular motions are, to a large extent, dynamically decoupled from the lower frequency external optic modes or lattice modes. There is much experimental evidence

that the intramolecular contribution constitutes the major portion of the
internal optic mode vibrational potential energy. In molecular crystals the
gas phase frequencies of the constituent molecules show only small frequency
shifts upon condensation to the crystalline phase [15]. In addition, the
internal mode frequencies of polyatomic ions usually show little variation
between different crystal lattices [16]. For these branches the phonon
coordinate (Eq. (48)) might be written to a reasonable degree of approximation as

$$Q(\underset{\sim}{k}j) = \frac{1}{\sqrt{N}} \sum_{\tau} q(\tau j) \exp[-i\underset{\sim}{k} \cdot \underset{\sim}{r}(\tau)] \qquad (49)$$

in a crystal containing only one molecular unit in the primitive cell. The
coordinate $q(\tau j)$ is a molecular normal coordinate corresponding to an intramolecular vibrational mode labeled by j in the τth unit cell. In the case of a
multiply occupied unit cell, $q(\tau j)$ is replaced by an appropriately symmetrized
linear combination of molecular normal coordinates. However, the important
point is the picture suggested by Eq. (49). In many crystals an internal
optic mode may be thought of as a collective motion of all the translationally
equivalent molecular units in the crystal vibrating in one molecular normal
mode, with a phase relation between adjacent unit cells imposed by the
requirements of translational symmetry and characterized by the wave vector
$\underset{\sim}{k}$. Even further simplification is possible since almost all spectroscopic
studies involve the interaction of an electromagnetic wave with an internal
optic mode in the vicinity of the Brillouin zone center ($\underset{\sim}{k} \cong 0$) due to the
conservation of wave vector selection rules. Therefore Eq. (49) becomes

$$Q(\underset{\sim}{0}j) = \frac{1}{\sqrt{N}} \sum_{\tau} q(\tau j) \qquad (50)$$

in the case of a simple molecular crystal and the picture is one of all
molecular units vibrating essentially in phase. The notation $\underset{\sim}{0}$ indicates the
limit of very small but finite $\underset{\sim}{k}$ so that the direction of $\underset{\sim}{k}$ is preserved, and
will be used to explicitly designate Brillouin zone center (BZC) quantities.
Because of the retention of molecular characteristics in the crystalline phase
and the dominance of intramolecular motion in the dynamics of most internal
optic modes, it might be expected that an appropriate analysis of the internal
mode frequencies might yield molecular parameters in these crystals. A
formalism which adopts this viewpoint (the Molecular Dipole Model) has been
developed and will be described in a later section.

B. Harmonic Oscillator Model

Undoubtedly the model most widely used to describe optical constants is based on a classical dispersion model first given by Kettler [17] for the refractive index and subsequently modified by Helmholtz [18] to include a damping term in order to describe absorption effects. This was used by Rubens in early infrared reflectivity studies of quartz, sodium chloride, potassium chloride, and fluorite [19,20]. Many versions of this dispersion formula have been used since then by numerous investigators in a wide variety of systems. One familiar form of the classical dispersion model writes the σ component of the complex dielectric constant following Seitz [21] as

$$\hat{\varepsilon}_{\sigma\sigma}(\omega) = \varepsilon_{\sigma\sigma}(\infty) + \sum_j \frac{\Omega_j^2 S_{j\sigma}}{\Omega_j^2 - \omega^2 + i\gamma_j\omega} \tag{51}$$

where the sum is over the infrared active modes. At this level of approximation the resonance frequency Ω_j is taken as a transverse optic mode frequency at the Brillouin zone center. The strength factor $S_{j\sigma}$ is a measure of the interaction of the jth vibrational mode with a light wave polarized in the σ direction while γ_j is the damping constant.

The harmonic oscillator dispersion equation can be re-derived in a form in which its utility to a molecular spectroscopist interested in obtaining molecular parameters in crystals is more apparent [3]. The description of the optical constants of a crystal in terms of the underlying vibrational dynamics is obtained by solving a phonon coordinate equation of motion which includes the interaction of the phonon mode with the electric field of an electromagnetic wave. The interaction potential energy between a unit cell with vibrationally induced dipole moment $\underset{\sim}{\mu}$ and the electric field $\underset{\sim}{E}$ of the light wave is

$$V_{int} = -\underset{\sim}{\mu}^\dagger \underset{\sim}{E}. \tag{52}$$

Here the vibrationally induced dipole moment originates in the dynamic distortion of the charge distribution resulting from the vibrational motion of the nuclei. The translational symmetry-based selection rule (conservation of wave-vector) requires that only Brillouin zone center (BZC) phonon modes need to be considered; consequently, the interaction of a single unit cell with the electromagnetic wave is examined. This treatment neglects local field effects which will be discussed in a later section, and for simplicity considers only an isotropic, uniaxial, or orthorhombic crystal.

In the principal dielectric axes system of the crystal, the σ component of the vibrationally induced dipole moment, μ_σ, may be expanded as a function of the BZC phonon coordinates Q_j,

$$\mu_\sigma = \mu_\sigma^{(o)} + \sum_j \mu_{j\sigma} Q_j + \ldots \tag{53}$$

where $\mu_{j\sigma}$ is the σ component of the dipole moment derivative with respect to Q_j, $\mu_{j\sigma} \equiv (\partial\mu_\sigma/\partial Q_j)_o$, and $\mu_\sigma^{(o)}$ is the permanent dipole moment in the σ direction. The higher order terms in (53) result from electrical anharmonicity and contribute to the breakdown of the harmonic oscillator selection rules. The vibrational kinetic and potential energy contributions with the crystal in an external electric field in the σ direction are

$$T = \frac{1}{2} \sum_j \dot{Q}_j^2 \tag{54a}$$

$$V = \frac{1}{2} \sum_j \Omega_j^2 Q_j^2 - \sum_j \sum_\sigma \mu_{j\sigma} Q_j E_\sigma \tag{54b}$$

where the interaction of the electric field with the permanent dipole moment components has been neglected without loss of generality. As before, the frequencies Ω_j are the BZC transverse optic mode frequencies. The resulting equation of motion for the coordinate Q_j, assuming a simple harmonic solution $Q_j = Q_j^{(o)}\exp(-i\omega t)$ and a non-vanishing dipole moment derivative for the σ component is

$$\ddot{Q}_j + \Omega_j^2 Q_j - \mu_{j\sigma}E_\sigma = 0 \tag{55}$$

which results in

$$Q_j = \mu_{j\sigma}E_\sigma/(\Omega_j^2 - \omega^2). \tag{56}$$

The electric polarization $\underset{\sim}{P}$ may be written as the induced dipole moment of the unit cell divided by the volume of the cell. There are two contributions to the induced moment; the vibrationally induced moment μ and the electronic polarization moment, $\underset{\approx}{\alpha}\underset{\sim}{E}$, where $\underset{\approx}{\alpha}$ is the electronic polarizability tensor. The σ component of the electric polarization is therefore

$$P_\sigma = \frac{1}{v} [\alpha_{\sigma\sigma} E_\sigma + \mu_\sigma], \tag{57}$$

noting that in the principal axes system of the crystals considered here, α is diagonal. After substituting Eq. (53) for μ_σ and using Eq. (56),

$$P_\sigma = \frac{1}{v} \alpha_{\sigma\sigma} E_\sigma + \frac{1}{v} \sum_j \mu_{j\sigma}^2 E_\sigma/(\Omega_j^2 - \omega^2). \tag{58}$$

From the definition of the electric displacement $\underset{\sim}{D}$ as $\underset{\sim}{D} = \underset{\sim}{E} + 4\pi\underset{\sim}{P}$ and Eq. (4b), the σ component of the dielectric tensor is

$$\varepsilon_{\sigma\sigma}(\omega) = 1 + \frac{4\pi}{v} \alpha_{\sigma\sigma} + \frac{4\pi}{v} \sum_j \mu_{j\sigma}^2/(\Omega_j^2 - \omega^2). \tag{59a}$$

Naturally in these crystal systems ε is also diagonal. The first two terms in (59a) are usually written as the high frequency contribution to the dielectric tensor originating in electronic polarization effects, $\varepsilon_{\sigma\sigma}(\infty)$, so that

$$\varepsilon_{\sigma\sigma}(\omega) = \varepsilon_{\sigma\sigma}(\infty) + \frac{4\pi}{v} \sum_j \mu_{j\sigma}^2/(\Omega_j^2 - \omega^2). \tag{59b}$$

However, the dielectric constant described by (59b) is unsatisfactory in several respects. It is a real quantity rather than a complex quantity, and the lack of an imaginary term implies that there is no absorption of electromagnetic radiation. Furthermore, there is a singularity at $\omega = \Omega_j$.

The presence of anharmonic interactions between the mode j and the other modes dissipates the energy absorbed by the mode j in a fundamental transition. This effect can be simulated through the addition of a phenomenological damping term which is proportional to the velocity, $\gamma_j \dot{Q}_j$, in the equation of motion (55). Lax has shown that in a real crystal γ_j must be frequency-dependent and have both a real and an imaginary part which obey a Kramers-Kronig relation [22]. Again assuming a harmonic time dependence of Q_j leads to

$$\hat{\varepsilon}_{\sigma\sigma}(\omega) = \varepsilon_{\sigma\sigma}(\infty) + \frac{4\pi}{v} \sum_j \mu_{j\sigma}^2/(\Omega_j^2 - \omega^2 - i\gamma_j\omega). \tag{60}$$

This result has been obtained by assuming that the interaction energy Eq. (52) is given by the inner product of the unit cell dipole moment $\underset{\sim}{\mu}$ and external field, that is, the electric field of the light wave. However, the external field will also polarize the crystal, and the electric field resulting from the polarization moment will also interact with the vibrationally induced moment. Therefore Eq. (52) should contain an additional factor to account for this effective field effect. In lieu of that, Eq. (60) is usually written in a manner formally identical with Eq. (51) by combining the effective field factor, the square of the dipole moment derivative, and $\frac{4\pi}{v}$ as the strength factor $S_{j\sigma}$ multiplied by Ω_j^2.

If the phonons are well-separated in frequency, then in the vicinity of a mode of interest, say Ω_i, Eq. (51) can be written

$$\hat{\varepsilon}_{\sigma\sigma}(\omega) = \varepsilon_{\sigma\sigma}(\infty) + \sum_{j\neq i} S_{j\sigma}\Omega_j^2/(\Omega_j^2 - \omega^2 - i\gamma_j\omega)$$

$$+ S_{i\sigma}\Omega_i^2/(\Omega_i^2 - \omega^2 - i\gamma_i\omega). \tag{61a}$$

Since all of the terms except the last are slowly varying functions of frequency, these are usually set equal to a real constant $\varepsilon'_{\sigma\sigma}$ (neglecting the small contributions from the imaginary parts) and the equation as it is usually applied becomes

$$\hat{\varepsilon}_{\sigma\sigma}(\omega) = \varepsilon'_{\sigma\sigma} + S_{i\sigma}\Omega_i^2/(\Omega_i^2 - \omega^2 - i\gamma_i\omega). \tag{61b}$$

The quantity $\varepsilon'_{\sigma\sigma}$ consists of an electronic polarization part plus contributions from all other infrared active modes which are still able to follow the exciting field, that is, modes whose frequencies are higher than Ω_i,

$$\varepsilon'_{\sigma\sigma} \cong \varepsilon_{\sigma\sigma}(\infty) + \sum_j S_{j\sigma}, \tag{62}$$

for $\Omega_j > \Omega_i$. The dielectric constant parameters $\varepsilon'_{\sigma\sigma}$, $S_{j\sigma}$, Ω_j, and γ_j are not true molecular parameters, although they can be related to molecular parameters through several models which will be discussed shortly.

The behavior of the real and imaginary parts of the dielectric constant are illustrated in Fig. 1a for the damped harmonic oscillator model of Eq. (61b). The parameters used in the calculation are $S_{i\sigma} = 0.373$, $\Omega_i = 1350$ cm^{-1}, $\gamma_i = 20$ cm^{-1}, and $\varepsilon'_{\sigma\sigma} = 2$. By way of comparison, the real and imaginary parts of the refractive index are shown in Fig. 1b as calculated using Eq. (20) and the damped harmonic oscillator parameters of Fig. 1a. From Eq. (61b) it is apparent that the transverse optic modes appear as poles of the complex dielectric function while the longitudinal modes occur as zeros. Within the approximation $\gamma_i/\Omega_i \ll 1$, the maximum of ε_2 occurs at Ω_i, which is the BZC transverse optic mode frequency.

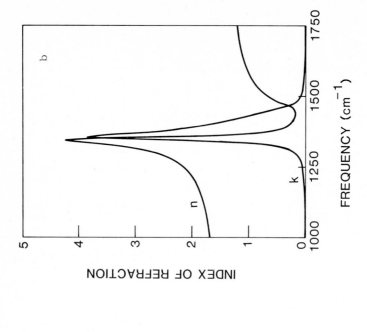

FIG. 1. (a) The real and imaginary parts of the complex dielectric constant based on the damped harmonic oscillator model. The dielectric constant parameters are: resonance frequency Ω_i = 1350 cm^{-1}, damping constant γ_i = 20 cm^{-1}, strength factor $S_{i\sigma}$ = 0.373, and $\varepsilon'_{\sigma\sigma}$ = 2. (b) The real and imaginary parts of the index of refraction calculated with the same parameters as in (a).

C. Green's Function Model

Absorption effects in the optical constants originate in the anharmonic coupling of the phonon of interest to other phonons in the crystal. In the family of classical dispersion models based on the harmonic oscillator, inclusion of absorption effects is accomplished through the addition of a phenomenological damping term to the equation of motion. An alternative technique of calculating optical constants in an anharmonic crystal uses the methods of quantum field theory to treat the anharmonic interactions between the normal modes as perturbations in a many body system. One commonly used formalism obtains thermodynamic Green's functions [23,24] through a perturbation procedure described by Cowley [25]. The dielectric properties of an anharmonic crystal have been examined with Green's function methods by Maradudin and Wallis [26] and others [27]. In a convenient notation the σ component of the dielectric tensor in principal dielectric axes resulting from the one phonon contributions is [25]

$$\varepsilon_{\sigma\sigma}(\omega) = 1 + \frac{1}{Nvh} \sum_j \frac{8\pi\omega(0j) \, M_{j\sigma} M_{j\sigma}}{\omega^2(0j) - \omega^2 + 2\omega(0j)[\Delta(0j,\omega,T) - i\Gamma(0j,\omega,T)]} \qquad (63)$$

where $M_{j\sigma}$ is the linear coefficient of the dipole moment expansion in the phonon coordinates. The quantity $\Delta(0j,\omega,T)$ is the shift term of the transverse optic mode with frequency $\omega(0j)$ resulting from the anharmonic interactions and is both temperature and frequency dependent. The inverse of the lifetime of the mode j is $\Gamma(0j,\omega,T)$ and in Eq. (63) describes the absorption of electromagnetic radiation by the crystal. This quantity written in terms of the anharmonic coupling coefficients to the lowest order is

$$\Gamma(0j,\omega,T) = \frac{18\pi}{\hbar^2} \sum_{j'} \sum_{j''} \sum_{\underset{\sim}{k}} |V^{(3)} (0j,kj',-kj'')|^2$$

$$\times \{[n(kj') + n(kj'') + 1] \, \delta(\omega - \omega(kj') - \omega(kj''))$$

$$+ [n(kj') - n(kj'')] [\delta(\omega + \omega(kj') - \omega(kj'')) - \delta(\omega - \omega(kj') + \omega(kj''))]\}$$

$$\qquad (64)$$

where $n(kj')$ is the Bose-Einstein factor

$$n(kj') = 1/\{\exp[\hbar\omega(kj')/kT] - 1\}. \qquad (65)$$

The damping constant of the mode j is seen to result from the anharmonic interactions with all other phonons in the crystal. Therefore $\Gamma(0j,w,T)$ is not a molecular parameter even though one might speak of the damping constant of an internal optic mode originating in a particular intramolecular vibrational motion. That same intramolecular motion might result in a very different value of Γ for the corresponding internal optic mode in another crystal lattice, since the underlying anharmonic interactions might be very different.

It is easily seen that the form of the dielectric constant (63) is equivalent to that of the classical dispersion oscillator model (51), providing the following identifications are made [28]:

$$(8\pi/hv)\ w(0j)M_{j\sigma}\ M_{j\sigma} = S_{j\sigma}\Omega_j^2 \tag{66a}$$

$$w^2(0j) + 2w(0j)\ \Delta(0j,w,T) = \Omega_j^2 \tag{66b}$$

$$2w(0j)\ \Gamma(0j,w,T) = \gamma_j w \tag{66c}$$

This equivalence requires that the resonance frequencies and damping parameters in the classical oscillator model are both temperature and frequency dependent.

D.   Pole and Zero Model

Since significant differences may exist between the longitudinal and transverse damping of an optic mode [25,29,30], an alternate description of the dielectric function can be given in which these quantities are assumed to be independent. In its most general form $\hat{\varepsilon}(w)$ is written as a function of poles and zeros in the complex frequency plane [31], where the transverse optic modes appear as poles and the longitudinal modes appear as zeros. The reflectivity data are then fit by adjusting the locations of the poles and zeros, rather than by fitting band parameters as in a classical dispersion analysis. In a convenient notation [32,33]

$$\hat{\varepsilon}_{\sigma\sigma}(w) = \varepsilon_{\sigma\sigma}(\infty)\ \Pi_j\ \frac{\Omega_{jL}^2 - w^2 + i\gamma_{jL}w}{\Omega_{jT}^2 - w^2 + i\gamma_{jT}w} \tag{67}$$

where the product is over all infrared active modes polarized in the σ direction. It should be noted that a resonance frequency is the modulus of the complex pole and not its real part. Four independent parameters are required to adequately describe the behavior of $\hat{\varepsilon}(\omega)$ in the vicinity of an isolated phonon mode, hence this model is also called the four parameter semi-quantum (FPSQ) model [28,32]. In the case of an isolated phonon mode, the classical dispersion oscillator and the pole and zero model are equivalent providing

$$S_{j\sigma} = \varepsilon_{\sigma\sigma}(\infty)\left[\frac{\Omega_{jL}{}^2 - \Omega_{jT}{}^2}{\Omega_{jT}{}^2}\right] \tag{68a}$$

with

$$\gamma_{jL} = \gamma_{jT} \tag{68b}$$

and

$$\Omega_j = \Omega_{jT}. \tag{68c}$$

The relationship between the pole and zero model and the form of $\hat{\varepsilon}(\omega)$ resulting from a Green's function formalism has been discussed [29,32]; the dielectric function can be written

$$\hat{\varepsilon}_{\sigma\sigma}(\omega) = \varepsilon_{\sigma\sigma}(\infty) \prod_j \frac{\omega_L{}^2(\underset{\sim}{Q}j) - \omega^2 + 2\omega_L(\underset{\sim}{Q}j)[\Delta\omega_L(\underset{\sim}{Q}j,T) + i\omega\bar{\gamma}_{jL}(\omega,T)]}{\omega_T{}^2(\underset{\sim}{Q}j) - \omega^2 + 2\omega_T(\underset{\sim}{Q}j)[\Delta\omega_T(\underset{\sim}{Q}j,T) + i\omega\bar{\gamma}_{jT}(\omega,T)]} \tag{69}$$

where the identification of $\omega_T(\underset{\sim}{Q}j)$ with Ω_{jT} in Eq. (67) is obvious. The quantity $\bar{\gamma}_j(\omega,T)$ is a slowly varying function of frequency in crystals with a regular distribution of branches and is related to the Green's function damping function (64) through

$$\bar{\gamma}_j(\omega,T) = \frac{1}{\omega} \Gamma(\underset{\sim}{Q}j,\omega,T). \tag{70}$$

IV. THEORY OF THE REFLECTION COEFFICIENT AND THE REFLECTIVITY

A. Overview

The process of molecular parameter deduction from reflectivity data is summarized in Fig. 2. From a set of measured reflectivities the dielectric constant $\hat{\varepsilon}$, or alternatively the refractive index $\hat{n}$, is calculated. For convenience, the analysis of the reflectivity data will be discussed in terms of

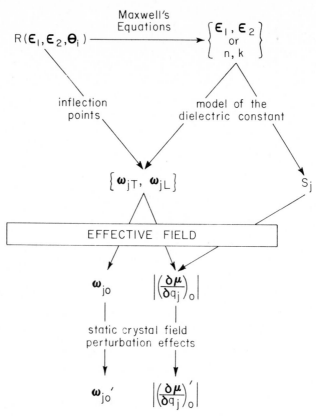

FIG. 2. The calculation of molecular parameters from infrared reflectivity data.

the dielectric constant, although it should be understood that the refractive index could as easily be used. The relationship between the reflectivity and the dielectric constant follows simply from a consideration of Maxwell's equations and the constitutive equations at the interface between two optically different media at which reflection of a plane electromagnetic wave occurs. No model is assumed. In the next step of the reduction, the dielectric constant data in the vicinity of an infrared active normal mode labeled by j are analyzed to give w_{jT}, the transverse optic mode frequency; w_{jL}, the longitudinal optic mode frequency; and S_j, the strength factor. This step requires a specific model of the dielectric constant in terms of the underlying vibrational dynamics. Usually the equation of motion of the jth normal mode is combined with an expression for the electric polarization written as a function of the jth normal coordinate, and the resulting expression yields the dielectric constant in terms of the normal mode vibrational parameters w_{jT}, w_{jL}, and S_j as was illustrated in Sec. III-B. It should be noted that these three parameters are not independent; the damped harmonic oscillator model uses ω_{jT} and S_j, while the pole and zero model utilizes ω_{jT} and ω_{jL}. Up to this point it has been necessary to use different notations for the BZC transverse optic mode frequency as derived from the various models for purposes of clarity and comparison. In this section that frequency will simply be written as ω_{jT} except in the discussion of classical dispersion analysis where the damped harmonic oscillator is explicitly assumed.

The resulting transverse and longitudinal mode frequencies of the jth mode may be solved for the molecular parameters ω_{jo} and $|(\partial\mu_\sigma/\partial q_j)_o|$, provided that the effective field factor can be calculated. The static crystal field frequency ω_{jo} is the molecular normal mode frequency of an individual molecule or polyatomic ion in the absence of dynamical coupling effects but in the presence of static crystal field interactions. The quantity $|(\partial\mu_\sigma/\partial q_j)_o|$ is a molecular dipole moment derivative and alternatively may be calculated directly from the strength factor S_j as indicated in Fig. 2.

A final step in the reduction is shown since the molecular parameters ω_{jo} and $|(\partial\mu_\sigma/\partial q_j)_o|$ contain perturbation effects from the static potential energy environment of the crystal. Therefore these quantities are not directly comparable to analogous gas phase values. However, correction for these perturbation effects is almost never attempted due to the complexity of the problem.

An additional pathway is suggested in Fig. 2 by which the vibrational parameters ω_{jT} and ω_{jL} can be directly obtained from the reflectivity bandshape without recourse to a model of the dielectric constant. It is easily shown that the inflection point on the low-frequency side of a reflectivity band occurs at a frequency close to the transverse optic mode frequency, while the high-frequency inflection point is in the vicinity of the longitudinal optic mode frequency. These inflection point frequencies are occasionally used as estimates of ω_{jT} and ω_{jL} in subsequent calculations.

Of the numerous methods for obtaining optical constants in crystals, three of the more commonly used will be very briefly described. These are based on the measurement and subsequent analysis of the reflectivity spectrum and may be classed as single frequency measurements of the specular reflectivity, the Robinson-Price technique or multiple frequency measurements, and classical dispersion analysis. Additional methods such as photoacoustic spectroscopy, calorimetric techniques, and extensions of reflection spectroscopy such as attenuated total reflection (ATR) spectroscopy will not be discussed here.

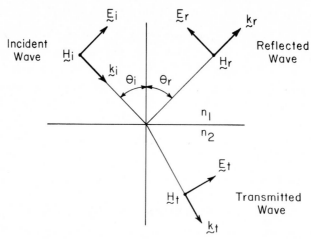

FIG. 3. The fields in the incident, transmitted, and reflected waves for a transverse magnetic (TM) reflection measurement at the interface between air ($n_1 = 1$) and an isotropic, homogeneous, absorbing medium.

B.   Reflectivity from an Absorbing Medium

The experimental geometry of a reflection measurement is described in Fig. 3 [34]. A plane, monochromatic, electromagnetic wave traveling through a nonabsorbing medium with index of refraction n_1 is incident on the boundary of an absorbing sample medium characterized by complex refractive index $\hat{n}_2$. Both media are assumed to be optically isotropic, homogeneous, and non-magnetic. Because of the latter condition, the magnetic permeabilities in each medium may be taken as unity. Part of the incident wave is reflected from the boundary at an angle of reflection θ_r, and part of the wave is transmitted through the sample medium at an angle θ_t. All angles including the angle of incidence θ_i are measured from the normal to the boundary. The electric field of the incident wave is

$$\underset{\sim}{E}_{inc}(\underset{\sim}{r},t) = \underset{\sim}{E}^o_{inc} \exp[i\underset{\sim}{k}_{inc} \cdot \underset{\sim}{r} - i\omega t] \qquad (71)$$

where $\underset{\sim}{k}_{inc}$ is the wave vector which describes the direction of incident wave propagation, $\underset{\sim}{E}^o_{inc}$ is the vector amplitude of the wave, and ω is the (circular) frequency of the wave. The magnitude of the incident wave vector may be written

$$k_{inc} = \frac{n_1 \omega}{c} \qquad (72)$$

with a similar equation for the magnitude of the reflected wave vector,

$$k_r = \frac{n_1 \omega}{c} . \tag{73}$$

However the transmitted wave vector is a complex quantity,

$$\hat{k}_t = \frac{\hat{n}_2 \omega}{c} . \tag{74}$$

Equations similar to Eq. (71) may be written for the electric fields of the transmitted and reflected wave.

In Fig. 3, $\underset{\sim}{H}$ is the magnetic vector and is related to the wave vector and the electric vector through

$$\underset{\sim}{H} = \frac{c}{\omega} \underset{\sim}{k} \times \underset{\sim}{E}. \tag{75}$$

The angle of incidence and the angle of reflection are easily shown to be equal through the application of appropriate boundary conditions at the interface between the two media. Similarly the relationship between the angle of incidence and the angle of the transmitted wave is given by Snell's law,

$$n_1 \sin \theta_i = \hat{n}_2 \sin \theta_t . \tag{76}$$

Note that both $\hat{n}_2$ and θ_t are complex.

Two independent reflection geometries can be described, dependent on the polarization of the incident electromagnetic radiation. If the electric vector of the incident radiation lies in the plane of incidence defined by the propagation direction of the incident wave and the normal to the boundary, the wave is a TM (transverse magnetic) wave or, equivalently, a plane polarized wave. This is the reflection geometry indicated in Fig. 3. The Fresnel reflection coefficient, defined as the ratio of the reflected wave amplitude to the incident wave amplitude, is

$$\hat{r}_{||} = \frac{\hat{n}_2 \cos \theta_i - n_1 \cos \theta_t}{\hat{n}_2 \cos \theta_i + n_1 \cos \theta_t} = \frac{\tan (\theta_i - \theta_t)}{\tan (\theta_i + \theta_t)} . \tag{77a}$$

The second form on the right hand side has been obtained using Snell's law. The Fresnel reflection coefficient is complex and may be also written

$$\hat{r}_{||} = \rho_{||} e^{i\phi_{||}} \tag{77b}$$

where $\phi_{||}$ is the phase change of the incident wave upon reflection at the boundary and $\rho_{||}$ is the magnitude of $\hat{r}_{||}$.

The reflectivity is

$$R_{||} = |\hat{r}_{||}|^2 = \rho_{||}^2 . \tag{78}$$

Note from Eqs. (77a) and (77b) that the reflectivity, which is the physical quantity measured in an experiment, may be written as an explicit function of the angle of incidence, the refractive index of the optically transparent medium through which the wave is incident on the boundary, and the optical constants of the second (sample) medium; $R_{||} = R_{||}(n_1,\theta_i,n_2,k_2)$. A similar set of equations can be written for the case of an incident wave with the electric vector perpendicular to the plane of incidence. This wave is a TE (transverse electric) wave or, equivalently, a perpendicularly polarized wave. The Fresnel reflection coefficient is

$$\hat{r}_{\perp} = \frac{n_1\cos\theta_i - \hat{n}_2\cos\theta_t}{n_1\cos\theta_i + \hat{n}_2\cos\theta_t} = -\frac{\sin(\theta_i - \theta_t)}{\sin(\theta_i + \theta_t)} \tag{79a}$$

and in the same manner as before

$$\hat{r}_{\perp} = \rho_{\perp} e^{i\phi_{\perp}} \tag{79b}$$

where $\phi_{\perp}$ and $\rho_{\perp}$ are the phase shift of the wave and the magnitude of $\hat{r}_{\perp}$, respectively. The reflectivity is

$$R_{\perp} = |r_{\perp}|^2 = \rho_{\perp}^2 \tag{80}$$

and can also be written as $R_{\perp} = R_{\perp}(n_1,\theta_i,n_2,k_2)$.

1.   Single Frequency Methods

It is apparent from Eqs. (77) through (80) that there are a number of methods which yield optical constants from reflectivity measurements at a single frequency. These require measuring $R_{\perp}$, $R_{||}$, or unpolarized reflectivity at two (or more) angles of incidence. Alternatively both $R_{\perp}$ and $R_{||}$ may be measured at one angle of incidence. An interesting variation is to measure either $R_\perp$ or $R_{||}$ at some angle of incidence and the Brewster angle, the angle for which $R_{||}$ is a minimum. Humphreys-Owen lists nine different pair combinations of reflectivity and Brewster angle measurements which are sufficient to determine the optical constants at a given frequency [35]. An analysis of the errors inherent in various techniques of reflectivity measurement and their effect on the calculated optical constants has been given by Humphreys-Owen and by Hunter [36].

Another widely used single frequency method is ellipsometry, which derives its name from the fact that plane polarized light undergoes a phase shift upon reflection from an absorbing medium and becomes elliptically polarized. A measurement of the phase and amplitude allows the calculation of the real and imaginary parts of the refractive index. Alternatively the thickness and refractive index of a transparent film on a substrate can be determined by such a measurement, although only the case of reflection from an absorbing medium will be discussed here. The basic equations were first derived by Drude [37] and later given by Winterbottom [38], among others. An excellent, if somewhat dated, overview is given by McCrackin et al. [39], which examines the theoretical basis, methods of measurement and computational problems of this technique.

Since the absolute change of amplitude and phase are difficult to measure accurately, the ellipsometric equations are derived for the relative shifts of the phase and amplitude between the parallel and perpendicular components. If Eq. (77b) is divided by Eq. (79b), there results

$$\frac{\hat{r}_{||}}{\hat{r}_\perp} = \hat{\rho} = \frac{\rho_{||}}{\rho_\perp} e^{i(\phi_{||} - \phi_\perp)} = \tan \psi e^{i\Delta} \qquad (81)$$

where Δ is the relative phase shift and $\tan \psi$ is the relative change in reflection amplitude. The complex index of refraction can be written

$$\hat{n}_2 = n_1 \tan \theta_i [1 - 4\hat{\rho}\sin^2 \theta_i/(\hat{\rho} + 1)^2]. \qquad (82)$$

Therefore a measurement of $\tan \psi$ and Δ at each frequency is sufficient to determine the real and imaginary parts of the refractive index.

2.   Robinson-Price Method

If the reflection coefficient at normal incidence is written as

$$\hat{r} = \rho e^{i\phi} = \frac{n + ik - 1}{n + ik + 1} \qquad (83)$$

it is possible to obtain the optical constants n and k of the reflecting surface by applying a Kramers-Kronig relationship to the amplitude and phase angle [40,41] and write

$$\phi(\omega) = \frac{2\omega}{\pi} \int_0^\infty \frac{\ln \rho(\omega')}{\omega'^2 - \omega^2} \, d\omega'. \tag{84}$$

It then follows that

$$n = \frac{1 - \rho^2}{1 + \rho^2 - 2\rho\cos\phi} \tag{85a}$$

and

$$k = \frac{2\rho}{1 + \rho^2 - 2\rho\cos\phi}. \tag{85b}$$

Since the values of $\rho(\omega)$ are not known for the entire range of frequencies required to perform the integration, the success of this method requires that techniques be developed for approximating the integral outside of the range for which experimental data are available. This has been discussed by a number of authors [42-44].

3.   Classical Dispersion Analysis

One of the more widely used methods of determining optical constants from reflectivity measurements is classical dispersion or classical oscillator analysis. The reflectivity at normal incidence is written in the usual manner as a function of the optical constants,

$$R = |\hat{r}|^2 = |\frac{\hat{\varepsilon}_{\sigma\sigma} - 1}{\hat{\varepsilon}_{\sigma\sigma} + 1}|^2, \tag{86}$$

and these in turn are written in terms of a damped harmonic oscillator model (or an equivalent form). Therefore the reflectivity in the vicinity of normal mode j is a function of the damped harmonic oscillator parameters, i.e., $R = R(\varepsilon_{\sigma\sigma}(\infty), S_{j\sigma}, \Omega_j, \gamma_j)$. Usually $\varepsilon_{\sigma\sigma}(\infty)$ is determined from the optical refractive index or a measurement of the reflectivity at a much higher frequency than Ω_j so that only $S_{j\sigma}$, Ω_j, and γ_j are adjusted to obtain the best fit to experimental reflectivity data. Classical dispersion theory has been reviewed by Wood [45] and techniques of fitting have been described by Spitzer et al. [46], among others. A flavor of the numerous applications of this method can be found in the studies of $BaTiO_3$ [47], $NaNO_3$ [48], and ZnO [49]. Recently this method of analysis has been modified by the substitution of a more realistic model of the dielectric constant, such as that predicted by a pole and zero model or by a Green's function formalism.

C. Anisotropic Media - Uniaxial Crystals

The problem of the reflectivity of an absorbing uniaxial crystal has been discussed by Mosteller and Wooten [50] and by Flournoy and Schaffers [51], who considered the case in which the optic axis is perpendicular to the reflecting crystal face. A calculation of the reflectivity in an absorbing uniaxial crystal in which the optic axis has a general orientation with respect to the plane of incidence and the reflecting face has been given by Berek [52] and later by Damany and Uzan [53]. The general solution in the latter two references is tedious and explicit equations are given only for a few selected geometries. The reflectivity equations for the general case have been re-derived in a more convenient form [54] in the context of a reflectivity study of sodium nitrate.

The experimental geometry of a general reflection measurement for an absorbing uniaxial crystal is shown in Fig. 4. The wave is propagating in the xz plane (plane of incidence) and is incident on the crystal face at an angle of incidence θ_i as measured from the z axis which is normal to the crystal face. The orientation of the optic axis with respect to the laboratory axes is given by the angles α and β. The two waves which can propagate in a uniaxial crystal as described in Sec. II-C are the ordinary wave and the extraordinary wave. The fields in an ordinary wave resulting from incident TE electromagnetic are described in Fig. 5. Basically the ordinary wave behaves as if the medium were isotropic. The electric field $\underset{\sim}{E}^o$ and the electric displacement $\underset{\sim}{D}^o$ are collinear and are perpendicular both to the direction of propagation given by wave vector $\underset{\sim}{k}^o$ and the principal plane defined by the optic axis $\underset{\sim}{A}$ and $\underset{\sim}{k}^o$. The direction of propagation is identical with the direction of energy flow and is governed by Snell's law. The field vectors in the extraordinary wave resulting from the incident TE wave are shown in Fig. 6. As previously described, the direction of the energy flow characterized by the Poynting vector $\underset{\sim}{S}^e$ does not coincide with the direction of wave propagation $\underset{\sim}{k}^e$. The electric displacement $\underset{\sim}{D}^e$ and the electric field $\underset{\sim}{E}^e$ are also not collinear but differ by the same angle as do $\underset{\sim}{S}^e$ and $\underset{\sim}{k}^e$. The vectors $\underset{\sim}{S}^e$, $\underset{\sim}{k}^e$, $\underset{\sim}{D}^e$, and $\underset{\sim}{E}^e$ all lie in the principal plane defined by $\underset{\sim}{k}^e$ and the optic axis.

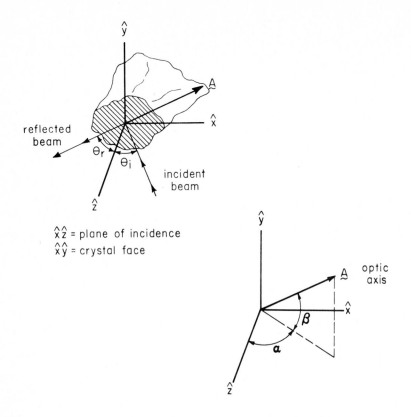

FIG. 4. Experimental geometry for the measurement of reflectivity from a uniaxial crystal face with general orientation of the optic axis $\underset{\sim}{A}$ with respect to the plane of incidence ($\hat{x}\hat{z}$ plane).

When the usual boundary conditions requiring continuity of field components across the optical boundary are applied to this problem, it is easily seen that both the ordinary wave field component and the extraordinary wave field component contribute to the TE and TM reflection coefficients in a complicated way. The detailed functional forms of the TE and TM reflectivities have been given elsewhere [54] and will be simply summarized here as R = $R(n_1, \theta_i, \alpha, \beta, \varepsilon_x, \varepsilon_z)$ where ε_x ($= \varepsilon_y$) and ε_z are the dielectric tensor components in the principal dielectric axis system of the crystal. An additional difficulty occurs in the general case since the transmitted wave is not propagating along one of the crystallographic axes. Consequently, the phonon resonances in the complex dielectric function are oblique phonon resonance whose frequencies depend on the angle between the wave vector and the direction of phonon polarization, further complicating the analysis of reflectivity data.

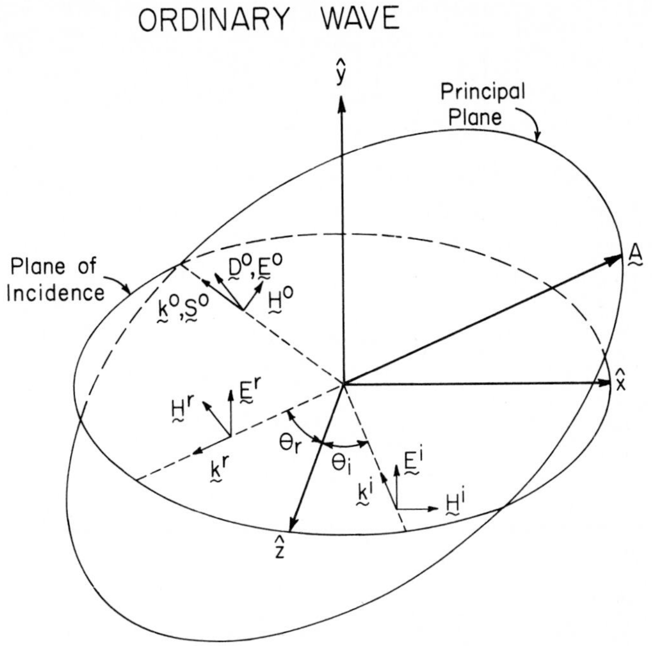

ORDINARY WAVE

FIG. 5. The fields in an ordinary wave propagating in a uniaxial crystal as a result of incident transverse electric (TE) electromagnetic radiation.

Considerable simplification can be achieved by judicious selection of the experimental reflection geometry. When the plane of incidence, defined by the incident and reflected wave normals, is perpendicular to the principal plane defined by the optic axis and the transmitted wave normal, then the TM reflectivity consists entirely of the contribution from the ordinary wave and the TE reflectivity results solely from the extraordinary wave. Conversely, when the principal plane and the plane of incidence coincide, the TM reflectivity contains only the extraordinary wave contribution while the TE reflectivity consists only of the ordinary wave component. A number of examples occur in the literature in which the reflectivity analysis is simplified by a careful choice of reflection geometry, but include effects of the angular dispersion of oblique phonons. These include studies of $NaNO_3$ [55], $LiIO_3$ [56], $KBrO_3$ [57], and calcite [58].

D. Anisotropic Media - Biaxial Crystals

The equations for the complex reflection coefficients in absorbing biaxial crystals were first given by Drude [59] and subsequently by Winterbottom

EXTRAORDINARY WAVE

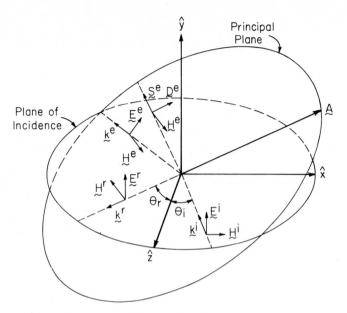

FIG. 6. The fields in an extraordinary wave propagating in a uniaxial crystal as a result of incident transverse electric (TE) electromagnetic radiation.

[60]. However, the calculation of optical constants in biaxial crystals from spectroscopic measurements is considerably more difficult than in uniaxial or isotropic crystals since the principal dielectric axes need not correspond with the crystallographic axes in monoclinic or triclinic crystals. Consequently the most complete studies of biaxial systems have been limited to orthorhombic crystals [61].

Belousov and Pavinich studied the infrared reflectivity from the ac face of diopside, $CaMgSi_2O_6$ [62], and spodumene, $LiAlSi_2O_6$ [63], both of which are monoclinic. Since one of the principal dielectric axes must coincide with the crystallographic b axis, the other two dielectric axes must lie in the ac plane. The orientation of these two axes was determined by a classical oscillator analysis in which the directions of the transition moments were varied until a satisfactory fit to the reflectivity data was found. Determination of the directions of the transition moments by analysis of reflectivity data has also been done in $Li_2SO_4 \cdot H_2O$ [64], and by transmission measurements of oriented plates in α-glycine-d_2 [65] and taurine and sarosine [66]. A method of obtaining optical constants in triclinic crystals has been given [67] which requires the measurement of polarized transmission spectra of three plates cut with different orientations in order to fully determine the elements of the dielectric tensor.

E. Polaritons

In the electrostatic approximation utilized thus far, a reflectivity measurement is viewed as a weak electromagnetic field (photons) probing a purely mechanical field (phonons). However, the vibrational motion of the particles involved in the phonon modes results in acceleration of charge, and this time-varying polarization gives rise to electromagnetic radiation fields. These radiation fields have a harmonic time dependence at the same frequency as the mechanical phonon vibration which created the field. The resulting interaction between the mechanical field and the electromagnetic field is a polariton and is the propagating excitation in a polarizable crystal. The description of polaritons is usually obtained by considering a retarded electromagnetic field, using the complete set of Maxwell's equations, which interacts with a vibrating lattice. This has been done by Huang for cubic crystals [68] and Loudon for uniaxial crystals [69]. A similar treatment has been given for orthorhombic crystals [70].

A convenient derivation and solution of the equations of motion for a polarizable crystal interacting with its own vibrationally induced radiation field begins with the equation of motion of a purely mechanical phonon (Eq. (55)),

$$\ddot{Q}_j + \Omega_j^2\, Q_j - \mu_{j\sigma}E_\sigma = 0.$$

The component of the polarization $\underset{\sim}{P}$ in the spectral region of normal mode j is

$$P_\sigma = \mu_{j\sigma}Q_j + \chi_{\sigma\sigma}(\infty)E_\sigma \qquad (87)$$

where $\chi_{\sigma\sigma}(\infty)$ is the $\sigma\sigma$ component of the dielectric susceptibility tensor which results from the electronic polarization. In the notation of Sec. III,

$$\varepsilon_{\sigma\sigma}(\infty) - 1 = 4\pi\chi_{\sigma\sigma}(\infty). \qquad (88)$$

If there are other infrared active modes higher in frequency which can also contribute to the dielectric constant as outlined in the development of Eqs. (61a and b), Eq. (87) and (88) can be easily modified. The wave equation for the electric field $\underset{\sim}{E}$ is obtained from Eq. (9) by writing $\varepsilon\underset{\sim}{E} = \underset{\sim}{\ddot{D}}$ and using the relationship between $\underset{\sim}{D}$ and $\underset{\sim}{P}$ to write

$$\nabla^2 \underset{\sim}{E} - \frac{1}{c^2} \ddot{\underset{\sim}{E}} = \frac{4\omega}{c^2} \ddot{\underset{\sim}{P}} .$$ (89)

Equations (55), (87), and (89) must be simultaneously solved for Q_j and the σ component of $\underset{\sim}{E}$ and $\underset{\sim}{P}$. Assuming the solutions are of the form

$$Q_j = Q_j^O \exp[i\underset{\sim}{k} \cdot \underset{\sim}{r} - i\omega t]$$ (90a)

$$P_\sigma = P_\sigma^O \exp[i\underset{\sim}{k} \cdot \underset{\sim}{r} - i\omega t]$$ (90b)

$$E_\sigma = E_\sigma^O \exp[i\underset{\sim}{k} \cdot \underset{\sim}{r} - i\omega t],$$ (90c)

substitution of these into the desired equations yields

$$[\Omega_j^2 - \omega^2]Q_j^O - \mu_{j\sigma}E_\sigma^O = 0$$ (91)

$$\mu_{j\sigma}Q_j^O - P_\sigma^O + \chi_{\sigma\sigma}(\infty)E_\sigma^O = 0$$ (92)

$$\frac{4\pi\omega^2}{c^2} P_\sigma^O - [k^2 - \frac{\omega^2}{c^2}]E_\sigma^O = 0.$$ (93)

In order for non-trivial solutions to exist for this set of coupled, linear, homogeneous equations in the unknowns Q_j^O, E_σ^O, and P_σ^O, the associated coefficient determinant must vanish, leading to the condition

$$[\Omega_j^2 - \omega^2] [k^2 - \frac{\omega^2}{c^2} - \frac{4\pi\omega^2}{c^2} \chi_{\sigma\sigma}(\infty)] - \frac{4\pi\omega^2}{c^2} \mu_{j\sigma}^2 = 0.$$ (94)

Using Eq. (88) and defining

$$\varepsilon_{\sigma\sigma}(0) = \varepsilon_{\sigma\sigma}(\infty) + 4\pi\mu_{j\sigma}^2/\Omega_j^2$$ (95)

the determinantal equation becomes

$$\omega^4 \varepsilon_{\sigma\sigma}(\infty) - \omega^2[\Omega_j^2 \varepsilon_{\sigma\sigma}(0) + c^2k^2] + c^2k^2\Omega_j^2 = 0.$$ (96)

Here $\varepsilon_{\sigma\sigma}(0)$ is understood to be the dielectric tensor component at a frequency lower than Ω_j but considerably above the transverse optic mode frequencies of any other infrared active normal modes. This derivation presupposes a crystal with widely space BZC frequencies, although the derivation can be easily modified if this is not the case. Eq. (96) is a quadratic equation in ω^2 whose roots are two polariton modes as shown in Fig. 7. The development up to this point has assumed $\underset{\sim}{\nabla} \cdot \underset{\sim}{E} = 0$ which implies that $\underset{\sim}{k}$ is perpendicular to $\underset{\sim}{E}$ from Eq. (90c). Thus the two polariton modes are trans-

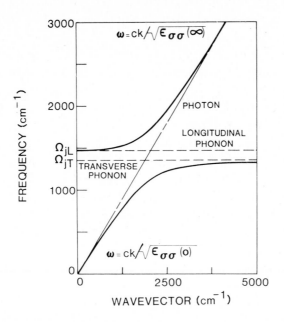

FIG. 7. Polaritons resulting from the interaction of the purely mechanical phonon field and the purely electromagnetic field due to the vibrational motion of the charges.

verse modes. A longitudinal solution can be found by removing the transversal condition ($\nabla \cdot \underset{\sim}{E} = 0$) in Eq. (8) which leads to an additional term $-\underset{\sim}{k}(\underset{\sim}{k} \cdot \underset{\sim}{E})$ added on to the left hand side of Eq. (89). Following the same procedure as in the transverse case leads to a determinantal equation yielding one root with no dispersion which will be written

$$\Omega_{jL}^{2} = \Omega_{jT}^{2} + 4\pi \, \mu_{j\sigma}^{2}/\varepsilon_{\sigma\sigma}(\infty). \tag{97}$$

The quantity Ω_{jT} is simply the quantity Ω_{j} which appears in the equations leading up to Eq. (97). Since these previous equations implicitly assumed a transverse solution (by writing $\nabla \cdot \underset{\sim}{E} = 0$), the additional subscript T will now be added to distinguish this case from the longitudinal case designated by Ω_{jL}. Combining Eq. (95) and Eq. (97) leads to

$$\frac{\Omega_{jL}^{2}}{\Omega_{jT}^{2}} = \frac{\varepsilon_{\sigma\sigma}(0)}{\varepsilon_{\sigma\sigma}(\infty)} \tag{98}$$

which is the well-known Lyddane, Sachs, and Teller relation [71]. This useful equation has been extended to more complex crystals [72,73].

The interaction of a weak external electromagnetic field with a polarizable crystal, such as occurs in a typical reflectivity measurement, is described in Fig. 7. In the frequency region well below the transverse optic mode Ω_{jT}, the dispersion in the polariton is that of a pure electromagnetic wave, $\omega = ck/\sqrt{\varepsilon_{\sigma\sigma}(0)}$, and the crystal is transparent to the probe field. Similarly at frequencies well above the longitudinal optic mode Ω_{jL}, the dispersion is also that of an electromagnetic wave, $\omega = ck/\sqrt{\varepsilon_{\sigma\sigma}(\infty)}$, and the crystal is again transparent. However, in the frequency range between Ω_{jT} and Ω_{jT} there is a band gap and the modes are a mixture of mechanical phonons and electromagnetic photons. Therefore the crystal will not admit the pure light-like field from the incident wave in a reflectivity measurement. The wave will be reflected back from the surface of the crystal and the crystal will have a high reflectivity in this spectral region. In a typical right angle Raman scattering experiment, the region of the polariton dispersion curve which is probed corresponds to a purely mechanical phonon at $\underset{\sim}{k} \cong 0$. The scale of the abscissa in Fig. 7 may be somewhat misleading, since the boundary of the first Brillouin zone (BZB) is inversely proportional to a unit cell dimension, which means that

$$|\underset{\sim}{k}(BZB)| \cong 10^8 \text{ cm}^{-1}.$$

Therefore the spectral region shown in the figure is indeed in the immediate vicinity of $\underset{\sim}{k} \cong 0$.

V.   THE MOLECULAR DIPOLE MODEL

The molecular dipole model of internal optic modes provides a useful formalism in which the relationship of molecular parameters to the optical constants and certain spectroscopic observables is quite transparent. The model is most appropriate for those internal modes in which the dominant intermolecular interaction governing the vibrational dynamics is the vibrationally induced dipolar coupling between molecular units. The earliest development and application of this model is a series of studies by Decius of the ν_2 out-of-plane bending modes of the nitrate and carbonate ion in crystals based on the aragonite structure [74]. In this classical treatment the interaction between nearest neighbor anions in the unit cell was described by a coupling constant. An alternative view was provided by Hexter who utilized vibrational excitons [75]. This work underlined the collective nature of the modes and the need to sum the dipolar interactions over the entire crystal. This view was reiterated by Decius et al. [76] in a continuation of the earlier

work on carbonate crystals with the aragonite lattice. The model was subsequently modified to include effects resulting from the electronic polarization
of the constituent molecules. One of the clearest discussions of the molecular
dipole model is a classic paper by Decius [77], using the expression for the
interaction energy of a lattice of polar and polarizable particles given by
Mandel and Mazur [78]. The model was further developed and applied to a
number of rhombohedral crystals including $CaCO_3$ (calcite), $MgCO_3$, $NaNO_3$,
and $LiNO_3$ by Frech and Decius [79]. Frech examined the relationship of the
molecular dipole model to the more familiar dynamical matrix method [80] and
exploited the versatility of the model in a calculation of dispersion curves
[81], an analysis of vibrational multiplet structure [82], and a calculation of
the angular dispersion of oblique phonons [83].

The description of the molecular dipole model presented here will follow
the development and notation used by Decius and Frech. An internal optic
mode frequency is written as an intramolecular term plus an additional term
resulting from the collective interaction of the vibrationally induced dipoles.
The intramolecular vibrational potential energy for a crystal of N unit cells,
each containing n molecular species with s molecular normal modes labeled by j
is

$$2V_{intramol} = \sum_{\tau=1}^{N} \sum_{\ell=1}^{n} \sum_{j=1}^{s} q_j(\tau\ell) \, w_{jo}^2(\tau\ell) \, q_j(\tau\ell). \qquad (99)$$

The quantity $q_j(\tau\ell)$ is the molecular normal coordinate associated with the
molecular normal mode frequency $w_{jo}(\tau\ell)$, where τ labels the unit cell and ℓ
specifies the molecular unit or site within a unit cell. The molecular normal
mode frequency $w_{jo}(\tau\ell)$ is actually a static crystal field frequency, that is,
the normal frequency of a molecular unit vibrating in the crystal in the
absence of all intermolecular dynamical coupling effects. The static crystal
field frequency may be written as $w_{jo}(\ell)$ since all translationally equivalent
molecules have the same static crystal field frequency.

The dipolar interaction energy is

$$2V_{dip} = \sum_{\tau\tau'} \sum_{\ell\ell'} \underset{\sim}{u}^{\dagger}(\tau\ell) \, \{T \, [I + \alpha T]^{-1}\}(\tau\ell, \tau'\ell') \, \mu(\tau'\ell') \qquad (100)$$

where $\underset{\approx}{I}$ is a unit matrix. The quantity $\underset{\approx}{\alpha}$ is a diagonal electronic polarizability tensor and arises from the contribution of the electronic polarization to
the total electric moment $\underset{\sim}{p}$,

$$\underset{\sim}{p}(\tau\ell) = \underset{\sim}{\mu}(\tau\ell) + \underset{\approx}{\alpha}(\tau\ell,\tau\ell) \, \underset{\sim}{E}(\tau\ell). \qquad (101)$$

Here $\mu(\tau\ell)$ is a column vector of the vibrationally induced dipole moments. The electric field $F(\tau\ell)$ produced at site $\tau\ell$ by a dipole $p(\tau'\ell')$ at site $\tau'\ell'$ is described through the dipole field tensor $\underset{\approx}{T}(\tau\ell,\tau'\ell')$ by

$$F(\tau\ell) = -\underset{\approx}{T}(\tau\ell,\tau'\ell') \, p(\tau'\ell'). \tag{102}$$

An element of the dipole field tensor may be written

$$T_{\sigma\sigma'}(\tau\ell,\tau'\ell') =$$

$$\frac{\delta_{\sigma\sigma'}}{|r(\tau\ell) - r(\tau'\ell')|^3} - \frac{3[r(\tau\ell) - r(\tau'\ell')]_\sigma \, [r(\tau\ell) - r(\tau'\ell')]_{\sigma'}}{|r(\tau\ell) - r(\tau'\ell')|^5} \tag{103}$$

where $r(\tau\ell)$ is the vector from the origin to site ℓ in unit cell τ. The vibrationally induced dipole moment is now expanded in molecular normal coordinates q_j

$$\mu(\tau\ell) = \mu^{(o)}(\tau\ell) + \sum_j \mu_j(\tau\ell) \, q_j(\tau\ell) + \ldots \tag{104}$$

where $\mu_j(\tau\ell)$ is a column vector of the dipole moment derivatives with respect to the q_j,

$$\mu_{j\sigma}(\tau\ell) \equiv (\partial\mu_\sigma(\tau\ell)/\partial q_j(\tau\ell))_o. \tag{105}$$

Both the permanent dipole moment $\mu^{(o)}$ and the higher order terms in the expansion will be neglected. The dipole interaction energy Eq. (100) is now

$$2V_{dip} = \sum_{jj'} \sum_{\tau\tau'} \sum_{\ell\ell'} q_j(\tau\ell)\mu_j^\dagger(\tau\ell) \, \{\underset{\approx}{T}[\underset{\approx}{I} + \alpha\underset{\approx}{T}]^{-1}\}(\tau\ell,\tau'\ell')\mu_{j'}(\tau'\ell')q_{j'}(\tau'\ell'). \tag{106}$$

The transformation to translationally symmetrized coordinates is

$$W_j(k\ell) = \frac{1}{\sqrt{N}} \sum_\tau q_j(\tau\ell) \, \exp[ik \cdot r(\tau)]. \tag{107}$$

Under this transformation the dipole energy becomes

$$2V_{dip} = \sum_{jj'} \sum_{\underset{\sim}{k}} \sum_{\ell\ell'} W_j(\underset{\sim}{k}\ell)\ \mu_j^\dagger(\ell)\ \{D[I + \alpha D]^{-1}\}(\underset{\sim}{k};\ell,\ell')\mu_{j'}(\ell)W_{j'}(\underset{\sim}{k}\ell') \qquad (108)$$

using the translational invariance of the lattice to write $\mu_j(\tau\ell) = \mu_j(\ell)$. The transformed matrices are necessarily diagonal in the index $\underset{\sim}{k}$. After noting from Eq. (103) that $T_{\sigma\sigma'}(\tau\ell,\tau'\ell') = T_{\sigma\sigma'}(\tau-\tau';\ell,\ell')$ and again invoking the translational invariance of the lattice, one can write an element of the dipole sum matrix $D(\underset{\sim}{k};\ell,\ell')$

$$D_{\sigma\sigma'}(\underset{\sim}{k};\ell,\ell') = \sum_{\tau'} T_{\sigma\sigma'}(\tau\ell,\tau'\ell')\ \exp\{i\underset{\sim}{k} \cdot [\underset{\sim}{r}(\tau\ell) - \underset{\sim}{r}(\tau'\ell')]\} \qquad (109)$$

which is easily identified as the Fourier component of the dipole field propagation matrix. This quantity should not be confused with the dynamical matrix of the crystal which appears in Sec. III. The problem of calculating the dipole sum implied by the equation will be deferred for the moment. The intermolecular vibrational energy retains its simple form under this transformation,

$$2V_{intramol} = \sum_{j} \sum_{\underset{\sim}{k}} \sum_{\ell} W_j(\underset{\sim}{k}\ell)\ w_{jo}^2(\ell)\ W_j(\underset{\sim}{k}\ell). \qquad (110)$$

In the general case of a multiply occupied unit cell, the symmetry of the cell may be utilized to further simplify the problem. Unit cell symmetrized coordinates $Q_j(\underset{\sim}{k}\gamma)$ are constructed using the usual projection operator techniques. The transformation to this basis is

$$Q_j(\underset{\sim}{k}\gamma) = \sum_{\ell} B(\underset{\sim}{k}\gamma,\ell)W_j(\underset{\sim}{k}\ell). \qquad (111)$$

The index γ designates a particular irreducible representation of the factor group of the crystal. Under this transformation the intramolecular contribution still retains its simple form

$$2V_{intramol} = \sum_{j} \sum_{\underset{\sim}{k}} \sum_{\gamma} Q_j(\underset{\sim}{k}\gamma)\ w_{jo}^2(\ell)\ Q_j(\underset{\sim}{k}\gamma). \qquad (112a)$$

In the usual case there is only one kind of molecular species present at the several sites in the unit cell labelled by ℓ. Therefore $w_{jo}(\ell) = w_{jo}$ and $\mu(\ell) = \mu_j$, leading to

$$2V_{intramol} = \sum_{j} \sum_{\underset{\sim}{k}} \sum_{\gamma} w_{jo}^2\ Q_j^2(\underset{\sim}{k}\gamma) \qquad (112b)$$

and

$$2V_{dipolar} = \sum_{jj'} \sum_{\underset{\sim}{k}} \sum_{\gamma} Q_j(\underset{\sim}{k}\gamma) \; \mu_j^\dagger \{\underset{\approx}{\mathfrak{D}}[I + \alpha\underset{\approx}{\mathfrak{D}}]^{-1}\}(\underset{\sim}{k}\gamma) \; \mu_{j'} Q_{j'}(\underset{\sim}{k}\gamma) \tag{113}$$

where $\underset{\approx}{\mathfrak{D}}(\underset{\sim}{k}\gamma)$ is the dipole sum matrix in the unit cell symmetrized basis

$$\underset{\approx}{\mathfrak{D}}_{\sigma\sigma'}(\underset{\sim}{k}\gamma) = \sum_{\ell\ell'} B(\underset{\sim}{k}\gamma,\ell) \; D_{\sigma\sigma'}(\underset{\sim}{k};\ell,\ell') \; B^{-1}(\underset{\sim}{k}\gamma,\ell'). \tag{114}$$

The vibrational energy is the sum of (112b) and (113)

$$2V_{vib} = \sum_{jj'} \sum_{\underset{\sim}{k}} \sum_{\gamma} Q_j(\underset{\sim}{k}\gamma) [\omega_{jo}^2 \, \delta_{jj'} + \mu_j^\dagger \{\underset{\approx}{\mathfrak{D}}[I + \alpha\underset{\approx}{\mathfrak{D}}]^{-1}\}(\underset{\sim}{k}\gamma)\mu_{j'}] Q_{j'}(\underset{\sim}{k}\gamma)$$

$$= \sum_{jj'} \sum_{\underset{\sim}{k}} \sum_{\gamma} Q_j(\underset{\sim}{k}\gamma) \, \Lambda_{jj'}^{\;2}(\underset{\sim}{k}\gamma) \, Q_{j'}(\underset{\sim}{k}\gamma). \tag{115}$$

It is important to note from Eq. (115) that the unit cell symmetrized coordinates Q are not the normal coordinates of the internal optic modes in the crystal. The presence of bilinear terms in jj' allows mixing of modes of the same symmetry γ but originating in different kinds of intramolecular motion, j and j'. This has been examined in $NaClO_3$ and $NaBrO_3$ [84].

Usually the internal optic phonon branches are well-separated in frequency and Eq. (115) can be simplified, since the different intramolecular motions are then dynamically decoupled, leading to

$$2V_{vib} = \sum_{j} \sum_{\underset{\sim}{k}} \sum_{\gamma} Q_j(\underset{\sim}{k}\gamma) \, [\omega_{jo}^2 + \mu_j^\dagger \{\underset{\approx}{\mathfrak{D}}[I + \alpha\underset{\approx}{\mathfrak{D}}]^{-1}\}(\underset{\sim}{k}\gamma)\mu_j] Q_j(\underset{\sim}{k}\gamma)$$

$$= \sum_{j} \sum_{\underset{\sim}{k}} \sum_{\gamma} Q_j(\underset{\sim}{k}\gamma) \, \omega_j^{\;2}(\underset{\sim}{k}\gamma) \, Q_j(\underset{\sim}{k}\gamma). \tag{116}$$

The unit cell symmetrized coordinates may then be viewed as the internal optic mode normal coordinates with the corresponding normal mode frequencies as indicated. In crystals in which the molecular unit, and hence the vibrationally induced dipole, occupies a site of symmetry belonging to one of the point groups C_n with $n \geq 3$, D_n, C_{nv}, C_{nh} with $n \geq 3$, D_{nh}, D_{nd}, or S_n, the dipole sum matrix $\underset{\sim}{D}$ and hence $\underset{\approx}{\mathfrak{D}}$ are diagonal in the Cartesian indices σ [77]. The frequency of a BZC phonon mode with polarization only in the σ direction would be simply

$$\omega_j^{\;2}(\underset{\sim}{0}\gamma) = \omega_{jo}^2 + \mu_{jo}\{\underset{\approx}{\mathfrak{D}}[I + \alpha\underset{\approx}{\mathfrak{D}}]^{-1}\}_{\sigma\sigma}(\underset{\sim}{0}\gamma) \, \mu_{jo}. \tag{117}$$

The use of the equations described in this section requires the calculation of the dipole sums described by Eq. (109). The matrix element

$D_{\sigma\sigma'}(\underset{\sim}{k};\ell\ell')$ may be viewed as the σ component of the dipole field at the set of translationally equivalent sites labeled with ℓ, produced by a plane lattice wave of unit dipoles oriented in the σ' direction and distributed over the set of sites ℓ'. Because of the plane wave summation required to evaluate the sums, the result is dependent on the direction of the wave vector with respect to the polarization direction of the dipole. Following Born and Huang [4], $D_{\sigma\sigma'}(\underset{\sim}{k};\ell,\ell')$ is written in the limit of small $\underset{\sim}{k}$ as

$$\lim_{\underset{\sim}{k}\to 0} D_{\sigma\sigma'}(\underset{\sim}{k};\ell,\ell') = d_{\sigma\sigma'}(\ell,\ell') + \frac{4\pi}{v}\frac{k_\sigma k_{\sigma'}}{|\underset{\sim}{k}|^2} \tag{118}$$

where v is the unit cell volume. As Decius and Hexter [3] have pointed out, $d_{\sigma\sigma'}$ depends on the crystal structure and not on $\underset{\sim}{k}$, while the second term depends on the direction of $\underset{\sim}{k}$, but not its magnitude or the crystal structure. In particular, if the lattice wave propagates in a direction perpendicular to the polarization direction σ', then $k_{\sigma'} = 0$ and the dipole sum is

$$D^T_{\sigma\sigma'}(\underset{\sim}{0};\ell,\ell') = d_{\sigma\sigma'}(\ell,\ell'). \tag{119}$$

Here, as before, the notation $\underset{\sim}{0}$ indicates the limit of a very small but finite $\underset{\sim}{k}$ so that the direction of $\underset{\sim}{k}$ is preserved. This type of wave is obviously a transverse wave and dipole sums obtained in this manner carry a superscript T and would be the proper ones to use in the calculation of BZC transverse internal optic phonon frequencies. In that case, Eq. (117) would more properly be written

$$\omega_{jT}^{2}(\underset{\sim}{0}\gamma) = \omega_{jo}^{2} + \mu_{jo}\{\mathfrak{A}[I + \alpha\mathfrak{A}]^{-1}\}^T_{\sigma\sigma'}(\underset{\sim}{0}\gamma)\mu_{jo'}. \tag{120}$$

However, if the dipole wave propagates in the same direction as the vibrational polarization moment, then the dipole sum resulting from this longitudinal polarization wave carries a superscript L and is

$$D^L_{\sigma\sigma}(\underset{\sim}{0};\ell,\ell') = d_{\sigma\sigma}(\ell,\ell') + \frac{4\pi}{v}. \tag{121}$$

The corresponding BZC longitudinal internal optic phonon frequency is then

$$\omega_{jL}^{2}(\underset{\sim}{0}\gamma) = \omega_{jo}^{2} + \mu_{jo}\{\mathfrak{A}[I + \alpha\mathfrak{A}]^{-1}\}^L_{\sigma\sigma}(\underset{\sim}{0}\gamma)\mu_{jo}. \tag{122}$$

A straightforward summation at the Brillouin zone center over the sites of a dipolar lattice is an inappropriate way to calculate the dipole sums of Eq.

(109) due to the slow convergence of the sum. In addition, such a procedure ignores the problems caused by the conditional convergence of these sums. One of the more widely used techniques which circumvents these problems is the Ewald-Kornfeld method, originally developed by Ewald [85] to calculate Coulombic sums in an ionic lattice and subsequently extended to dipolar lattices by Kornfeld [86]. This method essentially consists of replacing the potential due to point charges in a lattice by two potentials; the potential due to a Gaussian distribution of charge situated at each lattice point, and a lattice of point charges with a Gaussian distribution of charge of opposite sign superposed on the point charges. By proper choice of the width of the two Gaussian charge distributions, rapid convergence of the total potential is assured, even though the total potential is independent of this width parameter. One then takes advantage of the fact that the electric field of a dipolar array is simply the second derivative of the electrostatic potential with respect to the appropriate coordinate. An early description of this method can be found in a calculation of cubic lattice sums by Born and Bradburn [87]. Another technique for calculating dipolar sums is the planewise summation method of Nijboer and deWette [88] which has been further developed by deWette and Schacher [89]. Both the Ewald-Kornfeld method and the planewise summation method have been elegantly described by Decius and Hexter (Appendix XI of Ref. [3]) in a useful notation. An alternative method given by Benson and Mills [90] based on earlier work of Mackenzie [91] obviates the use of the reciprocal lattice and convergence parameters by converting the dipole sum to an expression involving a modified Bessel function of order two.

The question of retardation effects in the dipole sum calculations was examined by Mahan, who derived an Ewald sum for a lattice of retarded dipolar interactions in a simple crystal with one molecular unit per unit cell [92]. This treatment was subsequently extended to more complex crystals by Philpott and Lee, who concluded that retardation corrections to the static dipolar sums are negligible [93].

The utility of the molecular dipole model in obtaining molecular parameters from spectroscopic observables is most evident from Eqs. (120) and (122). There both the transverse and longitudinal BZC internal optic mode frequencies appear as an explicit function of the static crystal field frequency ω_{jo} and the molecular dipole moment derivatives, $\mu_{j\sigma} = (\partial\mu_\sigma/\partial q_j)_0$. It should be noted that the dipole moment derivative always occurs as a squared factor in this model, resulting in the familiar sign ambiguity. When a molecular unit occupies a crystal site of symmetry compatible with the irreducible representation structure of the molecular normal modes, the degeneracies present in the isolated unit are preserved, and a static crystal field frequency differs from the corresponding gas phase frequency only through the perturbation effects of the equilibrium field of the crystal. In this situation the molecular parameters calculated in the crystal with the molecular dipole model may be directly compared with their gas phase counterparts. This is not to say that the values are necessarily expected to be in close agreement, since the static crystal field effects may be appreciable. However, the comparison is simple and direct.

The calculation of w_{jo} and $(\partial \mu_\sigma / \partial q_j)_o$ requires that both $w_{jL}(\underset{\sim}{0}\gamma)$ and $w_{jT}(\underset{\sim}{0}\gamma)$ be measured. These are obtained either from the Raman spectrum or a measurement of the near-normal incidence reflectivity spectrum as has been previously discussed. The difference between the (squared) longitudinal and transverse frequencies is proportional to the square of the dipole moment derivative,

$$w_{jL}{}^2(\underset{\sim}{0}\gamma) - w_{jT}{}^2(\underset{\sim}{0}\gamma) =$$

$$u_{jo}(\{ \mathfrak{D}[I + \alpha\mathfrak{D}]^{-1} \}_{\sigma\sigma}^{L}(\underset{\sim}{0}\gamma) - \{ \mathfrak{D}[I + \alpha\mathfrak{D}]^{-1} \}_{\sigma\sigma}^{T}(\underset{\sim}{0}\gamma)) \mu_{jo}. \qquad (123)$$

The quantity in the outer parentheses can be calculated exactly if the dipole sums can be evaluated and the electronic polarizability components are known. Equation (123) is the origin of the statement that "the reflectivity bandwidth is proportional to the dipole moment derivative." The first calculation of dipole moment derivatives in crystals using an equation of this form was a study of several cubic and uniaxial crystals by Haas and Hornig [94], although in their work the factor in the outer parentheses was replaced by a quantity based on the Lorentz effective field.

In the case of a more complicated crystal, a number of vibrational modes may be observed in a spectral region where only a corresponding single transition is observed in the isolated molecular species. It is possible to distinguish several different kinds of effects responsible for this multiplet structure.

a.   Factor Group Splitting (correlation field splitting).   A nondegenerate vibration in an isolated species may exhibit a multiplet structure in a multiply occupied unit cell due to correlation effects. The vibrational motions of each molecular species in a cell are coupled through intermolecular forces, and the collective modes originating in a particular kind of intramolecular motion belong to the various irreducible representations of the unit cell group. Obviously, degenerate modes are similarly correlated but the "splitting" is perhaps most dramatic in the case of nondegenerate vibrations.

b. Site Group Splitting. A potential energy environment of lowered symmetry in a crystal may lift the degeneracy in the normal mode of an isolated species and lead to a multiplet structure. This may be expected to occur whenever the dimensionality of the irreducible representation of the degenerate mode is greater than the dimensionality of the irreducible representations of the site group of the site on which the molecular species is located.

c. Multiple Site Splitting. The static crystal field frequency of a particular vibrational mode may differ in species occupying nonequivalent sites with different potential energy environments. In the case of a degenerate vibration, the different site group splitting effects may lead to different multiplet structures.

The molecular dipole model not only provides a useful framework for individually viewing these different effects, but is also most helpful in the general case when several of these effects occur simultaneously. However, it must be realized that the molecular dipole model is only one possible description of the multiplet structure. An alternative view originally developed by Dows writes the intermolecular interactions which couple the vibrational motion in terms of atom-atom potentials [95]. Although the potential used in that study of methyl chloride consisted of a hydrogen atom repulsion part and a dipole-dipole contribution between neighbors, Dows concluded that the hydrogen atom repulsion term dominated the dynamical splitting in most of the bands studied. Numerous calculations of factor group splittings in other molecular crystals, especially alkanes, based on this model occur in the literature. Nevertheless, the molecular dipole model provides a particularly useful view since the role of the molecular parameters is most clear.

Consider a degenerate normal mode in an isolated molecular species which is located on a site of low symmetry in a crystal. If the site symmetry is incompatible with the symmetry of the normal mode, then the degeneracy may be lifted through the perturbing static crystal field. Each resulting static crystal field frequency component can then be labeled with a site group irreducible representation index s. Therefore the frequency of a BZC internal optic mode might be written

$$\omega_j^2(\underset{\sim}{0}\gamma) = \omega_{jo}^2(s) + \mu_{j\sigma}(s)\{ \boldsymbol{\beth}[I + \alpha\boldsymbol{\beth}]^{-1}\}_{\sigma\sigma}(\underset{\sim}{0}\gamma)\mu_{j\sigma}(s). \qquad (124)$$

It is important to note that the molecular dipole moment derivative also carries the site group irreducible representation label.

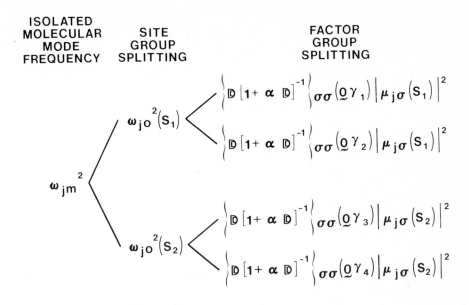

FIG. 8. Vibrational multiplet structure arising from both site group splitting and factor group correlation effects written in a molecular dipole formalism.

One interesting example of vibrational multiplet structure is illustrated in Fig. 8 for a hypothetical crystal with two molecular units in a primitive cell. The degenerate molecular mode ω_{jm} is split by site group perturbations into $\omega_{jo}(s_1)$ and $\omega_{jo}(s_2)$. The vibrational motion of each of these site group components is then correlated through the intermolecular interactions within the cell, resulting in two factor group components designated by γ_i. A measurement of $\omega_j(\underset{\sim}{0}\gamma_1)$ and $\omega_j(\underset{\sim}{0}\gamma_2)$ is sufficient to determine $\omega_{jo}(s_1)$ and $|\mu_{jo}(s_1)|$, while a measurement of $\omega_j(\underset{\sim}{0}\gamma_3)$ and $\omega_j(\underset{\sim}{0}\gamma_4)$ permits the calculation of $\omega_{jo}(s_2)$ and $|\mu_{jo}(s_2)|$. In this manner the site group splitting, that is, the difference between $\omega_{jo}(s_1)$ and $\omega_{jo}(s_2)$, can be obtained. An analysis of the site group splitting and factor group splitting of the sulfate ion modes in a number of crystals has been given by Wu and Frech [16]. In an alternative formulation of the molecular dipole model based on vibrational excitons, Hexter has extensively examined vibrational multiplet structure in a number of crystals, including methyl chloride [75], tetragonal alkali azides [96], and carbon dioxide [97]. He has also discussed the role of site group splitting and factor group splitting in the second part of Ref. [75], although a much more thorough discussion of this topic is to be found in Chapter 6, Section 3, of the Decius and Hexter book [3]. In that section is also a detailed comparison between molecular dipole theory and the models based on atom-atom potentials.

VI. THE INTERNAL FIELD IN CRYSTALS

During a spectroscopic measurement, the total electric field experienced by a species at a lattice site consists of the electric field of the external electromagnetic probe plus an internal field originating in the charge distribution of dipoles within the crystal. The terms "local field," "effective field," and "internal field" have been used interchangeably in the literature, however, it seems sensible to use either local field or effective field to describe the total field at a site and to reserve "internal field" to describe only that portion of the field resulting from the polarization of the medium.

The σ component of the internal field at site ℓ in a non-polarizable crystal due to a lattice wave of unit dipoles oriented in the σ' direction and distributed over the sublattice defined by ℓ' has already been calculated as $D_{\sigma\sigma'}(\underset{\sim}{k};\ell,\ell')$ in Eq. (118). Because of the nature of the infrared and Raman selection rules for fundamental transitions, only the BZC dipole sums need be considered, i.e., $D_{\sigma\sigma'}(\underset{\sim}{0};\ell,\ell')$. It will prove useful to develop results for a cubic crystal, since much of the earlier work in the area of effective and internal fields has been done for cubic systems. It is easily shown that, for a lattice dipole wave propagating in the z direction in a cubic crystal,

$$\underset{\approx}{D}(\underset{\sim}{0};\ell,\ell') = \frac{4\pi}{3v} \begin{bmatrix} -1 & 0 & 0 \\ 0 & -1 & 0 \\ 0 & 0 & 2 \end{bmatrix}, \tag{125}$$

provided that ℓ is a site of cubic symmetry with respect to the sublattice defined by ℓ'. Fox and Hexter [98] have shown that the same result can be obtained from a classical electrostatic calculation of the internal field in a cubic lattice of point dipoles for a slab-shaped crystal, which is the sample geometry most often encountered in a spectroscopic experiment. In the same article they have also given an excellent discussion of the shape dependence of dipole sums.

In a real crystal the electric fields from the dipoles induce additional moments through the polarizability of the constituent ions or molecules. These induced moments also create fields which further polarize the medium.

This polarization interaction must be calculated in a self-consistent way. In the presence of an external field $\underset{\sim}{E}$, the total or effective field at a lattice site is

$$\underset{\sim}{F}(\tau\ell) = \underset{\sim}{E}(\tau\ell) - \sum_{\tau'\ell'} \underset{\approx}{T}(\tau\ell,\tau'\ell') \ \underset{\sim}{p}(\tau'\ell') \tag{126}$$

which is a simple extension of Eq. (102) and, of course, the total electric moment at that site is (Eq. (101))

$$\underset{\sim}{p}(\tau\ell) = \underset{\sim}{\mu}(\tau\ell) + \underset{\approx}{\alpha}(\tau\ell,\tau\ell) \ \underset{\sim}{F}(\tau\ell).$$

Eliminating the electric moment between these two equations leads to

$$\underset{\sim}{F}(\tau\ell) = \sum_{\tau'\ell'} \{[\underset{\approx}{I} + \underset{\approx}{\alpha}\underset{\approx}{T}]^{-1}\}^{\dagger}(\tau\ell,\tau'\ell') \ \underset{\sim}{E}(\tau'\ell')$$

$$- \sum_{\tau'\ell'} \{\underset{\approx}{T}[\underset{\approx}{I} + \underset{\approx}{\alpha}\underset{\approx}{T}]^{-1}\}^{\dagger}(\tau\ell,\tau'\ell') \ \underset{\sim}{\mu}(\tau'\ell'). \tag{127}$$

This is the formal expression for the effective field at a site in the lattice due to vibrational polarization effects as well as those from the external electric field. If Eq. (127) is now transformed to a translationally symmetrized basis and the $\underset{\sim}{k} \cong 0$ component is considered, there results

$$\underset{\sim}{F}(0\ell) = \sum_{\ell'} \{[\underset{\approx}{I} + \underset{\approx}{\alpha}\underset{\approx}{D}]^{-1}\}^{\dagger}(\underset{\sim}{0};\ell'\ell') \ \underset{\sim}{E}^{O}$$

$$- \sum_{\ell'} \{\underset{\approx}{D}[\underset{\approx}{I} + \underset{\approx}{\alpha}\underset{\approx}{D}]^{-1}\}^{\dagger}(\underset{\sim}{0};\ell,\ell') \ \underset{\sim}{\mu}(\ell'). \tag{128}$$

At $\underset{\sim}{k} \cong 0$ the external field is essentially constant over all sites and is written as $\underset{\sim}{E}^{O}$. Numerous calculations of the effective field based on the first part of Eq. (128) have appeared in the literature for a variety of non-cubic crystals. These calculations have been done both in specific studies focused on the effective fields themselves and also in the context of investigations of optical and electronic phenomena. Examples of the former include Mueller's study in a primitive tetragonal lattice [99] and a calculation of the effective field in complex rhombohedral lattices by Frech and Decius [79].

The relationship between the dielectric constant and the polarization of the crystal due to the internal field can be described by eliminating the effective field $\underset{\sim}{F}$ between Eqs. (126) and (101), yielding

$$\underset{\sim}{p}(\tau\ell) = \underset{\tau'\ell'}{\Sigma} \; [I + \alpha T]^{-1} \; (\tau\ell,\tau'\ell') \; \alpha(\tau'\ell',\tau'\ell') \; \underset{\sim}{E}(\tau'\ell')$$

$$+ \underset{\tau'\ell'}{\Sigma} \; [I + \alpha T]^{-1}(\tau\ell,\tau'\ell') \; \mu(\tau'\ell'). \tag{129}$$

This is the formal expression for the total electric moment at a site, including the induced polarization effects from both vibrational moments and the external electric field. Again, in the limit of a long wavelength external field, the BZC component is

$$\underset{\sim}{p}(\underset{\sim}{0};\ell) = \underset{\ell'}{\Sigma} \; [I + \alpha D]^{-1}(\underset{\sim}{0};\ell'\ell') \; \alpha(\ell',\ell') \; \underset{\sim}{E}^{o}$$

$$+ \underset{\ell'}{\Sigma} \; [I + \alpha D]^{-1}(\underset{\sim}{0};\ell,\ell') \; \mu(\ell'). \tag{130}$$

The electric polarization is then written as the induced dipole moment within a unit cell divided by the volume of the cell.

$$\underset{\sim}{P} = \frac{1}{v} \underset{\ell}{\Sigma} \; \underset{\sim}{p}(\underset{\sim}{0},\ell) = \frac{1}{v} \underset{\ell\ell'}{\Sigma} \; [I + \alpha D]^{-1}(\underset{\sim}{0};\ell,\ell') \; \alpha(\ell',\ell') \; \underset{\sim}{E}^{o}$$

$$+ \underset{\ell\ell'}{\Sigma} \; [I + \alpha D]^{-1}(\underset{\sim}{0};\ell,\ell') \; \mu(\ell'). \tag{131}$$

The separate contributions of the electronic polarization and the vibrational polarization are easily seen, including the appropriate effective field factors $[I + \alpha D]^{-1}$. An expression for the dielectric constant can be easily derived from Eq. (131) by first writing the interaction energy of the vibrationally induced dipoles with the external electric field as given by Mandel and Mazur [78],

$$V_{int} = - \underset{\tau\tau'}{\Sigma} \underset{\ell\ell'}{\Sigma} \; \mu^{\dagger}(\tau'\ell') \; [I + T\alpha]^{-1}(\tau'\ell',\tau\ell) \; \underset{\sim}{E}(\tau\ell). \tag{132}$$

This is simply an extension of Eq. (52) with the proper effective field correction. Considering the long wavelength limit after transformation to the usual translationally symmetrized basis leads to

$$V_{int} = - \underset{\ell\ell'}{\Sigma} \; \mu^{\dagger}(\ell') \; [I + D\alpha]^{-1}(\underset{\sim}{0};\ell',\ell) \; \underset{\sim}{E}^{o}. \tag{133}$$

Following the same procedure as in Sec. III, the electric moment is expanded in terms of the normal coordinates of the crystal,

$$\mu(\ell') = \sum_j \mu_j(\ell')Q_j, \tag{134}$$

and these two equations are then utilized in the equation of motion for the coordinate Q_j to finally write

$$Q_j = \sum_{\ell\ell'} \mu_j^\dagger(\ell') \ [I + D\alpha]^{-1}(Q;\ell,\ell') \ E^o/[\Omega_j^2 - \omega^2 - i\gamma_j\omega]. \tag{135}$$

When this is substituted into Eqs. (134) and (131), the electric polarization is

$$P = \frac{1}{v} \sum_{\ell\ell'} \ [I + \alpha D]^{-1}(Q;\ell,\ell') \ \alpha(\ell',\ell') \ E^o$$

$$+ \frac{1}{v} \sum_j \{ \sum_{\ell\ell'} [I + \alpha D]^{-1}(Q;\ell,\ell') \ \mu_j(\ell')\}\{ \sum_{\ell\ell'} [I + \alpha D]^{-1}(Q;\ell\ell')$$

$$\times \mu_j(\ell')\}^\dagger \ E^o/[\Omega_j^2 - \omega^2 - i\gamma_j\omega]. \tag{136}$$

Finally from Eq. (31) and the definition $D = E^o + 4\pi P$ one can write

$$\varepsilon = I + \frac{4\pi}{v} \sum_{\ell\ell'} \ [I + \alpha D]^{-1}(Q;\ell\ell') \ \alpha(\ell',\ell')$$

$$+ \frac{4\pi}{v} \sum_j \{ \sum_{\ell\ell'} [I + \alpha D]^{-1}(Q;\ell,\ell') \ \mu_j(\ell')\} \{ \sum_{\ell\ell'} [I + \alpha D]^{-1}(Q,\ell,\ell')$$

$$\times \mu_j(\ell')\}^\dagger/[\Omega_j^2 - \omega^2 - i\gamma_j\omega]. \tag{137}$$

This equation is somewhat misleading as written, for as the development of Eq. (136) shows, the polarization arises in response to an external electromagnetic field which is naturally a transverse field. Therefore the dipole sum matrix elements in Eq. (137) must be the transverse dipole sum values. As in the case of the electric polarization, Eq. (137) illustrates the role of the effective field factors in both the electronic and the vibrational contribution to the dielectric constant. Comparison of Eq. (136) with Eq. (51) leads to the identifications

$$\varepsilon(\infty) = \underset{\approx}{I} + \frac{4\pi}{v} \underset{\ell\ell'}{\Sigma} [\underset{\approx}{I} + \alpha D]^{-1} (\underset{\sim}{0};\ell,\ell') \; \alpha(\ell',\ell) \tag{138}$$

and

$$S_{j\sigma}\Omega_j^{2} = \frac{4\pi}{v} \{ \underset{\ell\ell'}{\Sigma} [\underset{\approx}{I} + \alpha D]^{-1}(\underset{\sim}{0};\ell,\ell') \; \mu_j(\ell')\}_\sigma \{ \underset{\ell\ell'}{\Sigma} [\underset{\approx}{I} + \alpha D]^{-1}(\underset{\sim}{0};\ell\ell')$$

$$\times \; \mu_j(\ell')\}_\sigma^\dagger. \tag{139}$$

The derivation of Eqs. (138) and (139) closely follows that given by Decius and Hexter in Chapter 5 of their text [3].

Equations (128) and (137) yield a rather familiar result when the high frequency limit is considered in a simple cubic crystal with only one species per unit cell. In that case there is no vibrational polarization since the infrared active optic modes are unable to follow the external field, and the equations reduce to

$$\underset{\sim}{F} = \{[\underset{\approx}{I} + \alpha D]^{-1}\}^\dagger \; \underset{\sim}{E}^o \tag{140}$$

and

$$\underset{\approx}{\varepsilon} = \underset{\approx}{I} + \frac{4\pi}{v} [\underset{\approx}{I} + \alpha D]^{-1} \; \alpha \tag{141}$$

The redundant site indices ℓ and ℓ' have been suppressed, as well as $\underset{\sim}{k} \cong 0$. Using the transverse dipole sums for $\underset{\approx}{D}$ in this system

$$\underset{\sim}{D} = - \frac{4\pi}{3v} \; \underset{\approx}{I} \tag{142}$$

results in

$$\underset{\sim}{F} = [1 - \frac{4\pi\alpha}{3v}]^{-1} \; \underset{\sim}{E}^o \tag{143}$$

and

$$\varepsilon \underset{\approx}{I} = \{1 + \frac{4\pi\alpha}{v} [1 - \frac{4\pi\alpha}{3v}]^{-1}\} \; \underset{\approx}{I} \tag{144}$$

since in a simple cubic crystal both $\underset{\approx}{\varepsilon}$ and $\underset{\approx}{\alpha}$ are constant diagonal matrices. Rearrangement of Eq. (144) leads to

$$\frac{1}{3} [\varepsilon + 2] = [1 - \frac{4\pi\alpha}{3v}]^{-1} \tag{145}$$

which, when substituted into Eq. (143), results in an expression for the Lorentz effective field,

$$\underset{\sim}{F} = \frac{1}{3} [\varepsilon + 2] \underset{\sim}{E}^o. \tag{146}$$

This relationship was originally derived by Lorentz [100], who calculated the field in a virtual spherical cavity in the interior of a uniformly polarized medium. Carlson and Decius compared dipole moment derivatives calculated with the correct effective field and using the Lorentz effective field in a series of rhombohedral crystals and in an orthorhombic crystal [101]. Their conclusion was that the Lorentz field provided an adequate approximation for the effective field in a number of crystals.

The Lorentz effective field has been modified by Guertin and Stern to describe delocalization effects in cubic crystals [102]. Their treatment is equivalent to replacing Eq. (143) by

$$\underset{\sim}{F} = [1 + (\gamma - 1) \frac{4\pi\alpha}{3v}] [1 - \frac{4\pi\alpha}{3v}]^{-1} \underset{\sim}{E}^o \tag{147}$$

where γ is a parameter which varies between 0 and 1. The tight binding limit is described by $\gamma = 1$, where the Lorentz field is obtained. In the case of extreme charge delocalization, $\gamma = 0$ and there is no effective field correction [103].

A well-known and useful relationship between the high frequency dielectric constant and the polarizability can be derived using Eq. (145), although it is naturally restricted to those systems in which the Lorentz effective field is valid. From Eq. (145), with the dipole sum replaced by the cubic value (Eq. (142)),

$$\underset{\sim}{P} = \frac{1}{v} [1 - \frac{4\pi\alpha}{3v}]^{-1} \alpha_m \underset{\sim}{E}. \tag{148}$$

Notice that α_m is a molecular polarizability, which contains both a vibrational polarization contribution and an electronic polarization part.

Using $[\varepsilon - 1] \underset{\sim}{E} = 4\pi \underset{\sim}{P}$ and the Lorentz effective field factor (Eq. (145)) results in

$$\frac{\varepsilon - 1}{\varepsilon + 2} = \frac{4\pi\alpha_m}{3v} . \tag{149a}$$

For a simple cubic crystal, the number of species per unit volume, N, is $\frac{1}{v}$. In its more general form this equation is written as a function of the number density of species with polarizability α_m

$$\frac{\varepsilon - 1}{\varepsilon + 2} = \frac{4\pi}{3} N\alpha_m \qquad\qquad\qquad (149b)$$

and is the Clausius-Mossotti equation. As Böttcher [104] has pointed out, this equation is appropriate at frequencies where the vibrational and electronic polarization can follow the external field. At sufficiently high frequencies where only the electronic polarization can contribute, the dielectric constant is replaced by the optical refractive index and the equation becomes

$$\frac{n^2 - 1}{n^2 + 2} = \frac{4\pi}{3} N\alpha_{elec}. \qquad\qquad\qquad (149c)$$

This is the Lorentz-Lorentz equation. Equations (149) are another example of the relation of optical constants to a molecular parameter (the polarizability) through a model of the effective field.

REFERENCES

1. E. B. Wilson, J. C. Decius, and P. C. Cross, Molecular Vibrations, McGraw-Hill, New York, 1955, Ch. 7.

2. Reprinted in Chem. Eng. News, April 5, 1982.

3. J. C. Decius and R. M. Hexter, Molecular Vibrations in Crystals, McGraw-Hill, New York, 1977.

4. M. Born and K. Huang, Dynamical Theory of Crystal Lattices, Oxford University Press, London, 1954.

5. A. A. Maradudin, E. W. Montroll, G. H. Weiss, and I. P. Ipatova, Theory of Lattice Dynamics in the Harmonic Approximation, 2nd Ed., Academic Press, New York and London, 1971.

6. L. D. Landau and E. M. Lifshitz, Electrodynamics of Continuous Media, Pergamon Press, New York, 1960.

7. M. Born and E. Wolf, Principles of Optics, Pergamon Press, Oxford, 1959.

8. H. A. Kramers, Atti del Congresso Internazionale dei Fisici, 1927, 11-20 Settembre; Como-Pavia-Roma, (Published by Nicola Zanichelli, Bologna) 2, 545 (1928).

9. R. de L. Kronig, J. Opt. Soc. Am., 12, 547 (1926).

10. H. Fröhlich, Theory of Dielectrics, Oxford University Press, London and New York, 1949.

11. F. Stern, Sol. State Phys., 15, 299 (1963).

12. B. Crawford, Jr., A. C. Gilby, A. A. Clifford, and T. Fujiyama, Pure Appl. Chem., 18, 373 (1969).

13. e.g. see the discussion in Ref. 11, Sec. 23.

14. G. Weinreich, Solids: Elementary Theory for Advanced Students, Wiley, New York, 1965.

15. K. Nakamoto, Infrared Spectra of Inorganic and Coordination Compounds, Wiley, New York, 1963.

16. G. J. Wu and R. Frech, J. Chem. Phys., 66, 1352 (1977).

17. E. Kettler, Wied. Ann. Physik, 30, 299 (1887); 31, 322 (1887).

18. H. V. Helmholtz, Wied. Ann. Physik, 48, 389 (1893); 48, 723 (1893).

19. H. Rubens, Ann. Physik. Chem., 53, 267 (1894).

20. H. Rubens, Ann. Physik. Chem., 54, 476 (1895).

21. F. Seitz, Modern Theory of Solids, McGraw-Hill, New York, 1940, Ch. 17.

22. M. Lax, Phys. Chem. Solids, 25, 487 (1964).

23. A. I. Alekseev, Usp. Fiz. Nauk., 73, 41 (1961). Translation: Soviet Physics Uspekhi, 4, 23 (1961).

24. L. P. Kadanoff and G. Baym, Quantum Statistical Mechanics, Benjamin, New York, 1962.

25. R. A. Cowley, Adv. Phys., 12, 421 (1963).

26. A. A. Maradudin and R. F. Wallis, Phys. Rev., 123, 777 (1961); 125, 1277 (1962).

27. V. S. Vinogradov, Fig. Tverdogo Tela., 4, 712 (1962). Translation: Soviet Physics-Solid State, 4, 519 (1962).

28. F. Gervais and B. Piriou, Phys. Rev., B10, 1642 (1974).

29. R. P. Lowndes, Phys. Rev., B1, 2754 (1970).

30. R. P. Lowndes, J. Phys. C.: Sol. State Phys., 4, 3083 (1971).

31. D. W. Berreman and F. C. Unterwald, Phys. Rev., 174, 791 (1968).

32. F. Gervais and B. Piriou, J. Phys. C.: Sol. State Phys., 7, 2374 (1974).

33. F. Gervais and B. Piriou, Phys. Rev., B11, 3944 (1975).

34. A detailed treatment of the reflection of electromagnetic waves can be found in any basic text on electromagnetic theory, e.g., J. Jackson, Classical Electrodynamics, John Wiley & Sons, Inc., New York, 1962; J. A. Stratton, Electromagnetic Theory, McGraw-Hill, New York, 1941.

35. S. P. F. Humphreys-Owen, Proc. Phys. Soc. (London), 77, 949 (1961).

36. W. R. Hunter, J. Opt. Soc. Am., 55, 1197 (1965).

37. P. Drude, Wied. Ann. Physik, 36, 532, 865 (1889); 39, 481 (1890).

38. A. B. Winterbottom, Trans. Faraday Soc., 42, 487 (1946).

39. F. L. McCrackin, E. Passaglia, R. R. Stromberg, and H. L. Steinberg, J. Res. Nat'l. Bur. Standards, 67A, 363 (1963).

40. T. S. Robinson, Proc. Phys. Soc., 65B, 910 (1952).

41. T. S. Robinson and W. C. Price, Proc. Phys. Soc., 66B, 969 (1953).

42. M. Gottlieb, J. Opt. Soc. Am., 50, 343 (1960).

43. F. C. Jahoda, Phys. Rev., 107, 1261 (1957).

44. G. Andermann, A. Caron, and D. A. Dows, J. Opt. Soc. Am., 55, 1210 (1965).

45. R. W. Wood, Physical Optics, MacMillan, New York, 1934, Ch. XV.

46. W. G. Spitzer, D. A. Kleinman, and D. Walsh, Phys. Rev., 113, 127 (1959).

47. W. G. Spitzer, R. C. Miller, D. A. Kleinman, and L. E. Howarth, Phys. Rev., 126, 1710 (1962).

48. M. V. Belousov, D. E. Pogarev, and A. A. Shultin, Soviet Physics-Solid State, 11, 2185 (1970).

49. R. J. Collins and D. A. Kleinman, J. Phys. Chem. Solids, 11, 190 (1959).

50. L. P. Mosteller, Jr. and F. Wooten, J. Opt. Soc. Am., 58, 511 (1968).

51. P. A. Flournoy and W. J. Schaffers, Spectrochim. Acta, 22, 5 (1966).

52. M. Berek, Z. Kristallogr., 76, 396 (1931); 89, 144 (1934); 93, 116 (1936).

53. H. Damany and E. Uzan, Opt. Acta, 17, 131 (1970).

54. R. Frech, Phys. Rev., B13, 2342 (1976).

55. J. A. A. Ketelaar, C. Haas, and J. Fahrenfort, Physica, 20, 1259 (1954).

56. W. Otaguro, E. Wiener-Avnear, C. A. Arguello, and S. P. S. Porto, Phys. Rev., B4, 4542 (1971).

57. B. Unger and S. Haussühl, Phys. Stat. Solid, B54, 183 (1972).

58. R. Frech and H. Nichols, Phys. Rev., B17, 2775 (1978).

59. P. Drude, Wied. Ann. Physik, 32, 584 (1887).

60. A. B. Winterbottom, Kgl. Norske Videnskab. Selskab Forh., 45, 28 (1955).

61. R. Jacobsen, J. Opt. Soc. Am., 54, 1170 (1964).

62. M. V. Belousov and V. F. Pavinich, Opt. Spectrosc., 45, 771 (1978).

63. V. F. Pavinich and M. V. Belousov, Opt. Spectrosc., 45, 881 (1978).

64. G. J. Wu and R. Frech, J. Chem. Phys., 66, 4557 (1977).

65. J. Herranz and J. M. Delgado, Spectrochim. Acta, 31A, 1255 (1975).

66. U. Stahlberg and E. Stenger, Spectrochim. Acta, 23A, 475 (1967).

67. J. Herranz and C. Gómez-Aleixandre, Spectrochim. Acta., 33A, 833 (1977).

68. K. Huang, Proc. Roy. Soc. (London), A208, 352 (1951).

69. R. L. Loudon, Adv. Phys., 13, 423 (1964).

70. C. K. Asawa, Phys. Rev., 2, 2068 (1970).

71. R. H. Lyddane, R. G. Sachs, and E. Teller, Phys. Rev., 59, 673 (1941).

72. W. Cochran, Z. Kristallogr., 112, 465 (1959).

73. A. S. Barker, Jr., Phys. Rev., A135, 1290 (1964).

74. J. C. Decius, J. Chem. Phys., 22, 1941, 1946 (1954); 23, 1290 (1955).

75. R. M. Hexter, J. Chem. Phys., 33, 1833 (1960); 36, 2285 (1962); 39, 1608 (1963).

76. J. C. Decius, O. G. Malan, and H. W. Thompson, Proc. Roy. Soc. (London), A275, 295 (1963).

77. J. C. Decius, J. Chem. Phys., 49, 1387 (1968).

78. M. Mandel and P. Mazur, Physica, 24, 116 (1958).

79. R. Frech and J. C. Decius, J. Chem. Phys., 51, 1536, 5315 (1969); 54, 2374 (1971).

80. R. Frech, J. Chem. Phys., 58, 5067 (1973).

81. R. Frech, J. Chem. Phys., 60, 2354 (1974).

82. R. Frech, J. Chem. Phys., 61, 5344 (1975).

83. R. Frech, J. Chem. Phys., 67, 952 (1977).

84. R. Frech, J. Chem. Phys., 76, 86 (1982).

85. P. P. Ewald, Wied. Ann. Phys., 54, 519, 557 (1917); 64, 253 (1921).

86. H. Kornfeld, Z. Phys., 22, 27 (1924).

87. M. Born and M. Bradburn, Proc. Camb. Phil. Soc., 39, 104 (1943).

88. B. R. A. Nijboer and F. W. deWette, Physica, 23, 309 (1957); 24, 422, 1105 (1958).

89. F. W. deWette and G. E. Schacher, Phys. Rev., A137, 78 (1965).

90. B. Benson and D. L. Mills, Phys. Rev., 178, 839 (1969).

91. J. K. Mackenzie, Thesis, Univ. of Bristol, Bristol, 1950.

92. G. D. Mahan, J. Chem. Phys., 43, 1569 (1965).

93. M. R. Philpott and J. W. Lee, J. Chem. Phys., 58, 595 (1973).

94. C. Haas and D. F. Hornig, J. Chem. Phys., 26, 707 (1957).

95. D. A. Dows, J. Chem. Phys., 32, 1342 (1960).

96. R. M. Hexter, J. Chem. Phys., 37, 1347 (1962).

97. R. M. Hexter, J. Chem. Phys., 41, 1125 (1964).

98. D. Fox and R. M. Hexter, J. Chem. Phys., 41, 1125 (1964).

99. H. Mueller, Phys. Rev., 47, 947 (1953).

100. H. A. Lorentz, The Theory of Electrons, Dover, New York, 1952, pp. 138, 306.

101. R. E. Carlson and J. C. Decius, J. Chem. Phys., 58, 4919 (1973).

102. R. F. Guertin and F. Stern, Phys. Rev., 134A, 427 (1964).

103. C. G. Darwin, Proc. Roy. Soc. (London), A146, 17 (1934).

104. C. J. F. Böttcher, Theory of Electric Polarization, Elsevier, Amsterdam, 1973.

Chapter 3

RECENT ADVANCES IN MODEL CALCULATIONS OF
VIBRATIONAL OPTICAL ACTIVITY

P. L. Polavarapu

Department of Chemistry
Vanderbilt University
Nashville, Tennessee

I. INTRODUCTION

The measurement of circular dichroism (CD) in electronic transitions of chiral molecules and the empirical rules that relate the circular dichroism to molecular stereochemistry are well known to scientists [1]. However the measurements of CD owing to the molecular vibrational transitions are relatively new. The differential vibrational absorption of the left and right circularly polarized infrared radiation is referred to as vibrational circular dichroism (VCD) whereas the differential vibrational Raman scattering of the left and right circularly polarized monochromatic visible light is referred to as Raman optical activity (ROA). Although VCD and ROA refer to the same molecular vibrations, phenomenologically significant differences exist between them.

Vibrational optical activity (VOA) representing VCD and ROA has become a research area of immense interest since 1970. As in electronic optical activity (EOA) studies [1], it is the information on absolute configuration and on different possible conformations of chiral molecules that is anticipated from VOA studies. However, it is hoped that the VOA studies would be more informative since they can provide stereochemical information in various segments of a chiral molecule. This is due to the fact that all vibrations of a chiral molecule are infrared and Raman active and support VOA. To obtain configurational information around a chosen atom or group, the optical activity associated with the vibrations localized at that center can be expected to provide the primary source of information. This type of selective information, when assembled together, offers the hope that the VOA experimental data might be used to derive the three-dimensional structure of the entire chiral molecule chosen. This optimistic prognosis however will be restricted by certain facts of nature. The molecular vibrations in general need not be localized at a chosen center. Further, whether or not the electronic charge distribution is considered to be localized, the nuclear motions in one part of the molecule can cause significant disturbances in the equilibrium electronic charge distribution of the entire molecule. Also it is possible that a fundamental vibration of interest can have resonance interactions with another of the same type or of multi-quanta nature. Increasing evidence for the high probability of such instances is emerging from recent experimental measurements. As a consequence, greater emphasis rests on developing the appropriate theoretical models by taking the aforementioned effects into consideration. For these reasons the current VOA research has taken some specific directions. VOA experiments are being conducted on a series of chemically similar molecules of known absolute configurations to look for trends that correlate with the known stereochemical features. Partially guided by such experimental observations, theoretical models are being improved to understand the origin of VOA in a given molecular structural fragment and to apply that knowledge for deriving stereochemical information in molecular systems where configurational or conformational detail is not available or uncertain.

In this review I hope to present a detailed picture of the theoretical models for VOA in terms of the recent advances, their interrelations and their ability to explain the existing experimental VOA data. From different perspectives, reviews have been written by Stephens, Nafie, Keiderling and Mason for VCD [2-6] and by Buckingham, Barron and Nafie for ROA [4,7-10]. In the present review, care has been taken to avoid repetition of the information given in earlier reviews. Besides these reviews, excellent sources of additional information in this field are the doctoral theses [11-16] of Faulkner, Marcott, Havel and Laux (all from the University of Minnesota) on VCD studies and by Diem and Cuony on ROA studies. In the following section, a brief survey of the existing theoretical models for VOA is presented. Detailed description of various theoretical models is presented in a later section.

II.   A BRIEF SURVEY OF THEORETICAL MODELS

A.   Vibrational Circular Dichroism

Although the effects of molecular vibrations on electronic optical rotation were considered by Fickett [17] in terms of Kirkwood's polarizability theory and by Hameka [18] from a quantum mechanical viewpoint, Cohan and Hameka were the first to consider the optical activity specifically in vibrational transitions. They have evaluated [19] the CD in vibrational transitions of CHDBrCl, CHDTBr and $C^{35}Cl^{37}ClBrF$ where bare nuclear charges alone are considered in the formulation. Later Deutsche and Moscowitz have calculated [20] the VCD in helical polymers by considering them to be consisting of point masses each with an effective partial charge. In order to facilitate the choice of molecules that are favorable for VCD experimental investigations, Holzwarth and Chabay have presented a coupled oscillator model [21] following the procedure adapted for a similar model in electronic circular dichroism [22]. In this model, the optical activity associated with the vibrations of an optically active dimer, composed of two identical diatomic molecules, is considered. The two coupled vibrations of such dimers are predicted to exhibit circular dichroism of equal intensity but opposite sign. Faulkner has also considered this coupled oscillator model and presented the VCD intensity expressions when the oscillators are degenerate as well as nondegenerate [11]. The coupled oscillator expression in a more general form [23] and the situation when two degenerate oscillators are located at the same center [24] have also been considered.

Schellman has developed [25] a general model which subsequently enjoyed a wide popularity due to its simplicity. For calculating the magnetic dipole transitions moments, that are required to obtain VCD intensities, Schellman has assumed that each atom is associated with an effective charge and that this charge rigidly moves with the nuclei during vibrations. Reflecting these assumptions, this model is referred to as the fixed partial charge (FPC) model. In a collaborative effort, the research groups of Moscowitz, Holzwarth and Mosher have employed this model to predict the circular dichroism in C-H and C-D stretching vibrations of 2,2,2-trifluoro-1-phenylethanol and neopentyl-1-d-chloride and the results were compared [26,27] to the experimental observations. These calculations, more details of which are given in Faulkner's thesis [11], indicated that the predicted and experimental signs of VCD are in favorable agreement, but the predicted magnitudes of VCD intensities are well below the experimental magnitudes. Further comparison of the experimental results, obtained in subsequent years with the FPC model predictions, revealed some discrepancies in VCD signs as well. The failure of the FPC model is considered to be a consequence of ignoring the charge redistribution during molecular vibrations.

A procedure to incorporate the effects of charge redistribution into VCD model calculations was later presented by Abbate et al. [28]. These authors have expressed the electric and magnetic dipole moment components resulting from the charge redistribution in terms of bond moment derivatives, while retaining those resulting within the FPC concept. This model is appropriately termed as the charge flow model and has been applied [13,14] to a few molecular systems. Moskovits and Gohin [29] have also presented a formalism to incorporate the charge redistribution effects into the magnetic dipole transition moment in terms of bond currents. This procedure is different from that of Abbate et al. [28] in certain respects.

From a different perspective, one can view the charge redistribution as inducing the electric dipole moments. Then, for a given vibrational motion localized in a group or chiral moiety, electric dipole moments can be induced in other groups of the molecule as dictated by the respective polarizability tensors. A model, termed the dynamic polarizability model, underlying this concept was formulated by Barnett et al. [30,31] to explain the failure of the coupled oscillator model in accounting for the observed circular dichroism in the C-H stretching vibrations of (S)-(+)-9,10-dihydrodibenzo[c,g]phenanthrene. An identical concept was also developed independently by the Minnesota group and is detailed in the doctoral theses of Havel and Laux [13,14].

Earlier to these developments, Barron had suggested [10] a bond dipole theory for VCD intensities where bond moments and their derivatives are utilized in formulating the electric and magnetic dipole transition moments. Recently, a detailed formulation and a comparison of this model with the charge flow model has been presented by Polavarapu [32] where it was noted that the representations of the electric dipole transition moment and the charge flow contribution to the magnetic dipole transition moment are equivalent in both models. However, the fixed charge contribution to the magnetic dipole transition moment is represented differently in these models.

Utilizing the bond moment concept, Barron has given concise VCD formulas [10] for the vibrations of two bonds or groups that are chirally disposed. Also, Barron and Buckingham have developed simple VCD expressions [33] for the torsional vibration of a methyl group and suggested that the structure of chiral molecules can be probed by analyzing the optical activity associated with this vibration.

For treating the VCD intensities from a quantum mechanical viewpoint, the main problem is that the electronic contribution to the magnetic dipole transition moment vanishes in the Born-Oppenheimer approximation. Then, in order to include these electronic contributions, one has to go beyond the Born-Oppenheimer approximation or invoke this approximation at a different stage of formulation. While Walnut and Nafie [34] suggested the former approach and formulated the VCD intensity expression, the second approach was suggested by Nafie and Walnut [35]. Both approaches lead to an identical set of final equations and require that the molecular orbitals used for calculating the VCD intensities be localized. An algorithm for performing these localized molecular orbital (LMO) calculations within a semiempirical molecular orbital theory was developed and has been applied [36,37] to a few molecular systems. As the nature of these calculations requires repeating MO calculations for each molecular vibration, the computational time required, even for the modest size chiral molecular systems, is formidable. A computationally more efficient method, derived from the one used for infrared and Raman intensity calculations by Komornicki and McIver [38], was then formulated and model calculations performed [39,40]. In this efficient method, MO calculations are performed in the presence of an electric field perturbation to obtain the orbital displacement vectors for all nuclear displacements. Nafie and Freedman [41] also reported the same efficient method about the same time but did not report any calculations.

Considering the origin of VCD from a different perspective, Craig and Thirunamachandran [42] treated VCD as arising from the mixing, to a different extent, of initial and excited vibrational states with higher electronic states. Their approach is similar to the treatment of vibronic effects in electronic circular dichroism [43].

More recently two more important developments have taken place in this field. One is the interpretation of the VCD observed in a series of molecules containing a common $CH_2CH_2\overset{*}{C}H$ fragment that is inherently dissymmetric. In the C-H stretching region, several six membered ring molecules containing the aforementioned fragment are found to exhibit three VCD bands with alternating sign pattern. This sign pattern was correlated to the stereochemical arrangement of the $CH_2CH_2\overset{*}{C}H$ fragment by treating, through a perturbation scheme, three vibrations of this fragment as three oscillators that are coupled. The second development is the calculation of VCD resulting from an external magnetic field on the sample. A recent measurement of this magnetic vibrational circular dichroism (MVCD) [44] prompted the development of theoretical procedures for calculating MVCD. Both of the above aspects are detailed in the doctoral thesis of Laux [14].

B. Raman Optical Activity

The basic theoretical expressions [45] and a major portion of the model developments for ROA are given by Barron and Buckingham [7-10]. Simple expressions for conceptual model systems have greatly facilitated the understanding of ROA phenomenon. One model system consists of two identical bonds or groups, which are inherently achiral, oriented in such a way to constitute a chiral structure. This two group model [46] predicts, just as the coupled oscillator model [21] for VCD, oppositely signed ROA for the two vibrations associated with the two groups. Since the frequencies of the vibrations are expected to be slightly different due to perturbative interactions, a bisignate ROA couplet is expected to be observed. ROA expressions for a two group structure with a more general geometry have also been given [9,47]. The twisting vibrational motion of a chiral two group structure was also considered and a simple ROA expression was obtained for such motion. Gohin and Moskovits have utilized [48] the two group model to interpret the experimentally observed ROA features associated with the methylene deformation motions.

Another model system considered [49,50] is an achiral group with two degenerate normal modes of vibration, like a methyl group, situated in a chiral environment. The perturbation from the chiral environment lifts the degeneracy, and the separated normal modes of vibration are predicted to have oppositely signed ROA. Therefore it is thought that, if the frequency splitting is sufficient, the degenerate vibrations of the methyl group will exhibit bisignate ROA and might serve as a probe for the chirality of the environment.

Besides these conceptual models, three general calculational models, namely the bond polarizability model, the atom dipole interaction (ADI) model [51-53] and the orbital polarizability model [40], have been developed. The bond polarizability model is a generalization of the two group model [46] and the ADI model is an extension of the version used in electronic optical activity studies (vide infra). The orbital polarizability model [40] is formulated in terms of the localized molecular orbitals and is well suited for the adaption of the efficient MO approach [38] to vibrational intensities. A different procedure for the MO calculations of ROA has been presented by Cuony [16] in his doctoral thesis.

The experimental measurements of magnetic Raman optical activity [54], in ferrocytochrome-c and some metal complexes have been reported by Barron [55-59]. A theoretical method for calculating the magnetic resonance Raman optical activity, using vibronic theory and irreducible tensor methods, has been presented [60] recently by Barron and coworkers.

III.   THEORETICAL DETAILS

A.   Vibrational Circular Dichroism

1.   Formal Difficulties

For a fundamental vibrational one quantum transition associated with a normal mode of vibration Q_ℓ, the circular dichroism intensity is represented by rotational strength [61] R_ℓ, which is defined [11,19,20] as

$$R_\ell \equiv \text{Im} \langle \psi_{00} | \mu_\alpha | \psi_{01}^\ell \rangle \langle \psi_{01}^\ell | m_\alpha | \psi_{00} \rangle. \tag{1}$$

Here ψ_{00} is the total wavefunction for a molecule in the electronic and vibrational ground states; ψ_{01}^ℓ is that in an excited (by one quantum) vibrational state associated with the same ground electronic state where the superscript ℓ

identifies the single vibrational mode Q_ℓ that is excited; and μ_α and m_α represent, respectively, the components of the electric and magnetic dipole moment operators. There is a subtlety in expressing the operator and an associated formal difficulty in evaluating the electric dipole transition moment, $\langle\psi_{00}|\mu_\alpha|\psi_{01}^\ell\rangle$, when the Born-Oppenheimer approximation [62] is considered. One can express the electric dipole moment operator either in dipole length form or dipole velocity form and it has been shown [63] that the dipole length form is to be preferred. This is because if it is expressed in dipole velocity form the electronic contribution to the transition moment, within the Born-Oppenheimer approximation, vanishes [63,64]. This is not so when μ_α is expressed in dipole length form. Therefore whenever the Born-Oppenheimer approximation is considered the following dipole length form is chosen [64].

$$\mu_\alpha = \mu_\alpha^{nuc} + \mu_\alpha^{el} = \sum_A eZ_A R_{A\alpha} - \sum_K er_{K\alpha}. \tag{2}$$

Here Z_A is the bare nuclear charge on atom A, R_A and r_K are the positional vectors, respectively, of atom A and electron K, and e is the unit of electronic charge. All through this article, the molecules considered are implied to have no net charge. In the expressions we employ the Einstein summation convention.

The magnetic dipole moment operator component m_α is given as

$$m_\alpha = m_\alpha^{nuc} + m_\alpha^{el} = \frac{1}{2c} \sum_A \frac{eZ_A}{M_A} \varepsilon_{\alpha\beta\gamma} R_{A\beta} P_{A\gamma} - \frac{1}{2c} \sum_K \frac{e}{m} \varepsilon_{\alpha\beta\gamma} r_{K\beta} P_{K\gamma}. \tag{3}$$

Here M_A and m are the masses of atom A and electron K, respectively, P_A and P_K are the respective momenta and $\varepsilon_{\alpha\beta\gamma}$ is the three-dimensional Levi-Civita function.

In the Born-Oppenheimer approximation the total wavefunction is written as

$$\psi_{00} = \phi_0(r,R)\chi_{00}(R) \tag{4a}$$

$$\psi_{01}^\ell = \phi_0(r,R)\chi_{01}^\ell(R), \tag{4b}$$

where ψ_{01} is the total wavefunction in the ground electronic and excited vibrational states. The superscript ℓ, again, represents that a single vibrational mode Q_ℓ is excited. The terms in the parentheses for ϕ_0 in Eqs. (4) represent the dependence of the electronic wavefunction ϕ_0 on the electron coordinates r and nuclear coordinates R: it is implied that the dependence on

$\underset{\sim}{R}$ is parametrical. For the sake of simplicity the parentheses will be dropped hereafter. When Eqs. (2), (3) and (4) are substituted into Eq. (1), the resulting expression for the rotational strength will essentially contain four integrals with the operators μ_α^{nuc}, μ_α^{el}, m_α^{nuc} and m_α^{el}. By expanding the operators in normal coordinates, Q_ℓ, these integrals within the harmonic oscillator approximation can be shown, for nondegenerate electronic states, to be

$$\langle \phi_0 \chi_{00} | \mu_\alpha^{nuc} | \phi_0 \chi_{01}^\ell \rangle = \langle \chi_0^\ell | Q_\ell | \chi_1^\ell \rangle \sum_A eZ_A S_{A\alpha}^\ell \tag{5a}$$

$$\langle \phi_0 \chi_{00} | \mu_\alpha^{el} | \phi_0 \chi_{01}^\ell \rangle = - \langle \chi_0^\ell | Q_\ell | \chi_1^\ell \rangle \sum_K e\sigma_{K\alpha}^\ell \tag{5b}$$

$$\langle \phi_0 \chi_{01}^\ell | m_\alpha^{nuc} | \phi_0 \chi_{00} \rangle = \langle \chi_1^\ell | P_\ell | \chi_0^\ell \rangle \sum_A eZ_A \varepsilon_{\alpha\beta\gamma} R_{A\beta}^0 S_{A\gamma}^\ell \tag{5c}$$

$$\langle \phi_0 \chi_{01}^\ell | m_\alpha^{el} | \phi_0 \chi_{00} \rangle = 0. \tag{5d}$$

In these equations χ_0^ℓ and χ_1^ℓ are the initial and final harmonic oscillator functions for the normal mode of vibration Q_ℓ; $\underset{\sim}{S}_A = (\partial \underset{\sim}{R}_A / \partial Q_\ell)_0$ is the displacement vector of atom A and $\underset{\sim}{\sigma}_K^\ell$ is the displacement vector of the kth orbital centroid in the normal mode of vibration Q_ℓ; P_ℓ is the momentum conjugate to the normal coordinate Q_ℓ; and $\underset{\sim}{R}_A^0$ is the equilibrium positional vector of atom A. In Eq. (5b), the electronic wavefunction is considered as a single Slater determinant [65] of molecular orbitals and the proof for the relation (5d) is given by Hameka and Cohan [64] and Faulkner [11].

The formal difficulty in evaluating the rotational strength with the Born-Oppenheimer approximation is reflected in Eq. (5d) where it can be noticed that the electronic contribution to the magnetic dipole transition moment vanishes. Furthermore, if this vanishing contribution is taken for granted, the rotational strength calculated from Eqs. (1), (4) and (5) will be origin dependent [11,66]. That is, different choices for the origin of the molecular coordinate system will yield different values for R_ℓ.

As a consequence, two fundamental questions arise in evaluating the VCD intensities. One concerns the development of the theoretical methods that model the magnetic dipole transition moment satisfactorily and circumvent the origin dependency for rotational strength. The second one concerns the pathways to recover the electronic part of the magnetic dipole transition moment, to its fullest extent. The semiclassical and molecular orbital models developed so far deal with these points.

Before going into the details of various models, it may be noted that Cohan and Hameka [19] did not encounter the origin dependency problem in their calculations because they expressed the electric dipole moment operator in dipole velocity form. In such form the electronic contribution to the electric dipole transition moment, i.e., Eq. (5b), also vanishes. Then the rotational strength contains only the nuclear contributions arising from Eqs. (5a) and (5c) and can be given as

$$R_\ell = \frac{e^2 h}{8\pi c} \left(\sum_A Z_A S_{A\alpha}^\ell \right) \left(\sum_A Z_A \varepsilon_{\alpha\beta\gamma} R_{A\beta}^0 S_{A\gamma}^\ell \right). \tag{6}$$

This equation is however not satisfactory since VCD is predicted to arise from the motion of nuclei with bare charges and not to have any contribution from the motion of electrons.

2.   Semiclassical Models

a.   Fixed Partial Charge Model.   Schellman [25] has used the concept of effective atomic charge to arrive at an expression which gives a better representation of VCD. Here the basic idea is to assume that each atom is associated with an effective charge, $e\zeta_A$, and that this whole charge moves with the nuclei at all points during molecular vibrations. In this "perfect following" approximation, which was also used earlier by Deutsche and Moscowitz [20] and is referred to as the fixed partial charge (FPC) approximation [25], the electric and magnetic dipole moment operators are identical to μ_α^{nuc} and m_α^{nuc} in Eqs. (2) and (3), except that the bare nuclear charge eZ_A is replaced now by $e\zeta_A$:

$$\mu_\alpha = \sum_A e\zeta_A R_{A\alpha}, \tag{7a}$$

$$m_\alpha = \frac{1}{2c} \sum_A \frac{e\zeta_A}{M_A} \varepsilon_{\alpha\beta\gamma} R_{A\beta} P_{A\gamma}. \tag{7b}$$

Expanding these operators in normal coordinates Q_ℓ

$$\mu_\alpha = \mu_\alpha^0 + \sum_{A,\ell} e\zeta_A^0 S_{A\alpha}^\ell Q_\ell \tag{7c}$$

$$m_\alpha = \frac{1}{2c} \sum_{A,\ell} e\zeta_A^0 \varepsilon_{\alpha\beta\gamma} R_{A\beta}^0 S_{A\gamma}^\ell P_\ell \tag{7d}$$

where the superscript zero refers the quantities to equilibrium values. On substituting Eqs. (7) and (4) into Eq. (1), the expression for rotational strength will contain two integrals of the type given in Eqs. (5a) and (5c). Using the standard formulas [67] for these integrals,

$$\langle x_0^\ell | Q_\ell | x_1^\ell \rangle = (\frac{h}{4\pi w_\ell})^{1/2} \tag{8a}$$

$$\langle x_1^\ell | P_\ell | x_0^\ell \rangle = i(\frac{h\omega_\ell}{4\pi})^{1/2} \tag{8b}$$

where w_ℓ is the harmonic frequency for the vibrational mode Q_ℓ, the rotational strength is given as

$$R_\ell = \frac{e^2 h}{8\pi c} (\sum_A \zeta_A^0 \, S_{A\alpha}^\ell)(\sum_A \zeta_A^0 \, \varepsilon_{\alpha\beta\gamma} R_{A\beta}^0 \, S_{A\gamma}^\ell). \tag{9}$$

Owing to the simplicity in FPC formalism, this model became popular and several calculations were performed (vide infra). However several experimental observations are not explained satisfactorily by the FPC model. One reason for this is that the charge redistribution, which can occur during molecular vibrations, is not represented by either of the operators μ_α or m_α.

It is possible to write the operator μ_α, as in the equilibrium (atomic) charge-charge flux (ECCF) model for infrared intensities [68-70], with charge redistribution terms, i.e.,

$$\mu_\alpha = \mu_\alpha^0 + \sum_{A,\ell} e\zeta_A^0 \, S_{A\alpha}^\ell \, Q_\ell + \sum_{A,\ell} (\frac{\partial(e\zeta_A)}{\partial Q_\ell})_0 \, Q_\ell R_{A\alpha}^0. \tag{10}$$

However, unless the charge redistribution terms are incorporated appropriately into the magnetic dipole moment operator also, the rotational strength will become origin dependent. The development of procedures to incorporate the charge redistribution effects into μ_α and m_α, while assuring that the rotational strength remains origin independent, formed the basis for the charge flow model [28], bond current model [29], and bond moment model [10,32]. Since all these approaches depend on the bond moment concept, either in part or full, the bond moment model is detailed first.

b. Bond Moment Model. In the bond moment model [71], the electric dipole moment μ_α is considered to be the sum of individual bond moment vectors,

$$\mu_\alpha = \sum_j eq_j r_j U_{j\alpha}.$$ (11)

Here eq_j is the equilibrium charge on bond j, $\underset{\sim}{U}_j$ is the unit vector and r_j is the length of the bond j. For a given nuclear displacement $\underset{\sim}{X}_A$, the gradient of μ_α can be written as

$$\left(\frac{\partial \mu_\alpha}{\partial X_{A\lambda}}\right)_0 = \sum_j eq_j^0 r_j^0 \left(\frac{\partial U_{j\alpha}}{\partial X_{A\lambda}}\right)_0 + \sum_j eq_j^0 \left(\frac{\partial r_j}{\partial X_{A\lambda}}\right)_0 U_{j\alpha}^0 + \sum_j \left(\frac{\partial (eq_j)}{\partial X_{A\lambda}}\right)_0 r_j^0 U_{j\alpha}^0.$$ (12)

Again the superscript zeros refer the quantities to equilibrium values. If the bond j is between atoms B and C and the unit vector component is written as $U_{j\alpha} = (R_{B\alpha} - R_{C\alpha})/r_j$, then it is straightforward to show that

$$r_j^0 \left(\frac{\partial U_{j\alpha}}{\partial X_{A\lambda}}\right)_0 + U_{j\alpha}^0 \left(\frac{\partial r_j}{\partial X_{A\lambda}}\right)_0 = \delta_{\alpha\lambda}\Delta_{jA}$$ (13)

where $\Delta_{jA} = 1$ for $A = B$, $\Delta_{jA} = -1$ for $A = C$ and $\Delta_{jA} = 0$ for all other atoms; $\delta_{\alpha\lambda}$ is the Kronecker delta function. Then Eq. (12) can be simplified and written in normal coordinate space as

$$\left(\frac{\partial \mu_\alpha}{\partial Q_\ell}\right)_0 = \sum_j \left(\frac{\partial \mu_{j\alpha}}{\partial Q_\ell}\right)_0 = \sum_{j,A} \left(\frac{\partial \mu_{j\alpha}}{\partial X_{A\lambda}}\right)_0 S_{A\lambda}^\ell$$ (14a)

with

$$\left(\frac{\partial \mu_{j\alpha}}{\partial X_{A\lambda}}\right)_0 = eq_j^0 \Delta_{jA}\delta_{\alpha\lambda} + \left(\frac{\partial (eq_j)}{\partial X_{A\lambda}}\right)_0 r_j^0 U_{j\alpha}^0.$$ (14b)

Now the electric dipole moment operator, in the bond moment model, analogous to that in the ECCF model (Eq. 10), becomes

$$\mu_\alpha = \mu_\alpha^0 + \sum_{A,j,\ell} eq_j^0 \Delta_{jA}\delta_{\alpha\lambda}S_{A\lambda}^\ell Q_\ell + \sum_{j,\ell} \left(\frac{\partial (eq_j)}{\partial Q_\ell}\right)_0 Q_\ell r_j^0 U_{j\alpha}^0.$$ (15)

It is useful to note the following relations among the ECCF and bond moment models,

$$\sum_{A,j} eq_j^0 \Delta_{jA}\delta_{\alpha\lambda}S_{A\lambda}^\ell = \sum_A e\xi_A^0 S_{A\alpha}^\ell$$ (16a)

$$\sum_j \left(\frac{\partial(eq_j)}{\partial Q_\ell}\right)_0 r_j^0 U_{j\alpha}^0 = \sum_A \left(\frac{\partial(e\zeta_A)}{\partial Q_\ell}\right)_0 R_{A\alpha}^0. \tag{16b}$$

Because of these equalities, it does not matter whether the electric dipole transition moment is expressed in the bond moment model or in the ECCF model for calculating the rotational strength. However, with regard to the magnetic dipole transition moment one should be more circumspect. It is in the formulation of the magnetic dipole operator where the bond moment model [10,32], charge flow model [28] and bond current model [29] differ from each other.

In terms of the bond moments, the magnetic dipole moment operator may be written, for a neutral molecule with no intrinsic magnetic moments associated with the bonds, as

$$m_\alpha = \frac{1}{2c} \sum_{j,\ell} \varepsilon_{\alpha\beta\gamma} R_{j\beta}^0 \left(\frac{\partial\mu_{j\gamma}}{\partial Q_\ell}\right)_0 P_\ell \tag{17}$$

where $\partial\mu_{j\gamma}/\partial Q_\ell$ is given in Eq. (14) and R_j^0 is the positional vector of bond j at equilibrium from a chosen molecular origin. From Eqs. (17), (15) and (4), the rotational strength expression becomes,

$$R_\ell = \frac{e^2 h}{8\pi c} \left[\sum_A \sum_j \left\{q_j^0 \Delta_{jA}\delta_{\alpha\lambda} + \left(\frac{\partial q_j}{\partial X_{A\lambda}}\right)_0 r_j^0 U_{j\alpha}^0\right\} S_{A\lambda}^\ell\right] \times$$

$$\left[\sum_A \sum_j \varepsilon_{\alpha\beta\gamma} R_{j\beta}^0 S_{A\lambda}^\ell \left\{q_j^0 \Delta_{jA}\delta_{\gamma\lambda} + \left(\frac{\partial q_j}{\partial X_{A\lambda}}\right)_0 r_j^0 U_{j\gamma}^0\right\}\right]. \tag{18}$$

This equation is origin independent but has a different difficulty in the form of the bond position dependency. If a bond j is between atoms B and C, the positional vector of the bond R_j^0 can be chosen to be equal to that of either of the atoms B or C. This choice yields different values for R_ℓ as given in Eq. (18). To overcome this impasse one can consider that the change in the bond positional vector R_j, during a normal vibrational motion, also contributes to m_α. Then Eq. (17) can be rewritten as

$$m_\alpha = \frac{1}{2c} \sum_{j,\ell} \varepsilon_{\alpha\beta\gamma} \left[R_{j\beta}^0 \left(\frac{\partial\mu_{j\gamma}}{\partial Q_\ell}\right)_0 P_\ell + \left(\frac{\partial R_{j\beta}}{\partial Q_\ell}\right)_0 P_\ell \mu_{j\gamma}^0\right] \tag{19a}$$

$$m_\alpha = \frac{1}{2c} \sum_{A,j,\ell} \varepsilon_{\alpha\beta\gamma} eq_j^0 (R_{j\beta}^0 \Delta_{jA}\delta_{\lambda\gamma} + \delta_{R_{j\beta},R_{A\lambda}} r_j^0 U_{j\gamma}^0) S_{A\lambda}^\ell P_\ell$$

$$+ \frac{1}{2c} \sum_{A,j,\ell} er_j^0 \varepsilon_{\alpha\beta\gamma} (\frac{\partial q_j}{\partial X_{A\lambda}})_0 R_{j\beta}^0 U_{j\gamma}^0 S_{A\lambda}^\ell P_\ell. \tag{19b}$$

The first term on the right-hand side of Eq. (19b) represents the fixed charge contribution while the second one represents the charge redistribution contribution to m_α. From Eqs. (19), (15) and (4), the rotational strength becomes

$$R_\ell = \frac{e^2 h}{8\pi c} [\sum_A \sum_j \{q_j^0 \Delta_{jA} \delta_{\alpha\lambda} + (\frac{\partial q_j}{\partial X_{A\lambda}})_0 r_j^0 U_{j\alpha}^0 \} S_{A\lambda}^\ell] \times$$

$$[\sum_A \sum_j \varepsilon_{\alpha\beta\gamma} \{R_{j\beta}^0 q_j^0 \Delta_{jA}\delta_{\gamma\lambda} + \delta_{R_{j\beta},R_{A\lambda}} q_j^0 r_j^0 U_{j\gamma}^0$$

$$+ R_{j\beta}^0 (\frac{\partial q_j}{\partial X_{A\lambda}})_0 r_j^0 U_{j\gamma}^0 \} S_{A\lambda}^\ell] . \tag{20}$$

This equation, representing the bond moment model expression of VCD intensity, is origin independent, bond position independent and contains the charge redistribution contributions.

  c. Charge Flow Model. Abbate, Laux, Overend and Moscowitz [28] have formulated a different procedure for calculating the rotational strength. In formulating the magnetic dipole moment operator, they consider the current density to be arising from two sources, one of which is the motion of the fixed atomic charges that faithfully follow the nuclei. The magnetic dipole transition moment resulting from this source is identical to the one obtained in the FPC model. The second source is the change in bond charges due to nuclear displacements, which is essentially charge redistribution. The current generated due to the changes in bond charges is assumed to be localized along the line joining the two atoms of the bond. With these con-siderations the magnetic dipole moment operator is formulated as

$$m_\alpha = \frac{1}{2c} \{ \sum_{A,\ell} e\zeta_A^0 \varepsilon_{\alpha\beta\gamma} R_{A\beta}^0 S_{A\gamma}^\ell P_\ell + \sum_{A,j,\ell} er_j^0 (\frac{\partial q_j}{\partial X_{A\lambda}})_0 \varepsilon_{\alpha\beta\gamma} R_{j\beta}^0 U_{j\gamma}^0 S_{A\lambda}^\ell P_\ell \}. \tag{21}$$

The first term on the right hand side of this equation is equivalent to the one used in the FPC model (Eq. (7d)), while the second term is equivalent to the charge redistribution contribution in the bond moment model (Eq. (19b)). The electric dipole moment operator employed in the charge flow model is the same as that used in the bond moment model, i.e. Eq. (15). Therefore, the charge flow model and the bond moment model are equivalent in evaluating the electric dipole transition moment and the charge redistribution contribution to the magnetic dipole transition moment. A small difference, however, exists [32] between these two models in evaluating the fixed charge contribution to the magnetic dipole transition moment. Making use of Eqs. (21), (15) and (16), the rotational strength in the charge flow model is given as

$$R_\ell = \frac{e^2 h}{8\pi c} [\sum_A \zeta_A^0 S_{A\alpha}^\ell + \sum_{A,j} (\frac{\partial q_j}{\partial X_{A\lambda}})_0 r_j^0 U_{j\alpha}^0 S_{A\lambda}^\ell] \times$$

$$[\sum_A \zeta_A^0 \varepsilon_{\alpha\beta\gamma} R_{A\beta}^0 S_{A\gamma}^\ell + \sum_{A,j} (\frac{\partial q_j}{\partial X_{A\lambda}})_0 r_j^0 \varepsilon_{\alpha\beta\gamma} R_{j\beta}^0 U_{j\gamma}^0 S_{A\lambda}^\ell]. \tag{22}$$

This equation is origin independent and bond position independent.

d.   Bond Current Model.   For interpreting infrared absorption intensities, two different models are widely employed. One is the bond moment model [71] and the other is the effective (atomic) charge-charge flux (ECCF) model [68-70]. The basic difference in these models is one of concept as can be seen from Eqs. (10) and (15). In a previous section an expression for the rotational strength was formulated entirely within the bond moment concept (see Eq. (20)). The charge flow model utilizes the bond moment concept as well as the fixed atomic charge concept. Sometimes a question arises regarding whether or not the rotational strength can be formulated entirely within the ECCF concept.

The procedure formulated by Moskovits and Gohin [29] answers this query and is an elegant way of adapting the ECCF model of infrared intensities to the calculation of VCD intensities. In their formulation, Moskovits and Gohin considered the bond currents, that should have been created, to compensate for the changes in the effective atomic charges. The rate of change of the charge on atom A, in time, is given as

$$\dot{\zeta}_A = -\sum_B I_{AB} \tag{23}$$

where I_{AB} is the current generated in the chemical bond containing atoms A and B. During nuclear displacements these bond currents contribute to the magnetic dipole transition moment. The magnetic dipole moment operator is, therefore, given as

$$m_\alpha = \frac{1}{2c} \sum_{A,\ell} e\zeta_A^0 \, \varepsilon_{\alpha\beta\gamma} \, R_{A\beta}^0 \, S_{A\gamma}^\ell \, P_\ell \, +$$

$$\frac{1}{2c} \sum_{A,\ell} \sum_{B<A} \frac{e}{2} \varepsilon_{\alpha\beta\gamma} \, R_{A\beta}^0 \, R_{B\gamma}^0 \, (\frac{\partial I_{AB}}{\partial P_\ell})_0 \, P_\ell . \tag{24}$$

The first term on the right hand side of this equation represents the fixed charge contribution and is the same as that used in the FPC and charge flow models; the second term represents the charge redistribution contribution and is distinctly different from that used in the bond moment and charge flow models.

Using the electric dipole moment operator of the ECCF model (Eq. (10)) and the above expression for m_α, the rotational strength expression is obtained as

$$R_\ell = \frac{e^2 h}{8\pi c} [\sum_A \zeta_A^0 \, S_{A\alpha}^\ell + \sum_A R_{A\alpha}^0 \, (\frac{\partial \zeta_A}{\partial Q_\ell})_0] \times$$

$$[\sum_A \zeta_A^0 \, \varepsilon_{\alpha\beta\gamma} R_{A\beta}^0 \, S_{A\gamma}^\ell + \sum_A \sum_{B<A} (\varepsilon_{\alpha\beta\gamma}/2) \, R_{A\beta}^0 \, R_{A\gamma}^0 \, (\frac{\partial I_{AB}}{\partial P_\ell})_0]. \tag{25}$$

In order to evaluate the gradients of bond currents required in Eq. (25), Moskovits and Gohin have shown that

$$\sum_B (\frac{\partial I_{AB}}{\partial P_\ell})_0 = -(\frac{\partial \zeta_A}{\partial Q_\ell})_0 . \tag{26}$$

For a simple molecule containing a tetrahedral carbon attached to four different atoms, the above relation is quite simple since the summation on the left hand side runs only once for each end atom.

For the sake of simplicity in referring to the formalism of Moskovits and Gohin [29], we have taken the liberty to label it as the bond current model.

e. Polarizability Model. In the models considered so far, the effects of charge redistribution during molecular vibrations are represented in terms of either the bond charge derivatives or the atomic charge derivatives. The numerical values for these derivatives are generally transferred, either directly or with some judicial alterations, from smaller molecular systems of high symmetry. This invariably involves some amount of parameterization. It is quite common to encounter molecules with lone pairs of electrons and the contributions from these lone pairs may not have been embedded in the chosen parameterization. To account for such contributions, Laux presented a polarizability model in his doctoral thesis [14]. In this model each lone pair is considered to be a point dipole and is polarizable. In other words, an electric dipole moment can be induced at the location of the lone pair by the electric field emanating from the bond charges in the molecule. As the molecule vibrates, this effective electric field and hence the induced electric dipole moment at the lone pair also oscillate at the vibrational frequency. The variation of this induced electric dipole moment during a normal mode of vibration Q_ℓ, is given as

$$(\frac{\partial \mu_{p\alpha}}{\partial Q_\ell})_0 = \alpha_{p\alpha\beta}(\frac{\partial E_{p\beta}}{\partial Q_\ell})_0 \tag{27}$$

where μ_p is the induced electric dipole moment at lone pair p, $\underset{\sim}{E}_p$ is the electric field at p and $\underset{\sim}{\alpha}_p$ is the polarizability tensor for the lone pair. The variation of the polarizability of the lone pair itself is considered to be negligible. The electric field at the lone pair is represented as

$$E_{p\alpha} = \sum_j eq_j \; (\frac{R_{Bp,\alpha}}{R_{Bp}^3} - \frac{R_{Cp,\alpha}}{R_{Cp}^3}) \tag{28}$$

where $\underset{\sim}{R}_{Bp}$ and $\underset{\sim}{R}_{Cp}$ are the distance vectors directed, respectively, from the atoms B and C to the lone pair p; the atoms B and C constitute the chemical bond j with bond charge eq_j. By taking the differential of Eq. (28) and assigning a polarizability tensor $\underset{\sim}{\alpha}_p$ for the lone pair, Eq. (27) is evaluated. This lone pair contribution to the electric dipole transition moment is achieved by adding the term

$$\sum_\ell (\partial \mu_{p\alpha}/\partial Q_\ell)_0 Q_\ell,$$

to the operator μ_α. The analogous contribution to the magnetic dipole transition moment is incorporated by adding the term,

$$\frac{1}{2c} \sum_\ell \varepsilon_{\alpha\beta\gamma} \; R_{p\beta}^0 \; (\partial \mu_{p\gamma}/\partial Q_\ell) P_\ell$$

to the operator m_α. If more than one lone pair is present then these terms are summed over all lone pairs.

This model can be foreseen to have good potential in the calculation of VCD intensities in general. Although Eqs. (27) and (28) are specifically written for lone pairs, they are not restrictive and can be easily written for any polarizable group of atoms or bonds, with appropriate attention to the possible intrinsic magnetic moments. In fact, the dynamic polarizability model employed by Barnett et al. [30,31] in interpreting the circular dichroism in the C-H stretching vibrations of a dihydro-helicene closely resembles the above formalism.

3.   Quantum Mechanical Models

    a.   Localized Molecular Orbital Model.   In Sec III.A.1 we have summarized the formal difficulties associated with evaluating the VCD intensities in the Born-Oppenheimer approximation. The major problem noted there is the lack of electron contributions to the magnetic dipole transition moment. While evaluating the vibrational absorption intensities it was noted [64] that the operator in dipole velocity form should be converted to one in dipole length form before the molecular wavefunction is expressed in the Born-Oppenheimer approximation. Instead, if this sequence is reversed, then the dipole velocity operator for electrons yields a vanishing contribution to the electric dipole transition moment. To reinstate these contributions, which are of the same order as the nuclear contributions, it was found necessary to incorporate [64] some corrections to the Born-Oppenheimer wavefunction. These results imply that the Born-Oppenheimer wavefunctions have accurate correlations between nuclear and electron positions but no correlations between their velocities. Since the magnetic moment operator is known only in velocity form, the need for incorporating the velocity correlations into Born-Oppenheimer wavefunctions is apparent.

    In a series of two papers Walnut and Nafie have presented [34,35] the formalism to incorporate the velocity correlations into Born-Oppenheimer wavefunctions for calculating vibrational absorption and circular dichroism intensities. In their formulation, the corrections to the Born-Oppenheimer wavefunctions are obtained in terms of the so-called vibronic guaze function. An interesting feature of this function is that the dependence of the correction terms on higher electronic states, which generally results in perturbation theories as in, for example, the vibronic theory of Craig and Thirunamachandran (vide infra), is eliminated. By using the corrected wavefunctions and by representing the circularly polarized radiation perturbation in terms of a vector potential, they found that the troubled part of the rotational strength is now non-vanishing and can be written as

$$\langle\psi_{01}^{\ell}|m_{\alpha}^{el}|\psi_{00}\rangle = \langle\psi_{01}^{\ell}|-\frac{e}{2mc}\sum_{K}\varepsilon_{\alpha\beta\gamma}r_{K\beta}P_{K\gamma}|\psi_{00}\rangle \tag{29}$$

$$= \langle\chi_{01}^{\ell}|\sum_{A,\lambda} i\frac{P_{A\lambda}}{m_A}\{\langle\phi_0|u_{A\lambda}\sum_{K} - \frac{e}{2mc}\varepsilon_{\alpha\beta\gamma}r_{K\beta}P_{K\gamma}|\phi_0\rangle\}|\chi_{00}\rangle.$$

Here $u_{A\lambda}$ is the vibrational guaze function and other symbols have the meanings defined earlier. To obtain an expression for practical usage, they consider the ground state electronic wavefunction ϕ_o as a simple product of the molecular spin orbital functions. Each molecular spin orbital is represented by a Gaussian function and is considered to be ellipsoidal in form. Then the above equation is shown [34] to become,

$$\langle\psi_{01}^{\ell}|m_{\alpha}^{el}|\psi_{00}\rangle = \{-\frac{e}{2c}\sum_{K}\varepsilon_{\alpha\beta\gamma}\ r_{K0,\beta}\ \sigma_{K\gamma}^{\ell} + \Gamma_{K\alpha}^{\ell}\}\ \langle\chi_1^{\ell}|P_{\ell}|\chi_0^{\ell}\rangle \tag{30}$$

where r_{K0} is the centroid of the Kth occupied molecular spin orbital and

$$\Gamma_{K\alpha}^{\ell} = \frac{\Lambda_{K\beta\beta} - \Lambda_{K\gamma\gamma}}{2(\Lambda_{K\beta\beta} + \Lambda_{K\gamma\gamma})\Lambda_{K\beta\beta}\Lambda_{K\gamma\gamma}}\ (\frac{\partial\theta_{K\alpha}}{\partial Q_{\ell}})_0\ . \tag{31}$$

The components of tensor Λ_K determines the shape of the Kth molecular orbital. When the components $\Lambda_{K\alpha\alpha}$ in diagonal form of Λ_K are equal, the orbital is defined to have spherical shape and $\Gamma_{K\alpha}^{\ell}$ vanishes. The term $(\partial\theta_{K\alpha}/\partial Q_{\ell})_0$ represents the angle through which the orbital is rotated during a normal mode of vibration Q_{ℓ}. Therefore the term $\Gamma_{K\alpha}^{\ell}$ represents the rocking of a non-spherical orbital during molecular vibrations.

By making use of the expression for the electronic part of the magnetic dipole transition moment, as given by Eq. (30), and the expressions for the remaining contributions (given by Eqs. (5a), (5b) and (5c)), the rotational strength is obtained to be

$$R_{\ell} = \frac{e^2h}{8\pi c}\{\sum_{A} Z_A S_{A\alpha}^{\ell} - \sum_{K}\sigma_{K\alpha}^{\ell}\}\{\sum_{A} Z_A\varepsilon_{\alpha\beta\gamma}\ R_{A\beta}^0\ S_{A\gamma}^{\ell} -$$

$$\sum_{K}\varepsilon_{\alpha\beta\gamma}\ r_{K0\beta}\ \sigma_{K\gamma}^{\ell} + \sum_{K}\Gamma_{K\alpha}^{\ell}\}. \tag{32}$$

Here the summation over K extends over all occupied molecular spin orbitals.

The above approach, although somewhat involved, is more fundamental and provides very useful information. In particular, Eq. (31) shows that the rocking motions of non-spherical molecular orbitals contribute to the magnetic dipole transition moment. In practice the molecular orbitals obtained in Hartree-Fock calculations are linear combinations [72] of several atomic orbitals of the Slater type [65] or Gaussian type [73], and for such molecular orbitals it is not trivial to obtain the rocking contribution $r_{K\alpha}^{\ell}$. Therefore, in practical calculations, one assumes that if the molecular orbitals are well localized, $r_{K\alpha}^{\ell}$ will be small and can be ignored. If canonical orbitals are used, such orbitals are generally delocalized over the entire molecule and it is difficult to visualize them to be spherical. Thus, the canonical orbitals can introduce serious errors into the calculated rotational strength, unless Eq. (31) is evaluated to a good approximation.

An alternative approach has been suggested by Nafie and Walnut [35] and somewhat later in a more general form along with calculations by Nafie and Polavarapu [36,37]. In this approach the electron coordinates are first referred to the average positions of electrons. Then the electronic part of the magnetic dipole transition moment becomes

$$\langle \psi_{01}^{\ell} | m_{\alpha}^{el} | \psi_{00} \rangle = \sum_{K} - \frac{e}{2mc} \varepsilon_{\alpha\beta\gamma} \, r_{K0\beta} \, \langle \psi_{01}^{\ell} | P_{K\gamma} | \psi_{00} \rangle$$

$$+ \langle \psi_{01}^{\ell} | \sum_{K} - \frac{e}{2mc} \varepsilon_{\alpha\beta\gamma} \, \Delta r_{K\beta} \, P_{K\gamma} | \psi_{00} \rangle \tag{33}$$

where $\Delta r_{K\beta}$ is the difference between the instantaneous position $r_{K\beta}$ and the average position $r_{K0\beta}$. The second term in Eq. (33), which is generally referred to as the intrinsic term [35], is ignored on the basis that, if the molecular orbitals are considered localized, then $\Delta r_{\underset{\sim}{K}}$ will be much smaller than $r_{\underset{\sim}{K0}}$. Now the momentum operator is converted into the positional operator and the Born-Oppenheimer approximation is invoked. In the original formulation [35], the molecular wavefunction is considered to be a product of vibronic molecular spin orbitals and, in the latter formulation [36], the electronic wavefunction is represented by a Slater determinant [65] constructed from the localized molecular orbital functions.

The electronic contribution to the magnetic dipole transition moment obtained in Eq. (33) can be seen to be identical to that obtained in a fundamental approach, as given in Eq. (30), when the intrinsic term of Eq. (33) and the rocking term of Eq. (30) are ignored. Thus, both approaches lead to an identical result in the limit of spherical molecular orbitals. It is possible in principle to estimate the contribution of $\Gamma^{\ell}_{K\alpha}$, at least for Gaussian molecular orbitals [34], but it is not clear if a procedure even exists for calculating the intrinsic terms of Eq. (33).

Recently, Freedman and Nafie proposed [74] to replace $\underset{\sim}{r}_{K0}$ in Eq. (32) by nuclear positional vectors so as to obtain the VCD expression in terms of atomic polar tensors [75-76]. As no theoretical justification for such replacement is offered it is hard to validate the resulting VCD expression.

b.   <u>Vibronic Theory</u>.   Craig and Thirunamachandran have treated [42] circular dichroism in vibrational transitions in a manner similar to that of vibronic interactions in electronic transitions. They consider that a fundamental vibrational transition, within the ground electronic state, is capable of mixing the ground electronic state with a higher electronic state. By expressing the molecular wavefunction in an adiabatic approximation, they expand the electronic wavefunction about the equilibrium position as in the Herzberg-Teller approach.

$$\phi_0(r,Q) = \phi_0(r,Q_0) + \sum_{\ell} \sum_{u \neq 0} \frac{\langle \phi_u(r,Q_0)|A_\ell|\phi_0(r,Q_0)\rangle}{E_u - E_0} Q_\ell \phi_u(r,Q_0) \qquad (34)$$

where ϕ_u is a higher state electronic wavefunction with energy E_u, E_0 is the ground state energy and A_ℓ is the vibronic interaction operator. This expanded wavefunction is then used to calculate the rotational strength. Out of the four contributions to the rotational strength, the nuclear parts are nevertheless identical to those obtained within the Born-Oppenheimer approximation, i.e., to Eqs. (5a) and (5c). For electronic contributions to the electric dipole transition moment they obtain an expression involving higher state mixing. The electronic contribution to the magnetic dipole transition moment however is found to be zero. In addition to the adiabatic approximation, a crude adiabatic approximation is also considered, where again the electronic contribution to the magnetic dipole transition moment is predicted to be either zero or very small.

Although the procedures to evaluate the electric dipole transition moment integrals involving the vibronic interaction operator are known in the literature [77], no attempts have so far been made to calculate VCD intensities in this approach. This might be due to two points which are relevant to the discussion in earlier sections. Walnut and Nafie have concluded [34] that the VCD intensities do not have contributions from higher electronic states and that the electronic contributions to the magnetic dipole transition moment are significant and can be incorporated into the rotational strength expression.

The opposite view emerges from vibronic theory. The second point regards the origin dependency of the rotational strength value. In Sec. III.A.1 it was pointed out that if the electronic contribution to the electric dipole transition moment is included, and that to the magnetic dipole transition moment is not, then the rotational strength would become origin dependent. A similar situation is apparent in vibronic theory and this point was not addressed.

c. <u>Dynamic Polarization Model.</u> Barnett et al. have suggested [30,31] that the dynamic coupling model used in EOA calculations can be extended to VCD calculations. In this model the chiral molecule is divided into a symmetric vibrational chromophore and the substituent groups, which provide the chiral environment. The vibration under consideration, of the symmetric chromophore, is considered to induce an electric dipole moment in each of the substituent groups as dictated by their polarizabilities. This can result in a first order magnetic dipole transition moment which is represented by a vector product of the aforementioned induced electric dipole moments and the distance vectors between the vibrational chromophore and substituent groups. This point is conceptually equivalent to that in the polarization model described by Laux [14]. Operationally, however, the dynamic polarization model takes a different pathway and considers only the first order rotational strength. This is given by the dot product of the electric dipole transition moment of the vibration under consideration and the first order magnetic dipole transition moment as described above. In order to obtain the first order magnetic dipole transition moment, the ground and excited vibrational states of the chromophore are corrected with higher electronic and vibrational states of substituent groups. These corrections are considered to be arising from the dipolar interactions between the chromophore and substituents. After mathematical simplifications, the rotational strength for the chromophore transition of frequency ω_ℓ is found to be

$$R_\ell = \frac{\pi \omega_\ell}{2} D_\ell \sum_A \frac{1}{R_A^5} \{\sin 2\theta [R_{AX}(R_A^2 - 3R_{AZ}^2) \sin \chi$$

$$+ R_{AY}(R_A^2 - 3R_{AZ}^2) \cos \chi](\alpha_{AL} - \alpha_{AN})$$

$$+ (\alpha_{AM} - \alpha_{AN} \sin^2\theta - \alpha_{AL} \cos^2\theta) [6R_{AX}R_{AY}R_{AZ} \cos \chi$$

$$+ 3R_{AZ}(R_{AX}^2 - R_{AY}^2)]\} \qquad (35)$$

where D_ℓ is the dipole strength of the chromophore transition with the transition moment directed along the Z axis, $\underset{\sim}{R}_A$ are the equilibrium coordinates of the origin of substituent A whose principal components of the polarizability are α_{AL}, α_{AM}, and α_{AN}, and θ and χ are the angles relating the principal axes of the substituent with the chromophore axes. The summation in Eq. (35) is extended over all substituents A.

The dipole strength can be obtained from infrared absorption intensity and the polarizabilities of substituents from the literature. Thus, this model requires no parameterization. One notable difference in this model is that the rotational strength is dependent on the dipole strength. Therefore, the sign of the electric dipole transition moment is predicted to have no influence on the sign of VCD.

    d.   <u>Coupled Oscillator and Related Models.</u>   This conceptually simple model has been put forward by Holzwarth and Chabay [21] in the early stages of VCD research. In this model two identical oscillators ℓ and ℓ', separated by a distance $R_{\ell\ell'}$ and oriented at an angle in such a way as to make a chiral structure, are considered. If the ground state wavefunctions for the two oscillators are ψ_ℓ and $\psi_{\ell'}$ then the total ground state wavefunction ψ_0 is considered to be the product $\psi_\ell\psi_{\ell'}$. The excited state wavefunctions are given by $\psi'_\ell\psi_{\ell'}$ and $\psi_\ell\psi'_{\ell'}$ where the prime on the wavefunction represents a single excitation. Due to the perturbation interaction between the oscillators, the excited state wavefunctions can mix which results in two mixed functions ψ_+ and ψ_-. This development is similar in electronic and vibrational CD calculations. For example, the two oscillators in VCD calculations can be two C-H stretching vibrations in which case the mixed wavefunctions represent the symmetric and antisymmetric combinations of the two C-H stretching vibrations. The rotational strength for such vibrations is given [11,23] in a general form as

$$R^\pm = \frac{1}{2}\,\text{Im}(\mu_\alpha^\ell m_\alpha^\ell + \mu_\alpha^{\ell'} m_\alpha^{\ell'}) \pm \frac{1}{2}\,\text{Im}[(\mu_\alpha^\ell m_\alpha^{\ell'} + \mu_\alpha^{\ell'} m_\alpha^\ell) -$$

$$\pi\,\omega_\pm\,\varepsilon_{\alpha\beta\gamma}(R_{\ell'\alpha} - R_{\ell\alpha})\mu_\beta^{\ell'}\,\mu_\gamma^\ell] \qquad (36a)$$

where μ_α^ℓ, m_α^ℓ, $\mu_\alpha^{\ell'}$ and $m_\alpha^{\ell'}$ represent the electric and magnetic dipole transition moments of the oscillators ℓ and ℓ', and $\omega_\pm$ are the vibrational frequencies of two transitions designated by + and -. For groups lacking inherent chirality the first term vanishes. Then the coupled oscillator mechanism predicts oppositely signed CD for the vibrations of two oscillators.

The two degenerate stretching vibrations of a methyl group represent a special case of the above mechanism. In this case, the $R_{\ell\ell'}$ term vanishes since both vibrations are located on the same group and the first term of Eq.

(36a) vanishes due to local symmetry. Then, in a chiral environment, the degenerate stretching modes of a methyl group can have a non-vanishing rotational strength given [24] by the remaining term, i.e.,

$$R^{\pm} = \pm \frac{1}{2} \operatorname{Im}(\mu_\alpha^\ell m_\alpha^{\ell'} + \mu_\alpha^{\ell'} m_\alpha^\ell). \tag{36b}$$

Again, oppositely signed CD for the two vibrations, whose degeneracy can be lifted by a chiral perturbation, is predicted.

Recently Laux has extended [14] the coupled oscillator formulation to situations where three nondegenerate oscillators participate in exhibiting a recurring VCD sign pattern. A specific example is the two methylene antisymmetric stretching motions and a methine stretching motion in a $CH_2CH_2\overset{*}{C}H$ fragment. The ground state wavefunction of this fragment is considered to be the product of the ground vibrational states of the three oscillators. The excited states are formulated as linear combinations of the single excited states of the three oscillators with appropriate coupling coefficients. Using these wavefunctions the rotational strength is expressed in an extremely simple form to be

$$R_\ell = \frac{\pi}{\omega_\ell N_\ell} \sum_{\ell \neq \ell'} C_{\ell\ell'} \, \omega_\ell^0 \, \omega_{\ell'}^0 \, \varepsilon_{\alpha\beta\gamma}(R_{\ell'\alpha} - R_{\ell\alpha})\mu_\beta^{\ell'} \, \mu_\gamma^\ell \tag{37}$$

where $C_{\ell\ell'}$ are the coupling coefficients in the perturbed wavefunction, N_ℓ is a normalization factor, and ω_ℓ^0 and ω_ℓ are the vibrational frequencies of the oscillator ℓ when it is isolated and is in a whole molecular fragment.

B.   Raman Optical Activity

1.   Basic Expressions

The difference in scattered intensities for alternating left and right circularly polarized light incident on an optically active sample can be measured in forward, backward and 90° scattering directions. In the latter case the measurements with an analyzer oriented parallel and perpendicular to the scattering plane have different content [7]. For measurements in a 90° scattering direction with the analyzer oriented parallel to the scattering plane, the difference in scattered intensities $(I_p^L - I_p^R)$ is given as [7]

$$I_p^L - I_p^R = K \left[\frac{4\omega}{6c} \alpha_{\alpha\beta}\varepsilon_{\alpha\gamma\delta}A_{\gamma\delta\beta} - \frac{4\omega}{2}(3\alpha_{\alpha\beta}\beta_{\alpha\beta} - \alpha_{\alpha\alpha}\beta_{\beta\beta}) \right] \tag{38}$$

where $\alpha_{\alpha\beta}$ is the molecular electric dipole polarizability tensor, $\beta_{\alpha\beta}$ and $A_{\alpha\beta\gamma}$ are magnetic dipole and electric quadrupole optical activity tensors, ω is the angular frequency of incident light and K is a constant of the experiment. The fundamental formulas for these tensors are given as [9]

$$\alpha_{\alpha\beta} = \frac{4\pi}{h} \sum_{j\neq n} \frac{\omega_{jn}}{\omega_{jn}^2 - \omega^2} \, \text{Re}[<n|\mu_\alpha|j><j|\mu_\beta|n>] \tag{39a}$$

$$\beta_{\alpha\beta} = -\frac{4\pi}{h} \sum_{j\neq n} \frac{1}{\omega_{jn}^2 - \omega^2} \, \text{Im}[<n|\mu_\alpha|j><j|m_\beta|n>] \tag{39b}$$

$$A_{\alpha\beta\gamma} = \frac{4\pi}{h} \sum_{j\neq n} \frac{\omega_{jn}}{\omega_{jn}^2 - \omega^2} \, \text{Re}[<n|\mu_\alpha|j><j|\theta_{\beta\gamma}|n>] \tag{39c}$$

where μ_α, m_β and $\theta_{\beta\gamma}$, respectively, are the electric dipole moment, magnetic dipole moment, and electric quadrupole moment operators, n and j are the ground and excited molecular states, the mixing of which occurs due to the perturbation from the incident light of angular frequency ω, and $\omega_{jn} = \omega_j - \omega_n$.

In vibrational Raman scattering, the tensors $\underset{\sim}{\alpha}$, $\underset{\sim}{\beta}$ and $\underset{\sim}{A}$ in Eq. (38) are replaced by the Raman transition tensors. For a fundamental vibrational transition represented by normal mode Q_ℓ, the product $\alpha_{\alpha\beta}\beta_{\alpha\beta}$, for example, is replaced by

$$\alpha_{\alpha\beta}^\ell \beta_{\alpha\beta}^\ell <\chi_1^\ell |Q_\ell| \chi_0^\ell >^2$$

where

$$\alpha_{\alpha\beta}^\ell = (\partial\alpha_{\alpha\beta}/\partial Q_\ell)_0 \text{ and } \beta_{\alpha\beta}^\ell = (\partial\beta_{\alpha\beta}/\partial Q_\ell)_0$$

and the integral is one of the harmonic oscillator functions as defined in Eq. (8a). It is convenient to express ROA in terms of the chirality numbers q_p, where

$$q_p = 2(I_p^L - I_p^R)/(I_p^L + I_p^R)$$

which is analogous to the dissymmetry factor g defined in electronic and vibrational circular dichroism measurements. In this notation, the chirality numbers for four different ROA measurements are given as

$$q_p = -4\omega \left[\frac{3\alpha^\ell_{\alpha\beta} \beta^\ell_{\alpha\beta} - \alpha^\ell_{\alpha\alpha} \beta^\ell_{\beta\beta} - \frac{1}{3c} \alpha^\ell_{\alpha\beta} \varepsilon_{\alpha\gamma\delta} A^\ell_{\gamma\delta\beta}}{(3\alpha^\ell_{\alpha\beta} \alpha^\ell_{\alpha\beta} - \alpha^\ell_{\alpha\alpha} \alpha^\ell_{\beta\beta})} \right] \tag{40a}$$

$$q_s = -4\omega \left[\frac{7\alpha^\ell_{\alpha\beta} \beta^\ell_{\alpha\beta} + \alpha^\ell_{\alpha\alpha} \beta^\ell_{\beta\beta} + \frac{1}{3c} \alpha^\ell_{\alpha\beta} \varepsilon_{\alpha\gamma\delta} A^\ell_{\gamma\delta\beta}}{7\alpha^\ell_{\alpha\beta} \alpha^\ell_{\alpha\beta} + \alpha^\ell_{\alpha\alpha} \alpha^\ell_{\beta\beta}} \right] \tag{40b}$$

$$q_b = -16\omega \left[\frac{3\alpha^\ell_{\alpha\beta} \beta^\ell_{\alpha\beta} - \alpha^\ell_{\alpha\alpha} \beta^\ell_{\beta\beta} + \frac{1}{3c} \alpha^\ell_{\alpha\beta} \varepsilon_{\alpha\gamma\delta} A^\ell_{\gamma\delta\beta}}{7\alpha^\ell_{\alpha\beta} \alpha^\ell_{\alpha\beta} + \alpha^\ell_{\alpha\alpha} \beta^\ell_{\beta\beta}} \right] \tag{40c}$$

$$q_f = -8\omega \left[\frac{\alpha^\ell_{\alpha\beta} \beta^\ell_{\alpha\beta} + 3\alpha^\ell_{\alpha\alpha} \beta^\ell_{\beta\beta} - \frac{1}{3c} \alpha^\ell_{\alpha\beta} \varepsilon_{\alpha\gamma\delta} A^\ell_{\gamma\delta\beta}}{7\alpha^\ell_{\alpha\beta} \alpha^\ell_{\alpha\beta} + \alpha^\ell_{\alpha\alpha} \alpha^\ell_{\beta\beta}} \right] \tag{40d}$$

Here the subscripts f and b represent the measurements in forward and backward scattering arrangements while s and p represent the measurements in a 90° direction with the analyzer perpendicular and parallel to the scattering plane. To predict or compare with the experimental ROA intensities one seeks to evaluate the chirality numbers given by Eq. (40), and different theoretical models that are formulated for this purpose are described in the following sections.

In this section we have glossed over the basic theoretical expressions since our main motivation is to discuss the models employed for calculating ROA intensities. We refer to several informative reviews written by Barron [7-10] for further details. Nevertheless, from the above equations it may be noted that no formal difficulties such as those encountered in VCD formulation are present here. But one can anticipate that the quantum mechanical evaluation of the tensors $\underset{\sim}{\alpha}^\ell$, $\underset{\sim}{\beta}^\ell$ and $\underset{\sim}{A}^\ell$ would be quite involved due to the involvement of excited states in Eqs. (39).

An explanation of small differences in notation is in order. The magnetic dipole optical activity tensor is represented here by β and is related to the Barron's definition of G through the relation $c\beta = \underset{\sim}{G}'/\omega$. Secondly, the difference in scattered Raman intensities considered by Barron is $I^R - I^L$. To be in line with the definitions used in other optical activity measurements we consider the difference $I^L - I^R$ as suggested in the literature [78]. Therefore, the chirality parameter q should be noted to be equal to two times the circular intensity differential, defined [45] as $(I^R - I^L)/(I^R + I^L)$, with opposite sign.

2.   Semiclassical Models

a.   Bond Polarizability Model.   Although the bond polarizability model calculations for ROA intensities have been performed only recently [79-81], this model has its roots in the simple two group model developed by Barron and Buckingham in 1974 [46].   To obtain suitable formulas for optical activity tensors they considered the origin dependence of the β and A tensors.   If the origin is shifted from O to $O + a$ where a is a constant vector, for a neutral molecule the electric dipole moment operator μ_α is invariant but the magnetic dipole and electric quadrupole moment operators change from

$$m_\alpha \text{ to } m_\alpha - (1/2c)\, \varepsilon_{\alpha\beta\gamma} a_\beta \dot{\mu}_\gamma$$

and

$$\theta_{\alpha\beta} \text{ to } \theta_{\alpha\beta} - (3/2)(\mu_\alpha a_\beta + \mu_\beta a_\alpha) + \mu_\gamma a_\gamma \delta_{\alpha\beta}.$$

On substituting these expressions into Eq. (39), β and A tensors change [82] from

$$\beta_{\alpha\beta} \text{ to } \beta_{\alpha\beta} + (1/2c)\varepsilon_{\beta\gamma\delta} a_\gamma{}^\alpha{}_{\delta\alpha}$$

and

$$A_{\alpha\beta\gamma} \text{ to } A_{\alpha\beta\gamma} - (3/2)(a_\gamma{}^\alpha{}_{\beta\alpha} + a_\beta{}^\alpha{}_{\gamma\alpha}) + a_\lambda{}^\alpha{}_{\lambda\alpha} \delta_{\beta\gamma}.$$

Now consider a molecule with two groups j and j' separated by a distance and oriented at an angle to make a chiral structure as in the coupled oscillator model for VCD.   Choosing the molecular origin to be the same as the local origin of the group j', the optical activity tensors are written to be

$$\beta_{\alpha\beta} = \beta_{j\alpha\beta} + \beta_{j'\alpha\beta} - \frac{1}{2c}\, \varepsilon_{\beta\gamma\delta} R_{jj'\gamma}{}^\alpha{}_{j\delta\alpha} \tag{41a}$$

$$A_{\alpha\beta\gamma} = A_{j\alpha\beta\gamma} + A_{j'\alpha\beta\gamma} + \frac{3}{2} R_{jj'\beta}{}^\alpha{}_{j\gamma\alpha} + \frac{3}{2} R_{jj'\gamma}{}^\alpha{}_{j\beta\alpha} - R_{jj'\lambda}{}^\alpha{}_{\lambda\alpha} \delta_{\beta\gamma} \tag{41b}$$

where $R_{\sim jj'} = R_{\sim j} - R_{\sim j'}$, $\alpha_{\sim j}$ is the electric dipole polarizability of the jth group, and β_j and $A_{\sim j}$ are the intrinsic group optical activity tensors. It is important to note that the molecular as well as the group polarizabilities, $\alpha_{\sim}$ and $\alpha_{\sim j}$, are considered symmetric. These equations are now generalized to any molecule by considering each chemical bond to represent a group, and the molecular tensors are written as [46,9].

$$\alpha_{\alpha\beta} = \sum_j \alpha_{j\alpha\beta} \tag{42a}$$

$$\beta_{\alpha\beta} = \sum_j \beta_{j\alpha\beta} - \frac{1}{2c} \sum_j \varepsilon_{\beta\gamma\delta} R_{j\gamma} \alpha_{j\delta\alpha} \tag{42b}$$

$$A_{\alpha\beta\gamma} = \sum_j A_{j\alpha\beta\gamma} + \sum_j \{\frac{3}{2} (R_{j\beta}\alpha_{j\gamma\alpha} + R_{j\gamma}\alpha_{j\beta\alpha}) - R_{j\lambda}\alpha_{j\lambda\alpha}\delta_{\beta\gamma}\}. \tag{42c}$$

Here $R_{\sim j}$ represents the distance vector from a chosen molecular origin to the local bond origin. In order to obtain the chirality numbers, the normal coordinate gradients of the above tensors are required. Thus

$$\alpha^\ell_{\alpha\beta} = \sum_j (\frac{\partial\alpha_{j\alpha\beta}}{\partial Q_\ell})_0 \tag{43a}$$

$$\beta^\ell_{\alpha\beta} = \sum_j (\frac{\partial\beta_{j\alpha\beta}}{\partial Q_\ell})_0 - \frac{1}{2c} \sum_j \varepsilon_{\beta\gamma\delta}(R^0_{A_j\gamma} (\frac{\partial\alpha_{j\delta\alpha}}{\partial Q_\ell})_0 + S^\ell_{A_j\gamma} \alpha^0_{j\delta\alpha}) \tag{43b}$$

$$A^\ell_{\alpha\beta\gamma} = \sum_j (\frac{\partial A_{j\alpha\beta\gamma}}{\partial Q_\ell})_0 + \sum_j [\frac{3}{2}\{R^0_{A_j\beta} (\frac{\partial\alpha_{j\gamma\alpha}}{\partial Q_\ell})_0 + S^\ell_{A_j\beta} \alpha^0_{j\gamma\alpha}$$

$$+ R^0_{A_j\gamma} (\frac{\partial\alpha_{j\beta\alpha}}{\partial Q_\ell})_0 + S^\ell_{A_j\gamma} \alpha^0_{j\beta\alpha}\}$$

$$- R^0_{A_j\lambda} (\frac{\partial\alpha_{j\lambda\alpha}}{\partial Q_\ell})_0 \delta_{\beta\gamma} - S^\ell_{A_j\lambda} \alpha^0_{j\lambda\alpha} \delta_{\beta\gamma}] \tag{43c}$$

where again the superscript or subscript zero represents the values at equilibrium position. In the above equations, for the sake of simplicity we
assumed that the local origin in a bond is at one of the two atoms constituting
the bond; thus $S^\ell_{Aj\alpha} = (\partial R_{A\alpha}/\partial Q_\ell)_0$ where A is one of the two atoms of bond
j. As long as the local origin is chosen along the length of the bond, the
calculated ROA intensities will be independent of bond origin. To obtain ROA
intensities, appropriate products of $\underset{\sim}{\alpha}^\ell$, $\underset{\sim}{\beta}^\ell$ and $\underset{\sim}{A}^\ell$ are substituted into Eqs.
(40). Some simplifications can be incorporated if all the groups (or bonds) in
a molecule are considered to be axially symmetric and achiral. Then the
intrinsic optical activity tensors vanish and the following relations apply.

$$\alpha^\ell_{\alpha\alpha}\, \beta^\ell_{\beta\beta} = 0 \tag{44a}$$

$$\alpha^\ell_{\alpha\beta}\, \beta^\ell_{\alpha\beta} = \frac{1}{3c}\, \alpha^\ell_{\alpha\beta}\, \varepsilon_{\alpha\gamma\delta}\, A^\ell_{\gamma\delta\beta}. \tag{44b}$$

Then ROA intensities can be seen to depend on a single term $\alpha^\ell_{\alpha\beta}\, \beta^\ell_{\alpha\beta}$, which
in a simplified form is given as

$$\alpha^\ell_{\alpha\beta}\, \beta^\ell_{\alpha\beta} = \sum_{j'<j} - \frac{1}{2c}\, \varepsilon_{\beta\gamma\delta}\, R^0_{jj'\gamma}\, \left(\frac{\partial \alpha_{j'\alpha\beta}}{\partial Q_\ell}\right)_0 \left(\frac{\partial \alpha_{j\delta\alpha}}{\partial Q_\ell}\right)_0$$

$$+ \left[\sum_{j'} \left(\frac{\partial \alpha_{j'\alpha\beta}}{\partial Q_\ell}\right)_0\right]\left[\sum_{j} - \frac{1}{2c}\, \varepsilon_{\beta\gamma\delta}\, S^\ell_{A\,j\gamma}\, \alpha^0_{j\delta\alpha}\right]. \tag{45}$$

The chirality numbers are simply the normalized versions of this expression.
The first term in Eq. (45) represents the interaction between all pairs of
bonds and represents a generalization of the simple two group model [46].
The second term represents the changes in the positional vectors of the
bonds during normal modes of vibrations and is referred to as the inertial
term [9,10].
     Since the vectors $S^\ell_{\underset{\sim}{A}}$ are obtained from normal coordinate analyses, the
only quantities required for calculating the chirality numbers are the bond
polarizabilities and their normal coordinate derivatives. The bond polarizabilities used in Eqs. (41-43) are defined with respect to the chosen molecular coordinate system. In practice, however, the bond polarizabilities are
defined in local bond axes first and then transformed to the molecular axes.
This is done in two different ways.

If the bonds are assumed to be axially symmetric, then the bond polarizability in principal molecular axes can be written in a simplified form [79,80] as

$$\alpha_{j\alpha\beta} = \alpha_{js}\delta_{\alpha\beta} + (\alpha_{jp} - \alpha_{js})U_{j\alpha}U_{j\beta} \tag{46}$$

where α_{js} and α_{jp} are the bond polarizability components perpendicular and parallel to the principal bond axes. Then the bond polarizability derivatives in the principal molecular coordinate system are given as

$$(\frac{\partial\alpha_{j\alpha\beta}}{\partial Q_\ell})_0 = (\frac{\partial\alpha_{js}}{\partial Q_\ell})_0 \delta_{\alpha\beta} + [(\frac{\partial\alpha_{jp}}{\partial Q_\ell})_0 - (\frac{\partial\alpha_{js}}{\partial Q_\ell})_0] U_{j\alpha}^0 U_{j\beta}^0 +$$

$$\frac{(\alpha_{jp} - \alpha_{js})^0}{r_j^0} [\sum_A (\delta_{\alpha\lambda}U_{j\beta}^0 + \delta_{\beta\lambda}U_{j\alpha}^0 - 2U_{j\alpha}^0 U_{j\beta}^0 U_{j\lambda}^0) S_{A\lambda}^\ell \Delta_{jA}] \tag{47}$$

where the derivatives of the unit vectors are taken from Eq. (13). If the bond j of length r_j is between atoms B and C, and the unit vector is defined as $(R_{B\alpha} - R_{C\alpha})/r_j$ then $\Delta_{jA} = 1$ for $A = B$, $\Delta_{jA} = -1$ for $A = C$ and $\Delta_{jA} = 0$ for all other atoms.

In an alternate approach, without limiting the bonds to be axially symmetric and without limiting them to the principal molecular axes, the bond polarizabilities can be written in a general form as

$$\alpha_{j\alpha\beta} = T_{j\alpha\gamma} \alpha_{j\gamma\gamma}^b T_{j\beta\gamma} \tag{48a}$$

where $T_j = \begin{bmatrix} \dfrac{U_{jz}U_{jx}}{(U_{jx}^2 + U_{jy}^2)^{\frac{1}{2}}} & \dfrac{-U_{jy}}{(U_{jx}^2 + U_{jy}^2)^{\frac{1}{2}}} & U_{jx} \\[3mm] \dfrac{U_{jz}U_{jy}}{(U_{jx}^2 + U_{jy}^2)^{\frac{1}{2}}} & \dfrac{U_{jx}}{(U_{jx}^2 + U_{jy}^2)^{\frac{1}{2}}} & U_{jy} \\[3mm] -(U_{jx}^2 + U_{jy}^2)^{\frac{1}{2}} & 0 & U_{jz} \end{bmatrix}$ (48b)

Here T_j is the transformation matrix [83] between local axes of bond j and the chosen molecular axes, and $\alpha^b_{j\gamma\gamma}$ represents the diagonal bond polarizability tensor in the bond axes. The three components of $\underset{\sim}{\alpha}^b_j$ can be chosen to be different. Now the bond polarizability derivatives are given as

$$(\frac{\partial \alpha_{j\alpha\beta}}{\partial Q_\ell})_0 = (\frac{\partial T_{j\alpha\gamma}}{\partial Q_\ell})_0 \, (\alpha^b_{j\gamma\gamma})_0 \, T^0_{j\beta\gamma} + T^0_{j\alpha\gamma} \, (\frac{\partial \alpha^b_{j\gamma\gamma}}{\partial Q_\ell})_0 \, T^0_{j\beta\gamma}$$

$$+ \, T^0_{j\alpha\gamma} \, (\alpha^b_{j\gamma\gamma})_0 \, (\frac{\partial T_{j\beta\gamma}}{\partial Q_\ell})_0. \tag{49}$$

The derivatives of the transformation matrix $\underset{\sim}{T}_j$ can be obtained in a manner similar to that used in Eq. (47). We have used this procedure for ROA calculations in our laboratory. For axially symmetric bond polarizabilities and the principal molecular coordinate system, both approaches gave the same results.

Substitution of either Eqs. (48) and (49) or Eqs. (46) and (47) into Eqs. (40) and (43) will provide the necessary formulation for obtaining the chirality parameters.

b. Atom Dipole Interaction Model. The theory of the atom dipole interaction (ADI) model [84] has been used in the calculation of EOA [85], molecular polarizabilities [86,87], light scattering [88], collisional polarizability anisotropy [89] and Raman intensities [90]. The ADI model for ROA [51-53] is an extension of the formulation for other molecular properties. In this model the molecule is considered to be composed of polarizable atoms and each atom is assumed to have an isotropic polarizability α_A. In the presence of a uniform electric field E^0 of the incident light, an electric moment is induced at each atom. The interaction among the induced electric moments alters the electric field felt at each atom. The effective electric field, E_A, is

$$E_{A\alpha} = E^0_{A\alpha} - \underset{B \neq A}{\Sigma} T_{AB\alpha\beta} \, \mu_{B\beta}, \tag{50}$$

where μ_B is the induced electric dipole moment and $\underset{\sim}{T}_{AB}$ is the dipolar interaction tensor given as

$$
\underset{\sim}{T}_{AB} =
\begin{bmatrix}
\dfrac{1}{R_{AB}^3} - \dfrac{3R_{ABx}^2}{R_{AB}^5} & - \dfrac{3R_{ABx}R_{ABy}}{R_{AB}^5} & - \dfrac{3R_{ABx}R_{ABz}}{R_{AB}^5} \\[2ex]
& \dfrac{1}{R_{AB}^3} - \dfrac{3R_{ABy}^2}{R_{AB}^5} & - \dfrac{3R_{ABy}R_{ABz}}{R_{AB}^5} \\[2ex]
\text{sym} & & \dfrac{1}{R_{AB}^3} - \dfrac{3R_{ABz}^2}{R_{AB}^5}
\end{bmatrix}
\tag{51}
$$

with $\underset{\sim}{R}_{AB}$ representing the distance vector between atoms A and B. The induced electric dipole moment at each atom is given as

$$
\mu_{A\alpha} = \alpha_A\,\delta_{\alpha\beta}\Big(E_{A\alpha}^0 - \sum_{B\neq A} T_{AB\beta\gamma}\,\mu_{B\gamma}\Big)
$$

and can be rearranged to give

$$
E_{A\alpha}^0 = \sum_B A_{AB\alpha\beta}\,\mu_{B\beta}.
\tag{52a}
$$

The $\underset{\sim}{A}$ matrix, which should not be confused with the electric quadrupole optical activity tensor $A_{\alpha\beta\gamma}$, is explicitly shown to be

$$
\underset{\sim}{A} =
\begin{bmatrix}
\alpha_1^{-1} & 0 & 0 & T_{12}^{xx} & T_{12}^{xy} & T_{12}^{xz} & T_{13}^{xx} & T_{13}^{xy} & T_{13}^{xz} & - & - \\[1.5ex]
0 & \alpha_1^{-1} & 0 & T_{12}^{yx} & T_{12}^{yy} & T_{12}^{yz} & T_{13}^{yx} & T_{13}^{yy} & T_{13}^{yz} & - & - \\[1.5ex]
0 & 0 & \alpha_1^{-1} & T_{12}^{zx} & T_{12}^{zy} & T_{12}^{zz} & T_{13}^{zx} & T_{13}^{zy} & T_{13}^{zz} & - & - \\[1.5ex]
T_{21}^{xx} & T_{21}^{xy} & T_{21}^{xz} & \alpha_2^{-1} & 0 & 0 & T_{23}^{xx} & T_{23}^{xy} & T_{23}^{xz} & - & - \\[1.5ex]
T_{21}^{yx} & T_{21}^{yy} & T_{21}^{yz} & 0 & \alpha_2^{-1} & 0 & T_{23}^{yx} & T_{23}^{yy} & T_{23}^{yz} & - & - \\[1.5ex]
T_{21}^{zx} & T_{21}^{zy} & T_{21}^{zz} & 0 & 0 & \alpha_2^{-1} & T_{23}^{zx} & T_{23}^{zy} & T_{23}^{zz} & - & - \\[1.5ex]
- & - & - & - & - & - & - & - & - & - & -
\end{bmatrix}
\tag{52b}
$$

where T_{12}^{xy}, for example, are the appropriate elements of tensor $\underset{\sim}{T}_{AB}$ given by Eq. (51). Rearrangement of Eq. (52a) gives

$$\mu_{A\alpha} = \sum_B B_{AB\alpha\beta} E^0_{B\beta}, \tag{53}$$

where the $\underset{\sim}{B}$ matrix is known [84] as the relay tensor matrix. Assuming that the external electric field $\underset{\sim}{E}^0$ is uniform over molecular dimension, the atomic and molecular polarizabilities are obtained as

$$\alpha_{A\alpha\beta} = \sum_B B_{AB\alpha\beta} \tag{54a}$$

and

$$\alpha_{\alpha\beta} = \sum_A \alpha_{A\alpha\beta} = \sum_A \sum_B B_{AB\alpha\beta}. \tag{54b}$$

The central theme of the ADI model is to obtain the polarizabilities that have sampled the internal electric field of the molecule so that each atom 'knows' its environment. Thus, although the molecular polarizability is always symmetric, the atomic polarizabilities obtained from Eq. (54a) need not be symmetric.

To formulate the optical activity tensors, the procedure adapted in the ADI model is slightly different from that used in the bond polarizability model. The magnetic dipole moment operator given in Eq. (7b) is rewritten as

$$m_\beta = \frac{1}{2c} \sum_A \varepsilon_{\beta\gamma\delta} R_{A\gamma} \dot{\mu}_{A\delta} \tag{55}$$

where $\dot{\mu}_{A\delta}$ is the time derivative of $\mu_{A\delta}$, representing the oscillation in time of the electric dipole moments synchronous to that of the electric vector of the incident light wave. Then $\dot{\mu}_{A\delta}$ is equivalent to $\alpha_{A\delta\alpha}\dot{E}_\alpha$. Making use of the relation between the magnetic dipole moment $\underset{\sim}{m}$, the optical activity tensor $\underset{\sim}{\beta}$, and the oscillating electric field $\underset{\sim}{E}$, which is given as $m_\beta = -\beta_{\alpha\beta}\dot{E}_\alpha$ [9], the optical activity tensor β is formulated as

$$\beta_{\alpha\beta} = - \frac{1}{2c} \sum_A \varepsilon_{\beta\gamma\delta} R_{A\gamma} \alpha_{A\delta\alpha}. \tag{56}$$

Similarly, to formulate the electric quadrupole optical activity tensor $\underset{\sim}{A}$, the corresponding operator $\theta_{\beta\gamma}$, where

$$\theta_{\beta\gamma} = \sum_A \{ \frac{3}{2} (R_{A\beta}\mu_{A\gamma} + R_{A\gamma}\mu_{A\beta}) - R_{A\lambda}\mu_{A\lambda}\delta_{\beta\gamma}\}, \tag{57}$$

and the relation $\theta_{\beta\gamma} = A_{\alpha\beta\gamma}E_{\alpha}$ are used. Then the expression for $A_{\alpha\beta\gamma}$ is given to be

$$A_{\alpha\beta\gamma} = \sum_{A} \{\tfrac{3}{2} (R_{A\beta}{}^{\alpha}A_{\gamma\alpha} + R_{A\gamma}{}^{\alpha}A_{\beta\alpha}) - R_{A\lambda}{}^{\alpha}A_{\lambda\alpha}\delta_{\beta\gamma}\}. \tag{58}$$

The expressions for $\beta_{\alpha\beta}$ and $A_{\alpha\beta\gamma}$ are identical to those used in the bond polarizability model except that in Eqs. (56) and (58) the intrinsic optical activity tensors are omitted. This is justifiable since in the ADI model the polarizable units are essentially the point masses.

The normal coordinate gradients (analogous to Eqs. (43)) required to obtain ROA intensities are given as

$$\alpha_{\alpha\beta}^{\ell} = \sum_{A} \sum_{B} (\frac{\partial B_{AB\alpha\beta}}{\partial Q_{\ell}})_0 \tag{59a}$$

$$\beta_{\alpha\beta}^{\ell} = -\frac{1}{2c} \sum_{A} \varepsilon_{\beta\gamma\delta} \{R_{A\gamma}^0 \sum_{B} (\frac{\partial B_{AB\delta\alpha}}{\partial Q_{\ell}})_0 + S_{A\gamma}^{\ell} (\sum_{B} B_{AB\delta\alpha}^0)\} \tag{59b}$$

$$A_{\alpha\beta\gamma}^{\ell} = \sum_{A} \tfrac{3}{2} \{R_{A\beta}^0 [\sum_{B} (\frac{\partial B_{AB\gamma\alpha}}{\partial Q_{\ell}})_0] + S_{A\beta}^{\ell}(\sum_{B} B_{AB\gamma\alpha}^0) +$$

$$R_{A\gamma}^0 [\sum_{B} (\frac{\partial B_{AB\beta\alpha}}{\partial Q_{\ell}})_0] + S_{A\gamma}^{\ell}(\sum_{B} B_{AB\beta\alpha}^0)\}$$

$$- \delta_{\beta\gamma} \{R_{A\lambda}^0 [\sum_{B} (\frac{\partial B_{AB\lambda\alpha}}{\partial Q_{\ell}})_0] + S_{A\lambda}^{\ell} (\sum_{B} B_{AB\lambda\alpha}^0)\}. \tag{59c}$$

The appropriate combinations of the products of the above tensor gradients can be substituted into Eqs. (40) to obtain the chirality numbers. The simplifications (see Eqs. 44) that resulted from the symmetric nature of the bond polarizability tensors do not apply here since the atomic polarizabilities that have sampled the dipolar interactions need not be symmetric. In the ADI model calculation, the difficult part is the evaluation of the atomic polarizability derivatives, which are given as

$$\left(\frac{\partial \alpha_{A\alpha\beta}}{\partial Q_\ell}\right)_0 = \sum_B \left(\frac{\partial B_{AB\alpha\beta}}{\partial Q_\ell}\right)_0. \tag{60}$$

The differential of the relay tensor matrix $\underset{\sim}{B}$ is obtained by noting that $\underset{\sim}{B}\underset{\sim}{A} = 1$ and therefore $(\partial \underset{\sim}{B}/\partial Q)_0 = -\underset{\sim}{B}(\partial \underset{\sim}{A}/\partial Q)_0 \underset{\sim}{B}$ where $\underset{\sim}{A}$ is given in Eq. (52b). The differential of the $\underset{\sim}{A}$ matrix contains two parts. One is the gradient of the dipolar interaction tensor elements which can be obtained as

$$\left(\frac{\partial T_{AB}^{\alpha\alpha}}{\partial Q_\ell}\right)_0 = \left(-\frac{3}{R_{AB}^4} + \frac{15 R_{AB\alpha}^2}{R_{AB}^6}\right)\left(\frac{\partial R_{AB}}{\partial Q_\ell}\right)_0 - \frac{6 R_{AB\alpha}^0}{R_{AB}^5}\left(\frac{\partial R_{AB\alpha}}{\partial Q_\ell}\right)_0 \tag{61a}$$

$$\left(\frac{\partial T_{AB}^{\alpha\beta}}{\partial Q_\ell}\right)_0 = \frac{15}{R_{AB}^6}\left(\frac{\partial R_{AB}}{\partial Q_\ell}\right)_0 R_{AB\alpha}^0 R_{AB\beta}^0$$

$$- \frac{3}{R_{AB}^5}\left\{\left(\frac{\partial R_{AB\alpha}}{\partial Q_\ell}\right)_0 R_{AB\beta}^0 + R_{AB\alpha}^0\left(\frac{\partial R_{AB\beta}}{\partial Q_\ell}\right)_0\right\} \tag{61b}$$

with

$$\left(\frac{\partial R_{AB\alpha}}{\partial Q_\ell}\right)_0 = S_{A\alpha}^\ell - S_{B\alpha}^\ell \tag{61c}$$

and

$$\left(\frac{\partial R_{AB}}{\partial Q_\ell}\right)_0 = \frac{(R_{A\alpha}^0 - R_{B\alpha}^0)(S_{A\alpha}^\ell - S_{B\alpha}^\ell)}{R_{AB}^0}. \tag{61d}$$

The second part is the gradient of the diagonal elements of $\underset{\sim}{A}$, which can be given as $-(\alpha_A)^{-2}(\partial \alpha_A/\partial Q_\ell)_0$. The numerical values of α_A are obtained from the literature and those for $(\partial \alpha_A/\partial Q_\ell)_0$ are parameterized.

3.   Quantum Mechanical Models

a.   Orbital Polarizability Model.   To adapt the quantum mechanical methods for ROA calculations, two procedures can be envisioned. One is to evaluate Eqs. (39), using ab initio or semiempirical molecular orbital theories, at several different displaced geometries and obtain the numerical gradients of the $\underset{\sim}{\alpha}$, $\underset{\sim}{\beta}$ and $\underset{\sim}{A}$ tensors. This procedure has been adapted by Cuony in his doctoral thesis [16] for calculating ROA intensities of (R)-1,3-dideuteroallene. The second procedure first developed in our laboratory [40] is a molecular orbital analogue of the bond and atomic polarizability models. That is, a molecule can be envisioned as a collection of nuclei and localized molecular orbitals. The molecular polarizability can be partitioned into the component polarizabilities of the localized molecular orbitals. This can be seen as follows.

The electric dipole moment of a molecule can be written in terms of the localized molecular orbital centroids $\underset{\sim}{r}_{K0}$ and nuclear positions as

$$\mu_\alpha = \sum_A eZ_A R_{A\alpha} - 2e \sum_K^{n/2} r_{K0\alpha}. \tag{62}$$

Here we considered the molecule to contain n electrons and n/2 occupied molecular orbitals. In the presence of an electric field $\underset{\sim}{E}$, the induced electric dipole moment is proportional to the molecular polarizability, which is given as

$$\alpha_{\alpha\beta} = \frac{\partial \mu_\alpha}{\partial E_\beta} = \sum_K^{n/2} - 2e(\frac{\partial r_{K0\alpha}}{\partial E_\beta}) = \sum_K^{n/2} \alpha_{K\alpha\beta}. \tag{63}$$

Thus the molecular polarizability, as in the bond or atomic polarizability models, is a sum of the orbital polarizabilities $\alpha_{K\alpha\beta}$ which represent the relaxation of the average orbital positions in the presence of a weak electric field.

The optical activity tensors can be formulated in terms of the orbital polarizabilities, just as in the bond polarizability and ADI models. The normal coordinate gradients of the necessary tensors now become

$$\alpha_{\alpha\beta}^\ell = \sum_K (\frac{\partial \alpha_{K\alpha\beta}}{\partial Q_\ell})_0 \tag{64a}$$

$$\beta_{\alpha\beta}^\ell = \sum_K - \frac{1}{2c} \{\varepsilon_{\beta\gamma\delta} [r_{K0\gamma}(\frac{\partial \alpha_{K\delta\alpha}}{\partial Q_\ell})_0 + \sigma_{K\gamma}^\ell \alpha_{K\delta\alpha}^0]\} + \sum_K (\frac{\partial \beta_{K\alpha\beta}}{\partial Q_\ell})_0 \tag{64b}$$

$$A^\ell_{\alpha\beta\gamma} = \sum_K \frac{3}{2} \{r_{K0\beta}(\frac{\partial\alpha_{K\gamma\alpha}}{\partial Q_\ell})_0 + \sigma^\ell_{K\beta}\,\alpha^0_{K\gamma\alpha} + r_{K0\gamma}(\frac{\partial\alpha_{K\beta\alpha}}{\partial Q_\ell})_0 +$$

$$\sigma^\ell_{K\gamma}\,\alpha^0_{K\beta\alpha}\} - \delta_{\beta\gamma}\{r_{K0\lambda}(\frac{\partial\alpha_{K\lambda\alpha}}{\partial Q_\ell})_0 + \sigma^\ell_{K\lambda}\,\alpha^0_{K\lambda\alpha}\} + \sum_K A_{K\alpha\beta\gamma}. \qquad (64c)$$

Substitution of the appropriate products of these tensor elements into Eqs. (40) will provide the necessary formulation to obtain the chirality numbers. It may be noted that the intrinsic contributions to the optical activity tensors are present in this model as in the bond polarizability model. Since the orbitals considered in this model are assumed to be localized they can be considered to be inherently achiral and of high symmetry. Therefore, it is reasonable to assume that the intrinsic terms are small.

Recently, it was proposed [74] to replace the $r_{\sim K0}$ and $(\partial r_{\sim K0}/\partial X_{\sim A})_0$ terms by $R^0_{\sim A}$ and $(\partial R_{\sim A}/\partial X_{\sim A})_0$ so as to obtain the ROA expression in terms of atomic Raman tensors [91,92]. It is difficult to validate such a replacement unless a sound theoretical justification is available.

b. Two Group and Related Models. The two group model [46] for ROA considers the vibrations of two identical bonds or groups that are separated by a distance and constitute a chiral structure. The groups themselves are considered achiral. The ground vibrational state of the two group structure is considered to be the product of the vibrational states of the two groups, and the excited state is taken as the symmetric and antisymmetric combinations of singly excited individual group states. This concept is similar to that used in coupled oscillator models of electronic and vibrational circular dichroism. ROA associated with the two coupled vibrational transitions is predicted to be of opposite sign. Barron and Buckingham obtained [46] simple formulas for ROA associated with these transitions, which depend only on the separation of the two groups and the dihedral angle of the chiral structure. Gohin and Moskovits have presented [48] an interesting application of this model for the CH_2 scissoring vibrations of some terpenes.

A variant of this model considers two degenerate vibrations located on the same group, an example being the degenerate antisymmetric stretching motions of a methyl group. When a methyl group is attached to a chiral center, the vibrational degeneracy can be lifted and the two vibrations are predicted [49] to exhibit ROA with opposite signs.

Convenient formulas for ROA in the twisting mode of the two group structure [9] and in the methyl torsional mode [33] have also been developed.

We have not presented the equations involved in these models, as these are described several times in the literature. We refer to the reviews by Barron [8-10] for further details.

IV.   CALCULATIONAL DETAILS

A.   Electric Field Perturbative Calculations

In the orbital polarizability model calculations for ROA intensities [40], the primary information required is the displacement vectors for orbital centroids and the changes in orbital polarizabilities during molecular vibrations.   One way to obtain this information is to evaluate the orbital centroids and orbital polarizabilities at displaced nuclear positions on either side of the equilibrium position and take the numerical gradients.   This procedure is referred [40] to as the nuclear displacement gradients (NDG) scheme.   To obtain ROA intensities for all (3N - 6) vibrations in this procedure, the molecular orbital calculations need to be repeated at least 2 x (3N - 6) times. This can be prohibitively expensive even for semiempirical molecular orbital calculations.

Recently, ROA and VCD intensities have been evaluated [40] using the electric field perturbative procedure which was originally used [38] in the calculation of infrared and Raman intensities.   In this procedure, known as the electric field gradients (EFG) scheme, the forces on the atoms are evaluated in the presence of a weak electric field perturbation, and the variation of the forces with electric field strength are related to the displacement vectors of the orbital centroids and to the orbital polarizability changes during nuclear displacements.   The details of this procedure are summarized here.

In the presence of a weak external electric field $\underset{\sim}{E}$ the molecular electronic energy ε^{el} is given as $\varepsilon^{el}_0 - \mu^{el}_\alpha E_\alpha$ where ε^{el}_0 is the electronic energy in the absence of the electric field and μ^{el}_α is the electronic part of the dipole moment given in Eq. (62).   From the differentials of ε^{el} and Eq. (62), it can be shown that

$$(\frac{\partial \mu^{el}_\alpha}{\partial X_{A\beta}})_0 = (\frac{\partial \phi^{el}_{A\beta}}{\partial E_\alpha})_0 = -2e \sum_K (\frac{\partial r_{K0\alpha}}{\partial X_{A\alpha}})_0 \qquad (65)$$

where $\underset{\sim}{\phi}^{el}_A$ represents the electronic part of the force vector on atom A.   This force can be partitioned into the contributions from individual orbitals so that

$$\phi_{A\beta}^{el} = \sum_K \phi_{KA\beta}^{el} = - \sum_K \left(\frac{\partial \varepsilon_K^{el}}{\partial X_{A\beta}}\right)_0. \tag{66}$$

Here ε_K^{el} is the contribution from the Kth molecular orbital to the electronic energy. From Eqs. (65) and (66), the orbital displacement vectors can be obtained as

$$\sigma_{K\alpha}^\ell = \left(\frac{\partial r_{K0\alpha}}{\partial Q_\ell}\right)_0 = - \frac{1}{2e} \sum_A \left(\frac{\partial \phi_{KA\beta}}{\partial E_\alpha}\right)_0 S_{A\beta}^\ell. \tag{67}$$

Similarly, from the molecular polarizability relation given by Eq. (63), the polarizability derivatives become

$$\left(\frac{\partial \alpha_{\alpha\beta}}{\partial X_{A\gamma}}\right)_0 = \left(\frac{\partial}{\partial E_\beta} \left(\frac{\partial \mu_\alpha^{el}}{\partial X_{A\gamma}}\right)\right)_0. \tag{68}$$

Using the relations of Eqs. (65) and (66), the normal coordinate gradients of the polarizabilities become

$$\left(\frac{\partial \alpha_{K\alpha\beta}}{\partial Q_\ell}\right)_0 = \sum_A \left(\frac{\partial^2 \phi_{KA\gamma}}{\partial E_\beta \partial E_\alpha}\right)_0 S_{A\gamma}^\ell. \tag{69}$$

The orbital displacement vectors and polarizability gradients are related by Eqs. (67) and (69), respectively, to the first order and second order differentials of the orbital contributions to forces in the presence of the electric field. This is remarkable because the orbital displacements and polarizability gradients can be determined in the EFG scheme without ever displacing the nuclei. It is only necessary to perform MO calculations with varying electric field strengths to determine Eqs. (67) and (69) for all (3N - 6) vibrations of any given molecule. It is shown [40] that eighteen field perturbative calculations are sufficient for this purpose. Since the equilibrium orbital centroids (which can be determined in a normal MO calculation), and the orbital polarizabilities and their derivatives (Eqs. (63), (67) and (69)) are sufficient to determine both ROA and VCD intensities, the EFG scheme offers substantial savings in computational time.

The partition of the forces into individual orbital contributions in Eq. (66) requires that the molecular orbitals should not have inter-orbital interactions. This requirement is not met in practice; however, it is possible to minimize the inter-orbital interaction through localizing the orbitals using maximum self energy criterion [93]. In NDG calculations of ROA and VCD, the choice of the localization scheme is arbitrary. This choice is narrowed down to the energy localized orbitals [93] in the EFG scheme.

The EFG scheme has recently been employed for calculating ROA and VCD intensities. It is found that for calculating ROA and Raman intensities, the EFG scheme provides better numerical accuracy than the NDG scheme and that the VCD intensities obtained for NHDT in the two schemes match within a factor of two. This difference is attributed [40] to the residual interaction energy present in the localized orbitals.

B.   Sources of Information

For most of the model calculations the knowledge of normal mode descriptions is a prerequisite, and can be obtained using well documented procedures [67,94]. There are some intricate details regarding the redundancies in internal coordinate definitions and the associated problems in evaluating the force constants. The informative discussions by Marcott [12] and Overend et al. [95] should be consulted for details on this subject. Assuming that the descriptions of the normal modes of vibration are at hand, the information required for VCD and ROA calculations using semiclassical models centers around the charges, charge redistribution parameters, polarizabilities and polarizability gradients during nuclear displacements. The partial atomic charges are generally chosen to reproduce the molecular electric dipole moments [96]. Molecular orbital calculations have also been used to determine the net atomic charges. Similarly, the bond charges may be obtained from the documented bond moments [97]. The most difficult part, however, is to obtain the atomic charge or bond charge gradients. The best way of obtaining these values is through the analysis of infrared intensities using either the bond moment [71] model or the ECCF model [68-70]. While the bond moment model yields bond charge gradients, the ECCF model provides the atomic charge gradients. However, in most of the cases, it is well known that one can only evaluate a certain combination of charge redistribution parameters but not their individual values. This results in resorting to some approximations or transferring the combinations of parameters for a desired group from one molecule to another. Even for medium size molecules, the analysis of infrared intensities is somewhat tedious. Therefore, for chiral molecules encountered in optical activity studies it may only be possible to transfer the desired charge redistribution parameters from smaller molecular systems. Recent work [98-100] on infrared intensities may be consulted for obtaining atomic or bond charge gradients.

For ROA calculations the atomic or bond polarizabilities can be obtained from the literature [84,86,101]. The difficulties involved in obtaining the polarizability gradients are more acute. There is a limited amount of information regarding atomic or bond polarizability gradients [86,87,90,98,102] obtained from Raman intensities. Some information on bond polarizability gradients has been obtained using a delta function model [103,104].

V. COMPARISONS TO EXPERIMENTAL OBSERVATIONS

In the early stages of VCD research, the experimental measurements were limited to the 4000-2500 cm^{-1} region. Later the frequency range of measurements was extended to 900 cm^{-1} [105-107]. Certain vibrational bands appearing below this frequency range can also be significant for deriving stereochemistry. Therefore, in the past few months we have extended the VCD measurements up to 600 cm^{-1} in our laboratory [108]. The VCD spectrum of (+)-(3R)-methylcyclohexanone in the 1000-600 cm^{-1} region is shown in Fig. 1. The lower frequency limit for VCD measurements is likely to remain at 600 cm^{-1} until photoelastic modulators giving quarter wave retardation below this limit with sufficient transmission are developed. The situation is different in ROA measurements where the optical activity in transitions which appear down to 100 cm^{-1} can be measured. However, few ROA measurements in the C-H stretching region have been made. This is because most of the Raman bands in this region are polarized and are prone to artifacts [109] unless special precautions are taken. Thus VOA information for transitions below 600 cm^{-1} is now obtained solely from ROA measurements, while that for O-H, N-H and C-H stretching vibrations is obtained from VCD measurements. For transitions between 2000 to 600 cm^{-1}, VOA information is obtained from both ROA and VCD measurements.

Since not all vibrations of a molecule can easily relate the VOA observations to stereochemistry, the first problem is one of identifying the vibrations whose optical activity is most suitable for this purpose. To some extent, the coupled oscillator and two group models [21,46] facilitate this choice.

When a chiral perturbation splits the degenerate vibrational modes of a methyl group, these models predict that the optical activity associated with the two split modes will have opposite sign and will be sensitive to the sense of chiral perturbation. For some time it was thought that the optical activity

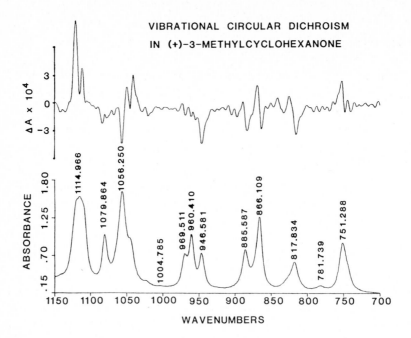

FIG. 1. Infrared vibrational absorption (bottom trace) and circular dichroism (top trace) in (+)-(3R)-methylcyclohexanone. These spectra are recorded for neat liquid with 100 micron path length on a Fourier transform infrared spectrometer (Nicolet 6000C).

in methyl group modes can be used to determine the configuration. Hug et al. observed a bisignate couplet centered at about 1450 cm^{-1} in the ROA spectrum of (+)-1-phenylethylamine [110] and, recently, of (+)-1-methyl indane [111] and attributed this to the degeneracy lifted methyl deformation modes. Barron also observed the bisignate couplets at about 1450 cm^{-1} in the ROA spectra of (+)-1-phenylethylamine, (+)-1-phenylethanol [49] and (+)-1-phenylethylisocyanate [112] and assigned them to the methyl deformation modes. The sign pattern of the couplet agreed with the relative stereo-chemistry of these molecules. However, Barron later expressed doubts about this assignment [50,113], because in the ROA spectrum of (+)-1-(p-bromo-phenyl)ethylamine, this couplet was not found. Barron noted [50] that a benzene ring mode which appears in the para-substituted amine at 1410 cm^{-1} could be near the methyl antisymmetric deformation modes in the afore-mentioned molecules. Thus, one part of the bisignate couplet could be arising from the benzene ring mode. This view is further supported in ROA studies on deuterated 1-methylindanes [114]. Su and Keiderling have measured VCD in the methyl deformation mode region of the aforementioned compounds and found [115] a single negative VCD band at 1450 cm^{-1}. In the para-substituted amine, a new VCD band of the same sign was found at 1405 cm^{-1}, and the VCD intensity at 1450 cm^{-1} decreased. This observation is in line with Barron's suggestion of phenyl ring mode participation in the observed ROA couplets at 1450 cm^{-1}. For the antisymmetric methyl stretching modes of these compounds, a single VCD band with positive sign was observed. Although these observations lead to lesser support for Eq. (36b) and its analogue in ROA, some supportive evidence is apparent in recent measurements. In the VCD spectrum of L-alanine-C*-d$_1$-N-d$_3$, the bisignate couplet near the antisymmetric methyl stretching mode absorption might be attributed [116-118] to the principle involved in Eq. (36b). In the VCD spectrum of dimethyl-trans-1,2-cyclopropane dicarboxylate a bisignate couplet was found [119] near the antisymmetric methyl deformation mode absorption. This bisignate couplet is persistent even when the ring hydrogen atoms at the 1 and 2 positions are deuterated. Therefore, this bisignate couplet was attributed [119] to the two antisymmetric methyl deformation modes. In the VCD spectrum of $C_6H_5CHDCD_3$, the CD_3 antisymmetric stretching modes were found to exhibit bisignate VCD bands [13]. Also, in the ROA spectrum of (+)-1-phenyltrifluoromethylethanol, a bisignate couplet centered at about 510 cm^{-1} has been attributed [50] to the antisymmetric deformation modes of the CF_3 group.

Some evidence for coupled oscillator phenomenon is also available despite some initial uncertainties. Sugeta et al. [23] measured the CD in the C-H stretching motions of tartaric acid-d$_4$ and concluded that VCD here arises from the first term in Eq. (36a), since no bisignate couplets corresponding to the second term of Eq. (36a) were found. In the following year, Keiderling and Stephens [120] observed two VCD bands of opposite sign and of nearly equal intensity in the O-H stretching region of dimethyl-d-tartrate. They considered this observation to be the first example of a coupled oscillator phenomenon, represented by the second term of Eq. (36a). However, Marcott et al. [121] felt that the two VCD bands in the O-H stretching region of dimethyl-d-tartrate are due to two possible conformations of this molecule. Later Su and Keiderling [122] found for the same molecule two VCD bands, again of opposite sign, in the carbonyl stretching region, suggesting the coupled oscillator behavior of the two carbonyl groups. Here the C=O absorption band appeared [122] to contain a shoulder on the high frequency side, giving a suspicion for the presence of a multiquantum transition. However, the absorption band is noted [123] to be perfectly symmetric which eliminates this doubt.

Although tartaric acid-d$_4$ does not show coupled oscillator behavior, the VCD sign of the C-H stretching mode has an interesting correlation with the configuration. The positive VCD in ℓ-tartaric acid-d$_4$, correlates with positive VCD in the methine stretching mode of L-alanine-C-d_3-N-d_3, L-lactic acid, L-serine, L-threonine and L-valine [117].

A successful application of the coupled oscillator phenomenon is found in the C-H stretching vibrations of six membered ring systems. Polavarapu et al. [124] reported that the VCD spectra of (+)-p-menth-l-ene, (+)-p-menth-l-en-9-ol and (+)-limonene show a common positive-negative-positive triplet feature in the C-H stretching region. They assigned the positive VCD around 2945 cm^{-1} to the methylene antisymmetric stretching mode, the negative VCD around 2925 cm^{-1} to the methine stretching mode, and the positive VCD around 2895 cm^{-1} to another methylene antisymmetric stretching mode. Recently, Moscowitz et al. [125] have examined the VCD spectra in this region for similar six membered ring systems and found that most of the compounds examined show this triplet and the sign order of the triplet relates to the chair form which the molecule is expected to prefer. Specifically, they found that the positive-negative-positive triplet corresponds to the CH_2-CH_2-C*H fragment in one chair form of the molecule and the negative-positive-negative triplet corresponds to this fragment in another chair form of the molecule. The observed triplets are considered to be characteristic of the three stretching vibrations of the CH_2-CH_2-C*H fragment. In order to support this interpretation, Laux developed Eq. (37) in his doctoral thesis [14] and performed calculations on the isolated CH_2-CH_2-C*H fragment. These calculations predicted the same sign pattern as observed, and supported the interpretation of the observed triplets.

Evidence for the coupled oscillator phenomenon is also apparent in the VCD spectra of some sugar acetates [13]. In aldohexose peracetates, the coupled oscillator theory applied to the C_1-H and C_5-H stretching motions predicted that the antisymmetric combination of these stretching motions results in positive VCD and the symmetric combination results in negative VCD. This prediction is in agreement with observations on some aldohexose peracetates where a positive VCD at 3010 cm^{-1} and a negative VCD at 2910 cm^{-1}, corresponding to the antisymmetric and symmetric stretching motions of C_1-H and C_5-H, were found. However, the C-H stretching region of glucose, allose, talose and idose did not show any measurable VCD bands, while mannose, galactose, altrose and gulose showed a single VCD band indicating that the coupled oscillator behavior is not apparent in these simple sugars. Nevertheless, Havel [13] found that the sign of the observed VCD in the C-H stretching region for simple sugars relates to the spatial arrangement of the C-H bonds at C_2, C_3 and C_4 as follows: when viewed along the bonds C_2-C_3 and C_3-C_4, if the three C-H bonds at C_2, C_3 and C_4 constitute a helical arrangement, then a net clockwise rotation corresponds to a positive VCD and a net counterclockwise rotation corresponds to a negative VCD; if

these three C-H bonds do not constitute a helix, then the VCD will be zero.
This is referred to as the "C-H bond helix rule" and is in agreement with
observations on several simple sugars.

Barnett et al. [31] found four VCD bands in the aliphatic C-H stretching region of (S)-(+)-9,10-dihydrodibenzo-[c,g]-phenanthrene and noted that
the coupled oscillator mechanism is not successful in explaining the observed
CD. Of the four observed VCD bands, two are believed to be associated with
the antisymmetric and symmetric CH_2 stretching modes and the other two to
be associated with the overtones of the CH_2 bending modes. Using the
dynamic polarization model (Eq. (35)), Barnett et al. have evaluated the VCD
in the antisymmetric and symmetric CH_2 stretching motions. The calculated
signs are noted to be in agreement with the observed ones. Somewhat
earlier, the same model was utilized in explaining the observed VCD in the
N-H stretching region of (R)-(+)-2,2'-diamino-1,1'-binaphthyl. Barnett et al.
[30] have calculated the VCD associated with the symmetric NH_2 stretching
mode using an equivalent of Eq. (35), and noted that the calculated and
observed VCD are in reasonable agreement.

Singh and Keiderling have compared [126] the VCD in the amide I and
amide II bands of poly-α-benzyl-L-glutamate with the theoretical predictions
obtained by an analogue of the electronic exciton coupling model [127]. It
was predicted [127] that bisignate VCD bands, due to the coupling of monomeric units, will be present at the amide I and amide II bands. The amide I
band is found to exhibit bisignate VCD bands with the sign order as predicted, but the amide II band did not show a bisignate couplet and is at
variance with the predictions.

So far we have concentrated on the comparison of experimental observations with the predictions by theoretical models that do not require the
explicit evaluation of the compositions of the normal modes of vibration. The
semiclassical models and the molecular orbital calculations, however, require a
detailed description of normal coordinates. A majority of the calculations in
this category have been performed using the FPC model for VCD and the ADI
and bond polarizability models for ROA. The first comparison of experimental
and theoretical results was presented by Faulkner and coworkers [11,27],
where the FPC model predictions for the C*-D stretching vibrations in
(R)-(-)-neopentyl-1-d-chloride, the C*-H and O-H stretching vibrations in
(S)-(+)-2,2,2-trifluoro-1-phenylethanol, and the C*-H stretching vibrations in
(S,S)-(-)-tartaric acid-d_4 were compared with experimental VCD measurements
in the appropriate regions. The magnitudes of the calculated VCD intensities
were found to be smaller than those observed, but the correct sign was
predicted for the C*-D vibration in neopentyl-1-d-chloride and the C*-H
vibration in trifluorophenylethanol. For the C*-H vibrations of tartaric

acid-d_4, however, the FPC model predicted two VCD bands of opposite sign, while the experimental spectrum showed a single positive VCD band. Keiderling and Stephens have compared [128] the FPC model predictions for the C-H stretching vibrations in (5S)-(+)-1,6-spiro[4,4]nonadiene with the experimental VCD spectrum in that region. They found a favorable comparison of magnitudes and signs for the antisymmetric methylene stretching modes; however, for the =C-H stretching motions, even a qualitative agreement was not found. For the O-H stretching vibration in (S)-(+)-1-dimethylamino-2-propanol the VCD predicted by the FPC model was found [12] to be two orders of magnitude short of the experimental value. To explain this, Marcott has investigated [12] the effect of including lone pairs of electrons, as pseudo atoms, in the FPC calculation and found that when these are included the calculated VCD magnitude is comparable to the experimental value. Calculations on (+)-(3R)-methylcyclohexanone also indicated [129-131] that the magnitudes of VCD predicted by the FPC model are much lower than the experimental values. This shortcoming is overcome by scaling the partial charges employed in the FPC calculation upwards and, thus, qualitatively good agreement is obtained for the VCD in the C-H stretching region. The sign pattern calculated using the FPC model for VCD bands in the C-H stretching region of [(trans-2S,6S)-^{2}H$_2$]cyclohexanone compared [132] well with that observed for the (-)enantiomer which indeed is known to have the (S,S) configuration.

Heintz and Keiderling have reported [119] VCD spectra of a series of trans-1,2-disubstituted cyclopropanes and identified the VCD due to the apex methylene stretching modes and the C-H stretching modes at the 1 and 2 positions. They stated that the FPC calculations using pseudo atoms for substituents qualitatively agree with the observed VCD sign pattern in the C-H stretching region. For the deformation modes they felt that the calculated signs do not agree well with the observed signs due to the approximate nature of the force field employed. Similar partial success was found [133] in the comparison of FPC model predictions with experimental observations on a series of terpenes.

The localized molecular orbital (LMO) model predictions and the experimental VCD observations on the C*-D stretching vibration in R-(-)-neopentyl-1-d-chloride and the ten C-H stretching vibrations of (+)-(3R)-methylcyclohexanone have been analyzed [36,37] by Polavarapu and Nafie. In these LMO calculations the canonical molecular orbitals obtained in the CNDO MO scheme [134] are localized using the maximum self energy criterion [93]. The sign and magnitude of the VCD for the C*-D vibration in neopentyl-1-d chloride are in good agreement with the observed ones. In methylcyclohexanone also, the VCD spectrum simulated from the calculated rotational strengths was found to be in good agreement with the experimental VCD spectrum in the C-H stretching region. The LMO predicted magnitudes in both molecules are higher than those predicted by the FPC model and are closer to the experimental magnitudes. The LMO predicted VCD signs are the same as those predicted by the FPC model for the C*-D stretching vibration in neopentyl-1-d-chloride and for ten out of the twelve C-H stretching normal modes in methylcyclohexanone. For one of the two modes where the LMO and FPC models are at variance, the VCD magnitude is small so the disagreement is less important. The VCD sign of the other mode is sensitive to the charges chosen in the FPC model calculation. Nevertheless, the overall spectral pattern, predicted by both models, is in favorable agreement with the observed one. For the C-H stretching motions in alanine-N-d_3, alanine-C*-d_1-N-d_3 and alanine-C-d_3-N-d_3, the LMO predictions were found [118] to be in good agreement with the observed VCD signs as well as the magnitudes. This favorable comparison is significant since the FPC model predictions were found to be at variance with the observed VCD signs as well as the magnitudes in alanine-C*-d_1-N-d_3 and alanine-C-d_3-N-d_3.

The above comparisons allow one to estimate the validity of the approximations involved in the LMO model. In the practical usage of Eq. (32), it is assumed that the localized orbitals will have small intrinsic magnetic moments and small orbital rocking contributions. As the predictions with this assumption compare favorably with the experimental observations, the omission of $\Gamma_{K\alpha}^{\ell}$ terms in Eq. (32) appears justified at least for the three molecules tested thus far.

For LMO calculations using the electric field perturbative procedure, the localization of MO's is required not only for minimizing the intrinsic orbital contributions to the magnetic dipole transition moment, but also for minimizing the inter orbital energy (see Eq. (66)). Thus the localization requirement is more stringent in EFG calculations than in NDG calculations. This is apparent in the calculations on NHDT [40] and alanine [74].

The bond moment and bond current models have been developed only recently and they have not been applied to molecules for which experimental data is available. The charge flow model has been utilized recently in the analysis of experimental observations. Havel has measured [13] the VCD associated with C-H and C-D stretching motions in a series of deuterated phenylethanes $C_6H_5CHDCH_3$, $C_6H_5CHDCD_3$, $C_6H_5CHDCH_2D$, and $C_6H_5CHDCHD_2$, and identified the bands due to the C-H, C-D, CH_3 and CD_3 groups. Using this experimental data, the FPC and charge flow models have been tested [13] for their predictive capabilities. It was noted that for a majority of the bands the VCD signs have been predicted correctly by the charge flow model but not by the FPC model. The magnitudes of the rota-

tional strengths predicted by the charge flow model were found to be somewhat smaller than those observed. This may be due to the fact that the
charge flow in the benzene ring was not incorporated into the calculation.
An attempt to incorporate these effects through the polarizability model (see
Sec. III.A.2e) did not prove to be successful.

All comparisons mentioned thus far have been limited to vibrations that
are highly localized like C-H, O-H and N-H stretching vibrations. When the
VCD associated with low frequency motions is compared with the experimental
observations one may not find an encouraging situation. In the VCD spectrum of (+)-(3R)-methylcyclohexanone, shown in Fig. 1, the absorption bands
at 947, 886, 866 and 818 cm^{-1} are seen to have negative VCD and the band at
751 cm^{-1} has a positive VCD. The same FPC model calculation [129], which
predicted correct VCD in the C-H stretching region, predicts incorrect signs
for the above bands. As the experimental VCD measurements in the 900 to
600 cm^{-1} region are very recent [108], other theoretical models are yet to be
tested in this region.

Turning to ROA calculations, very few calculations appropriate for
comparisons to experimental observations have been reported although
abundant ROA experimental data is now available. One basic reason for this
is that ROA measurements, as mentioned earlier, are carried out in the
2000-100 cm^{-1} region. Excluding the carbonyl stretching vibrations, all other
vibrations that belong to this region are those involving deformations, skeletal
modes and torsional modes, and are generally known to have extensive inter-
mode mixing. Therefore, simple correlations like those encountered in C-H
stretching VCD are hard to realize in this region. Calculations using the
ADI, bond polarizability and orbital polarizability models become imperative
where one faces insurmountable difficulties in the vibrational analysis due to
its lesser reliability in this region. As a consequence, a good comparison
between theoretical and experimental results is hard to achieve for low frequency vibrational modes, as seen earlier for VCD. Despite these un-
fortunate circumstances, some progress is being made in relating theory to
experiment. A recurring negative-positive-negative ROA pattern was noted in
the 1475-1400 cm^{-1} region for (+)-p-menth-1-ene and (+)-p-menth-1-en-9-ol.
The same pattern with opposite sign order is present in (-)-limonene
correlating with their configurations. These features were attributed [124] to
the three ring methylene deformation modes with some mixing from the ring
HCC deformations. A similar triad is present [135] in the ROA spectra of
(-)-menthol, (-)-menthylamine and (-)-menthylchloride between 1400 to 1275
cm^{-1}, which again is probably due to the ring CH_2 deformations, and
correlates with the absolute configurations [135] of these three molecules.
Assignments for several ROA bands between 1200-100 cm^{-1} have been

postulated [9,10] but some uncertainties exist especially in the assignment of the methyl torsional mode and the associated ROA.

There are a total of nine ROA calculations reported to date. Of these, the ADI and bond polarizability model calculations on bromochlorofluoromethane [53,79], the ADI model calculations on substituted haloethanes [24], the orbital polarizability model calculation on NHDT [40] and the molecular orbital calculation on deuterated allene [16] are of theoretical interest since the experimental ROA data for these molecules do not exist. The ADI model calculation on the C-Cl stretching vibration in 1-chloro-2-methylbutane [136] is informative but the experimental data on this molecule have not yet been verified for its reproducibility. Therefore, definite conclusions cannot be inferred from the comparison between the theoretical and experimental results on this molecule. This leaves the ADI and bond polarizability model calculations on (+)-(3R)-methylcyclohexanone [80,129] and the two group model calculation on terpenes [48] for comparison to experimental results.

In the ROA spectra of (-)myrtenol, (-)nopol and (-)myrtenal, Moskovits et al. [137] found bisignate couplets centered at about 1440 cm^{-1}. In (+) α-pinene also, this couplet is found with opposite sign order. Considering that this couplet arises from the methylene scissoring modes, Gohin and Moskovits [48] have applied the two group model for the scissoring vibrations of two methylene groups and found the resulting predictions to be in agreement with the observed sign order as well as the magnitudes.

In the ROA spectrum of (+)-(3R)-methylcyclohexanone [9] three bisignate couplets centered around 960, 500 and 400 cm^{-1} attracted attention. The couplet around 960 cm^{-1} is associated with the two rocking motions of the methyl group, the couplet around 500 cm^{-1} with one of the CCC ring motions and CCO in-plane deformation, and the couplet around 400 cm^{-1} with two CCC ring motions. For all these couplets the sign order is predicted correctly by the ADI model, but not by the bond polarizability model. The experimental spectrum shows a broad bisignate couplet centered at about 200 cm^{-1} which is not predicted correctly by both models.

Although the above comparison speaks more in favor of the ADI model, it is not completely correct to compare the two model calculations on (+)-(3R)-methylcyclohexanone for their performance. This is because in these calculations somewhat different force fields and geometries are employed; also different levels of approximations are used in giving numerical values to the polarizability derivatives. Both models require considerable effort in the proper parameterization of the polarizability derivatives.

VI. SUMMARY

The comparison of theoretical predictions with experimental observations, which was presented in the previous section, might leave mixed feelings regarding the success or failure of the existing theoretical models. However, some care is required in judging a given theoretical model because there are two distinct aspects that could be responsible for a favorable comparison between theoretical and experimental observations. While the principles underlying a given theoretical model is one factor, the second one is the normal mode composition employed in the calculations. In a broader sense, VOA bands can be classified into four categories. These are the VOA bands arising from: (I) localized vibrations without significant charge redistribution; (II) localized vibrations with charge redistribution dispersed to various other groups in the molecule; (III) delocalized vibrations without significant charge redistribution; and (IV) delocalized vibrations with dispersed charge redistribution. Type I bands are the simplest for theoretical treatment and might be successfully explained using simple concepts like those involved in the coupled oscillator, two group and FPC models. For type II bands, semiclassical models that consider charge redistribution effects for VCD and polarizability derivatives for ROA can be successful. For types I and II bands, the vibrations are considered localized and therefore the theoretical predictions will be more or less insensitive to small variations in the force field employed in the vibrational analysis. An appropriate example is the VCD in the C-H stretching region of (+)-(3R)-methylcyclohexanone, where three different FPC calculations [129-131] with varying force fields predicted qualitatively identical VCD features. Any disparities between theoretical and experimental observations on type II bands can then be traced to inadequate parameterization of charge fluxes in the case of VCD, and polarizability derivatives in the case of ROA.

Successful links between VOA band features and molecular stereochemistry are becoming apparent in these two categories of VOA bands. Evidence for the coupled oscillator and two group behavior, the recurring VCD pattern associated with $CH_2CH_2C^*H$ fragment, the C-H bond helix rule for carbohydrates, the dynamic polarization behavior, and the bisignate features in perturbed degenerate modes are encouraging examples.

Type III and IV VOA bands are hard to simulate by theoretical models. Here an accurate description of the normal coordinates is as crucial as the principles underlying the theoretical models. As a result, all the uncertainties that are associated with the determination of force constants will enter VOA calculations. To elaborate further, it is well known that different choices for force constants can give good agreement between the calculated and observed vibrational frequencies. These different choices, however, can lead to significant differences in the description of the normal coordinates for delocalized vibrations. Thus, for VOA bands of types III and IV, it is difficult to judge whether the disparities between the theoretical and experimental observations are due to incorrect normal mode descriptions or due to inadequate parameterization in the model employed. Pathways to overcome these difficulties remain to be investigated.

Coming back to the developments in theoretical models, it can be said that, after a period of vigorous activity in the last six years, the developments in semiclassical models now appear to have come to saturation. This is because, for ROA, the bond polarizability and ADI models are now well developed and understood; for VCD, the primary goal of incorporating the electronic contribution to the magnetic dipole transition moment has been achieved in the form of the bond moment, bond current and charge flow models. It can be anticipated that future VOA calculations will emphasize on improving the parameterization required in these semiclassical models. The situation for quantum mechanical calculations is different and several developments are anticipated in molecular orbital calculations. This is because the current calculations require that the molecular orbitals be well localized. Unfortunately, for larger cyclic molecules which are frequently the subjects of optical activity studies, the localization procedures are known to be less successful. In these cases the calculated VOA intensities are, at the best, approximate. Therefore a need to develop procedures that eliminate the localization requirement is clearly evident. This may be achieved by devising methods to calculate the rocking contributions from canonical molecular orbitals to the magnetic dipole transition moment and optical activity tensors.

ACKNOWLEDGEMENT

I would like to thank the Minnesota group, especially Dr. A. Moscowitz, for providing copies of recent Ph.D. theses from The University of Minnesota on VCD. Thanks are also due to Dr. T. A. Keiderling for providing recent VCD spectra of dimethyl-d-tartrate and a preprint of his VCD work. It is a pleasure to acknowledge my scientifically rewarding association with Dr. L. A. Nafie from 1978 to 1980. This work is supported by grants from NIH (GM29375), Research Corporation and Vanderbilt University. Acknowledgement is made to the donors of the Petroleum Research Fund administered by The American Chemical Society for partial support.

REFERENCES

1. (a) For theoretical details see: D. J. Caldwell and H. Eyring, The Theory of Optical Activity, Wiley Interscience, New York, 1971; (b) For a recent account on sector rules see: G. Snatzke in Optical Activity and Chiral Discrimination, (S. F. Mason, ed.) D. Reidel, Dordrecht, Holland, 1979.

2. P. J. Stephens and R. Clark in Optical Activity and Chiral Discrimination, (S. F. Mason, ed.) D. Reidel, Dordrecht, Holland (1979).

3. L. A. Nafie and M. Diem, Acct. Chem. Res., 12, 296 (1979).

4. L. A. Nafie in Vibrational Spectra and Structure, Vol. 10, (J. R. Durig, ed.) Elsevier, New York (1981).

5. T. A. Keiderling, Appl. Spectrosc. Rev., 17, 189 (1981).

6. S. F. Mason in Advances in Infrared and Raman Spectroscopy, Vol. 8, (R. J. H. Clark and R. E. Hester, eds.) Heyden, London (1981).

7. L. D. Barron and A. D. Buckingham, Ann. Rev. Phys. Chem., 26, 381 (1975).

8. L. D. Barron in Specialists Periodical Reports, Molecular Spectroscopy, Vol. 4, The Chemical Society, London (1976).

9. L. D. Barron in Advances in Infrared and Raman Spectroscopy, Vol. 4, (R. J. H. Clark and R. E. Hester, eds.) Heyden, London (1978).

10. L. D. Barron in Optical Activity and Chiral Discrimination, (S. F. Mason, ed.) D. Reidel, Dordrecht, Holland (1979).

11. T. R. Faulkner, On The Infrared Optical Activity of Small Molecular Systems, The University of Minnesota, 1976.

12. C. Marcott, Vibrational Circular Dichroism and the Structure of Chiral Molecular Systems, The University of Minnesota, 1979.

13. H. A. Havel, Vibrational Circular Dichroism Studies in Carbon-Hydrogen and Carbon-Deuterium Stretching Regions, The University of Minnesota, 1981.

14. L. A. Laux, The Optical Activity Associated with Molecular Vibrations, The University of Minnesota, 1982.

15. M. Diem, Raman Optical Activity and Vibrational Studies of Chiral Molecules, The University of Toledo, 1976.

16. B. Cuony, Calcul De L'activite Optique Vibrationelle Raman, The University of Fribourg, 1981.

17. W. Fickett, J. Am. Chem. Soc., 74, 4204 (1952).

18. H. F. Hameka, J. Chem. Phys., 41, 3612 (1964).

19. N. V. Cohan and H. F. Hameka, J. Am. Chem. Soc., 88, 2136 (1966).

20. C. W. Deutsche and A. Moscowitz, J. Chem. Phys., 49, 3257 (1968); 53, 2630 (1970).

21. G. Holzwarth and I. Chabay, J. Chem. Phys., 57, 1632 (1972).

22. W. Kuhn, Ann. Rev. Phys. Chem., 9, 417 (1958); J. A. Schellman, Acct. Chem. Res., 1, 144 (1968).

23. H. Sugeta, C. Marcott, T. R. Faulkner, J. Overend and A. Moscowitz, Chem. Phys. Lett., 40, 397 (1976).

24. L. A. Nafie, P. L. Polavarapu and M. Diem, J. Chem. Phys., 73, 3530 (1980).

25. J. A. Schellman, J. Chem. Phys., 58, 2882 (1973); 60, 343 (1974).

26. G. Holzwarth, E. C. Hsu, H. S. Mosher, T. R. Faulkner and A. Moscowitz, J. Am. Chem. Soc., 96, 251 (1974).

27. T. R. Faulkner, A. Moscowitz, G. Holzwarth, E. C. Hsu and H. S. Mosher, J. Am. Chem. Soc., 96, 252 (1974).

28. S. Abbate, L. Laux, J. Overend and A. Moscowitz, J. Chem. Phys., 75, 3161 (1981); 78, 609 (1983).

29. M. Moskovits and A. Gohin, J. Phys. Chem., 86, 3947 (1982).

30. C. J. Barnett, A. F. Drake and S. F. Mason, Chem. Comm., 43 (1980).

31. C. J. Barnett, A. F. Drake, R. Kuroda and S. F. Mason, Mol. Phys., 41, 455 (1980).

32. P. L. Polavarapu, Mol. Phys., (in press).

33. L. D. Barron and A. D. Buckingham, J. Am. Chem. Soc., 101, 1979 (1979).

34. T. H. Walnut and L. A. Nafie, J. Chem. Phys., 67, 1491 (1977); 67, 1501 (1977).

35. L. A. Nafie and T. H. Walnut, Chem. Phys. Lett., 49, 441 (1977).

36. L. A. Nafie and P. L. Polavarapu, J. Chem. Phys., 75, 2935 (1981).

37. P. L. Polavarapu and L. A. Nafie, J. Chem. Phys., 75, 2945 (1981).

38. A. Komornicki and J. W. McIver, Jr., J. Chem. Phys., 70, 2014 (1979).

39. P. L. Polavarapu and J. Chandrasekhar, Chem. Phys. Lett., 84, 587 (1981); 86, 326 (1982).

40. P. L. Polavarapu, J. Chem. Phys., 77, 2273 (1982).

41. L. A. Nafie and T. B. Freedman, J. Chem. Phys., 75, 4847 (1981).

42. D. P. Craig and T. Thirunamachandran, Mol. Phys., 35, 31 (1978).

43. R. T. Klingbliel and H. Eyring, J. Phys. Chem., 74, 4543 (1970).

44. T. A. Keiderling, J. Chem. Phys., 75, 3639 (1981).

45. L. D. Barron and A. D. Buckingham, Mol. Phys., 20, 1111 (1971).

46. L. D. Barron and A. D. Buckingham, J. Am. Chem. Soc., 96, 4769 (1974).

47. A. J. Stone, Mol. Phys., 29, 1461 (1975); 33, 293 (1977).

48. A. Gohin and M. Moskovits, J. Am. Chem. Soc., 103, 1660 (1981).

49. L. D. Barron, Nature, 255, 458 (1975).

50. L. D. Barron, J. Chem. Soc. Perkin II, 1970 (1977).

51. P. L. Prasad and D. F. Burow, J. Am. Chem. Soc., 101, 800 (1979).

52. P. L. Prasad and D. F. Burow, J. Am. Chem. Soc., 101, 806 (1979).

53. P. L. Prasad and L. A. Nafie, J. Chem. Phys., 70, 5582 (1979).

54. L. D. Barron and A. D. Buckingham, Mol. Phys., 23, 145 (1972).

55. L. D. Barron, Nature, 257, 372 (1975).

56. L. D. Barron, Chem. Phys. Lett., 46, 579 (1977).

57. L. D. Barron and C. Meehan, Chem. Phys. Lett., 66, 444 (1979).

58. L. D. Barron, C. Meehan and J. Vrbancich, Mol. Phys., 41, 945 (1980).

59. L. D. Barron, J. Vrbancich and R. S. Watts, Chem. Phys. Lett., 89, 71 (1982).

60. L. D. Barron, C. Meehan and J. Vrbancich, J. Raman Spectrosc., 12, 256 (1982).

61. Cohan and Hameka (Ref. 19) and Schellman (Ref. 25) used the word "rotatory strength" for R_ℓ. However, in other VCD papers R_ℓ is referred to as "rotational strength".

62. M. Born and R. Oppenheimer, Ann. Physik., 84, 457 (1927).

63. C. A. Mead and A. Moscowitz, Int. J. Quant. Chem., 1, 243 (1967).

64. N. V. Cohan and H. F. Hameka, J. Chem. Phys., 45, 4392 (1966).

65. J. C. Slater, Phys. Rev., 35, 509 (1930).

66. T. R. Faulkner, C. Marcott, A. Moscowitz and J. Overend, J. Am. Chem. Soc., 99, 8160 (1977).

67. E. B. Wilson, J. C. Decius and P. C. Cross, Molecular Vibrations, McGraw-Hill, New York, 1955.

68. S. Kh. Samvelyan, V. T. Aleksanyan and B. V. Lokshin, J. Mol. Spectrosc., 48, 47 (1973).

69. J. C. Decius, J. Mol. Spectrosc., 57, 348 (1975).

70. J. C. Decius and G. B. Mast, J. Mol. Spectrosc., 70, 294 (1978).

71. L. A. Gribov, Intensity Theory For Infrared Spectra of Polyatomic Molecules, Consultants Bureau, New York (1964).

72. C. C. J. Roothan, Rev. Mod. Phys., 23, 69 (1951).

73. S. F. Boys, Proc. Roy. Soc. (London), A200, 542 (1950).

74. T. B. Freedman and L. A. Nafie, J. Chem. Phys., 78, 27 (1983).

75. J. F. Biarge, J. Herranz and J. Morcillo, An. R. Soc. Esp. Fis. Quim., Ser. A57, 81 (1961).

76. W. B. Person and J. H. Newton, J. Chem. Phys., 61, 1040 (1974).

77. V. I. Baranov, L. A. Gribov and B. K. Novosadov, J. Mol. Struct., 70, 1 (1981).

78. W. Hug and H. Surbeck, Chem. Phys. Lett., 60, 186 (1979).

79. L. D. Barron and B. P. Clark, Mol. Phys., 46, 839 (1982).

80. L. D. Barron and B. P. Clark, J. Raman Spectrosc., 13, 155 (1982).

81. L. D. Barron, J. F. Torrance and J. Vrbancich, J. Raman Spectrosc., 13, 171 (1982).

82. A. D. Buckingham and H. C. Longuet-Higgins, Mol. Phys., 14, 63 (1968).

83. This particular transformation in Eq. (48) is frequently used in transforming the diatomic overlap integrals from the local diatomic coordinate system to the molecular coordinate system in molecular orbital calculations.

84. J. Applequist, J. R. Carl and K. -K. Fung, J. Am. Chem. Soc., 94, 2952 (1972).

85. J. Applequist, J. Chem. Phys., 71, 4324 (1979); 71, 4332 (1979); 58, 4251 (1973).

86. J. Applequist, Acct. Chem. Res., 10, 79 (1977).

87. K. Sundberg, J. Chem. Phys., 66, 114 (1977); 66, 1475 (1977).

88. B. M. Ladanyi and T. Keyes, J. Chem. Phys., 68, 3217 (1978).

89. A. D. Buckingham and D. A. Dunmur, Trans. Farad. Soc., 64, 1776 (1968).

90. J. Applequist and C. O. Quicksall, J. Chem. Phys., 66, 3455 (1977).

91. P. L. Polavarapu, J. Mol. Spectrosc., 93, 450 (1982).

92. M. P. Boggard and R. Haines, Mol. Phys., 41, 1281 (1980).

93. C. Edminston and K. Ruedenberg, Rev. Mod. Phys., 35, 457 (1963).

94. J. H. Schachtschneider, Technical Report No. 57-65 (Shell Development Co.), Emeryville, CA, 1966.

95. C. Marcott, S. D. Ferber, H. A. Havel, A. Moscowitz and J. Overend, Spectrochim. Acta, 37A, 241 (1981).

96. A. L. McClellan, Tables of Experimental Dipole Moments, Freeman, San Francisco, CA (1963).

97. J. W. Smith, Electric Dipole Moments, Butterworths, London, 1955.

98. M. Gussoni in Advances in Infrared and Raman Spectroscopy, Vol. 6, (R. J. H. Clark and R. E. Hester, eds.) Heyden, London (1980).

99. W. B. Person and D. Steele in Specialists Periodical Reports, Molecular Spectroscopy, Vol. 2, The Chemical Society, London, 1974.

100. P. L. Prasad, J. Chem. Phys., 69, 4403 (1978); J. Phys. Chem., 83, 1744 (1979).

101. M. J. Aroney, Angewand. Chemie., 16, 663 (1977); R. J. W. LeFevre, Advances in Physical Organic Chemistry, Vol. 3 (V. Gold, ed.) Academic Press, New York (1965); C. G. LeFevre and R. J. W. LeFevre, Rev. Pure and Appl. Chem., 5, 261 (1955).

102. For recent work on bond polarizability derivatives see Ref. [98] and also the papers by S. Montero; for example: F. Orduna, C. Domingo, S. Montero and W. F. Murphy, Mol. Phys., 45, 65 (1982) and references cited therein. For earlier work, see L. A. Woodward in Raman Spectroscopy. Theory and Practice, Vol. 1 (H. A. Szymanski, ed.) Plenum Press, New York (1967); G. W. Chantry in The Raman Effect, (A. Anderson, ed.) Marcel Dekker, New York (1971); R. E. Hester in Specialists Periodical Reports, Molecular Spectroscopy, Vol. 2, The Chemical Society, London, 1976.

103. T. V. Long, Jr. and R. A. Plane, J. Chem. Phys., 43, 457 (1965).

104. B. Fontal and T. G. Spiro, Spectrochim. Acta, 33A, 507 (1977).

105. E. D. Lipp, C. G. Zimba and L. A. Nafie, Chem. Phys. Lett., 90, 1 (1982).

106. C. N. Su, T. A. Keiderling, K. Misura and W. J. Stec, J. Am. Chem. Soc., 104, 7343 (1982).

107. V. J. Heintz, W. A. Freeman and T. A. Keiderling, Inorg. Chem., (in press).

108. P. L. Polavarapu, Appl. Spectrosc., (in press).

109. W. Hug, Appl. Spectrosc., 35, 115 (1981).

110. W. Hug, S. Kint, G. F. Bailey and J. R. Scherer, J. Am. Chem. Soc., 97, 5589 (1975).

111. H. -J. Hansen, H. -R. Sliwka and W. Hug, Helv. Chim. Acta, 62, 1120 (1979).

112. L. D. Barron, M. P. Boggard and A. D. Buckingham, Nature, 241, 113 (1973).

113. L. D. Barron and B. P. Clark, J. Chem. Res(s), 36 (1979).

114. W. Hug, A. Kamatari, K. Srinivasan, H. -J. Hansen and H. -R. Sliwka, Chem. Phys. Lett., 76, 469 (1980).

115. C. N. Su and T. A. Keiderling, Chem. Phys. Lett., 77, 494 (1981).

116. M. Diem, P. L. Polavarapu, M. Oboodi, T. B. Freedman and L. A. Nafie, J. Am. Chem. Soc., 104, 3329 (1982).

117. B. B. Lal, M. Diem, P. L. Polavarapu, M. Oboodi, T. B. Freedman and L. A. Nafie, J. Am. Chem. Soc., 104, 3336 (1982).

118. T. B. Freedman, M. Diem, P. L. Polavarapu and L. A. Nafie, J. Am. Chem. Soc., 104, 3343 (1982).

119. V. J. Heintz and T. A. Keiderling, J. Am. Chem. Soc., 103, 2395 (1981).

120. T. A. Keiderling and P. J. Stephens, J. Am. Chem. Soc., 99, 8061 (1977).

121. C. Marcott, C. C. Blackburn, T. R. Faulkner, A. Moscowitz and J. Overend, J. Am. Chem. Soc., 100, 5262 (1978).

122. C. N. Su and T. A. Keiderling, J. Am. Chem. Soc., 102, 511 (1980).

123. T. A. Keiderling, private communication.

124. P. L. Polavarapu, M. Diem and L. A. Nafie, J. Am. Chem. Soc., 102, 5449 (1980).

125. L. Laux, V. Paultz, S. Abbate, H. A. Havel, J. Overend, A. Moscowitz and D. A. Lightner, J. Am. Chem. Soc., 104, 4276 (1982).

126. R. D. Singh and T. A. Keiderling, Biopolymers, 20, 237 (1980).

127. J. Snir, R. A. Frankel and J. A. Schellman, Biopolymers, 14, 173 (1975).

128. T. A. Keiderling and P. J. Stephens, J. Am. Chem. Soc., 101, 1396 (1979).

129. P. L. Polavarapu and L. A. Nafie, J. Chem. Phys., 73, 1567 (1980).

130. C. Marcott, K. Scanlon, J. Overend and A. Moscowitz, J. Am. Chem. Soc., 103, 483 (1981).

131. R. D. Singh and T. A. Keiderling, J. Chem. Phys., 74, 5347 (1981).

132. P. L. Polavarapu, L. A. Nafie, S. A. Benner and T. H. Morton, J. Am. Chem. Soc., 103, 5349 (1981).

133. R. D. Singh and T. A. Keiderling, J. Am. Chem. Soc., 103, 2387 (1981).

134. J. A. Pople and D. L. Beveridge, Approximate Molecular Orbital Theory, McGraw Hill, New York (1970).

135. L. D. Barron and B. P. Clark, J. Chem. Soc. Perkin II, 1164 (1979).

136. P. L. Prasad, L. A. Nafie and D. F. Burow, J. Raman Spectrosc., 8, 255 (1979).

137. T. Brocki, M. Moskovits and B. Bosnich, J. Am. Chem. Soc., 102, 495 (1980).

138. L. D. Barron, Molecular Light Scattering and Optical Activity, Cambridge University Press, Cambridge (1982).

139. L. Laux, V. Paultz, C. Marcott, J. Overend and A. Moscowitz, J. Chem. Phys., 78, 4096 (1983).

140. D. W. Schlosser, F. Devlin, K. Jalkanen and P. J. Stephens, Chem. Phys. Lett., 88, 286 (1982).

141. L. A. Nafie and T. B. Freedman, J. Chem. Phys., 78, 7108 (1983).

142. L. A. Nafie, in Advances in Infrared and Raman Spectroscopy, (R. J. H. Clark and R. E. Hester, eds.), Heyden, London (in press).

143. L. D. Barron and J. Vrbancich, Mol. Phys., 48, 833 (1983).

NOTES ADDED IN PROOF

After this review was completed, a few more relevant contributions appeared in press. Barron has written a book [138] where an extensive account of light scattering and optical activity is presented. Both natural and magnetic optical activity aspects pertaining to vibrational, as well as electronic transitions, are discussed. The theoretical aspects of magnetic vibrational circular dichroism, which was referred earlier in the text to Laux's doctoral thesis [14], are now published [139]. Stephens and co-workers [140] have reported the VCD of matrix isolated molecules, demonstrating the feasibility of such experiments. Since the vibrational bands in matrix isolated infrared spectra have much narrower line widths, and the overlapping of neighboring bands is eliminated, the VCD spectra of matrix isolated molecules will provide more accurate estimates of the experimental magnitudes. This, in turn, provides a better evaluation of the theoretical predictions. An article [141] dealing with the vibronic theory of vibrational intensities and a review article on VOA [142] have been reported recently. Barron and Vrbancich have derived [143] a compact expression for the ROA associated with the in phase methyl torsion for (2S, 3S)-(-)-epoxy butane, using the bond polarizability model. The sign and magnitude of ROA predicted from this expression are in favorable agreement with the experimental observation [143].

Chapter 4

VIBRATIONAL EFFECTS IN SPECTROSCOPIC GEOMETRIES

L. Nemes

Research Laboratory for Inorganic Chemistry
Hungarian Academy of Sciences
Budapest, Hungary

I. INTRODUCTION

The joint application of spectroscopic and diffraction methods for the determination of molecular geometry is a rapidly advancing field (see, e.g., Refs. [1-3]). Such a combined treatment should ideally be based on geometric parameters already corrected for vibrational effects. The analysis of such effects is then very important in molecular structure research.

In this review an effort is made to summarize theoretical and methodical advances in the derivation and interpretation of empirical molecular structures, their reduction to the equilibrium representation, the various ways of extracting structural information from rotation, and rotation-vibration spectra induced by various intramolecular effects.

In the calculation of microwave substitution structures, the problem of small substitution coordinates is frequently encountered. These, as a rule, carry relatively large vibrational contributions. The developments in the assessment of small r_s structure coordinates are correspondingly of great significance in this summary.

The units applied presently are, in some cases, different from those commonly used in spectroscopy. They are, however, thought to be in accordance with the prescriptions of the Système International d'Unités. Bond lengths are given in pm units (1 pm = 10^{-12}m = 10^{-2} Å), whereas electric dipole moments are given in C·m units (1 C·m = 10^{21} c[m·s^{-1}] Debye = 2.99792458 x 10^{29} D). Moments of inertia are also given in u·pm^2 or u·nm^2 units in some cases (1 u·nm^2 = 10^2 u·Å^2; 1 u·pm^2 = 10^{-4} u·Å^2). The conversion factors between various spectroscopic quantities in this review were calculated from the latest values of fundamental constants, Ref. [4].

II. VIBRATIONAL AVERAGING EFFECTS - A SHORT SUMMARY

Vibrational effects manifest themselves in geometrical parameters through averaging processes. This is understood by considering that the quantities from which such parameters are calculated are derived from the overall molecular rotational motion that usually is much slower than vibrations. For instance, at 300K the mean rotational period for the H_2 molecule is roughly 1.5 x 10^{-13} s, whereas its vibrational period is about 7.6 x 10^{-15} s. The hydrogen molecule undergoes, therefore, about twenty vibrational periods within a single rotational cycle. Both types of motions are slower than the electronic motion by some magnitudes, and of course this provides the basis for the Born-Oppenheimer separation approximation.

The two main methods for gas-phase determinations of molecular geometries are electron diffraction and rotational spectroscopy, and they differ
significantly from each other with respect to what is being averaged in the
repeated vibrational cycles. Since the diffraction picture leads eventually to
distances within pairs of nuclei whereas the rotational spectrum yields the
position coordinates of nuclei in the molecular frame, one ought to distinguish
between vibrational averages of interatomic distances on the one hand, and
the distances derived from the average nuclear position coordinates on the
other. We shall return to this important distinction later.

The theory of vibrational averaging is inseparable from the theory of
vibration-rotation interactions, as is eminently exemplified by the series of
papers by Herschbach and Laurie [5-7]. Their general formalism is constructed, in turn, upon Wilson and Howard's perturbation approach for the
vibration-rotation Hamiltonian operator for polyatomic molecules, and Nielsen's
treatment of vibration-rotation energies [8]. For polyatomic molecules the
formulation of vibrational averages is carried out in terms of the effective
moments of inertia interpreted for given vibrational states. The notion of the
necessity of providing well-defined physical concepts for empirical, operationally defined geometrical structure representations appeared already in
Laurie and Herschbach's paper [5].

Vibrational averaging effects are treated nowadays in the formalism
mainly due to Japanese authors. As most empirical molecular parameters are
derived from the vibrational ground state, the latter has a special importance.
Oka [9], Morino et al. [10], and Toyama et al. [11] have further developed
Laurie and Herschbach's approach for the ground-state vibrational average
structures. The various spectroscopic and diffraction vibrational averages
and effective structures were reviewed by Kuchitsu and Cyvin [12], and
Kuchitsu has recently given a comprehensive theoretical summary [13] of
geometrical parameters. Presently, we only take the r_g and r_α structural
representations for exemplifying vibrational effects.

The r_g structure consists of vibrationally averaged interatomic distances,
whereas the r_α structure contains distances between vibrationally averaged
nuclear positions. The usual formalism for relating r_g and r_α structures to
the equilibrium r_e values is the following:

$$r_g = r_e + \langle \Delta z \rangle + (\langle \Delta x^2 \rangle + \langle \Delta y^2 \rangle)/2r_e, \tag{1}$$

and

$$r_\alpha = r_e + \langle \Delta z \rangle, \tag{2}$$

where $\langle\Delta z\rangle$ is the linear vibrational average of the difference of Cartesian displacements parallel to the internuclear direction, whereas the second term in Eq. (1) contains the quadratic averages of displacement differences perpendicular to the direction of the bond; $\langle\Delta x^2\rangle$, $\langle\Delta y^2\rangle$. Evidently, Eqs. (1) and (2) represent a diatomic bond picture applied to the polyatomic model. The linear $\langle\Delta z\rangle$ average is a function of vibrational anharmonicity, but the quadratic averages can be evaluated fully within the harmonic model.

As Mills [14] has shown, the formalism in Eqs. (1) and (2) can be replaced by one which is more familiar to the vibrational spectroscopist. The formalism utilizes the L tensor introduced by Hoy et al. [15], and Tsaune et al. [17] which describes the non-linear transformation from curvilinear internal coordinates to the normal coordinates, and which is a straightforward generalization of the usual L matrix of Wilson, Decius and Cross [16].

In the "diatomic bond" formalism, Eqs. (1) and (2) are written by Mills as

$$r_g = r_e + \sum_r L^r \langle Q_r\rangle + (1/2) \sum_s L^{ss} \langle Q_s^2\rangle + \ldots, \tag{3}$$

and

$$r_\alpha = r_e + \sum_r L^r \langle Q_r\rangle \underset{\sim}{\times} r_g - (1/2) \sum_s L^{ss} \langle Q_s^2\rangle. \tag{4}$$

The L^r elements of the L tensor are those of the usual L matrix [16] in the transformation of the internal coordinates $\underset{\sim}{R}$ to the dimensional normal coordinates $\underset{\sim}{Q}$:

$$\underset{\sim}{R} = L \underset{\sim}{Q}. \tag{5}$$

The L^{ss} elements are second derivatives of the curvilinear internal coordinates with respect to the normal coordinates, and are completely determined by the L matrix [15].

The equivalence of Eqs. (3) and (4) to Eqs. (1) and (2) is based upon the relationships:

$$\sum_r L^r \langle Q_r\rangle = \langle\Delta z\rangle, \tag{6}$$

$$\sum_s L^{ss} \langle Q_s^2\rangle = \langle\Delta x^2\rangle + \langle\Delta y^2\rangle. \tag{7}$$

The r_α structures of electron diffraction extrapolated to zero K are identical to the r_z structures of spectroscopy, and the difference $r_z - r_e$ for a given pair of chemically bound atoms is given by

$$r_z - r_e = \sum_r L^r \langle Q_r \rangle_o = -(1/4)(\frac{\hbar^2}{hc})^{1/2} \sum_r \sum_s L^r w_r^{-3/2} \phi_{rss}, \qquad (8)$$

where ϕ_{rss} is a cubic anharmonic force constant in cm^{-1} units, defined by the nuclear potential energy expansion in dimensionless q_r normal coordinates [15]:

$$V/hc = (1/2) \sum_r w_r q_r^2 + (1/6) \sum_r \sum_s \sum_t \phi_{rst} \; q_r q_s q_t + \cdots \qquad (9)$$

III.   DEVELOPMENTS IN ISOTOPIC SUBSTITUTION GEOMETRY

A dominant portion of modern spectroscopic research on gas-phase molecular geometry makes use of the Kraitchman-Costain method for isotopic substitution [18,19]. The method is described in a number of books on microwave spectroscopy [20-22]. For an appreciation of modern advances in this field, some of the fundamental aspects shall be reviewed here.

Before the application of any method for the calculation of geometries, it may be helpful to know in advance how many independent (non-redundant) structural parameters one should determine to build a unique geometry. It turns out that this number is specified by the number of totally symmetric normal modes in the point groups of molecules [1,23,24]. When the potential energy function has more than one minimum corresponding to more than one equilibrium nuclear arrangement, the counting of the totally symmetric vibrations is a little less straightforward in the point group classification, but no basic difficulties are encountered [24].

The mathematical procedures for the determination of molecular geometry from rotational spectra are based on the formulation of the moments of inertia in different axis systems, as described by Kraitchman [18].

According to Goldstein [25] the moment of inertia tensor is given in the dyadic form as

$$I = \sum_i m_i (r_i^2 \underset{\sim}{1} - \underset{\sim i \sim i}{r \, r}), \qquad (10)$$

when referred to the center of mass of the nuclei. In Eq. (10), $\underset{\sim}{1}$ is the unit dyadic; $\underset{\sim}{1} = \underset{\sim}{i}\,\underset{\sim}{i} + \underset{\sim}{j}\,\underset{\sim}{j} + \underset{\sim}{k}\,\underset{\sim}{k}$, expressed by unit vectors, where $\underset{\sim i \sim i}{r \, r}$ is the dyad of the position vector $\underset{\sim i}{r}$ of nucleus i.

When an arbitrary origin O is used to construct the position vectors $\underset{\sim}{r}_i$, Eq. (10) is modified according to the parallel axis theorem:

$$I' = I - M (R^2 \underset{\sim}{1} - \underset{\sim}{R} \underset{\sim}{R}), \tag{11}$$

where I' is analogous to I in Eq. (10), but is expressed by the $\underset{\sim}{r}_i' = \underset{\sim}{r}_i - \underset{\sim}{R}$ center-of-mass related vectors ($\underset{\sim}{r}_i$ and $\underset{\sim}{R}$ are the position vectors of the nucleus i and the center-of-mass, respectively, relative to the arbitrary origin O), and

$$M = \sum_i m_i,$$

the total mass of the nuclei. Note that I' refers to the center-of-mass.

Kraitchman [18] introduced another (second) moment of inertia; the planar dyadic, which he has defined for the case when the center-of-mass and the arbitrary origin coincide:

$$P = \sum_i m_i \underset{\sim}{r}_i \underset{\sim}{r}_i = \sum_i m_i \underset{\sim}{r}_i' \underset{\sim}{r}_i'. \tag{12}$$

For vectors expressed in a system with arbitrary origin;

$$P' = P - M \underset{\sim}{R} \underset{\sim}{R}. \tag{13}$$

The relationships among I and P, and I' and P', respectively, are then easily derived from Eqs. (10-12) and Eq. (13);

$$I = \sum_i m_i r_i^2 \underset{\sim}{1} - P, \tag{14}$$

and

$$I' = \sum_i m_i (r_i^2 - R^2) \underset{\sim}{1} - P'. \tag{15}$$

Eqs. (14) and (15) express the fact that the same principal axis transformation diagonalizes the matrix forms of I and P, or I' and P', respectively. The moment of inertia coefficients for P' and P were given for the first time explicitly by Kraitchman [18].

Another notation for expressing relations between tensors I and P was suggested by Watson [26] in which the planar dyadic P is denoted by $K_{\alpha\beta}$;

$$I_{\alpha\beta} = \delta_{\alpha\beta} K_{\gamma\gamma} - K_{\alpha\beta}, \tag{16}$$

and

$$K_{\alpha\beta} = (1/2)\delta_{\alpha\beta} I_{\gamma\gamma} - I_{\alpha\beta}, \tag{17}$$

where α, β and γ correspond to the Cartesian components x, y and z, and $\delta_{\alpha\beta}$ is the Kronecker delta symbol.

Equations (10-17) are rigorously valid only for the equilibrium geometry. Kraitchman's isotopic substitution equations are derived using the matrix forms of Eqs. (11) and (13) by expressing I or P of the isotopically substituted molecule using a coordinate system located at the center-of-mass of the parent molecule.

Kraitchman's original equations are valid for isotopic substitution of single atoms, or sequences of such steps, and utilize mainly the I tensor. It was only for the general, non-planar asymmetric top that he recommended the use of the planar dyadic P.

For nuclei that are equivalent with respect to C_n or S_n symmetry operations, Chutjian [27] extended Kraitchman's formulas for multiple isotopic substitution. For di-substitution these equations may be simplified as was shown by Nygaard [28] who replaced Chutjian's I elements by P elements.

A.   Inherent Problems with Substitution Coordinates

As is well known, Costain [19] expanded the use of Kraitchman's equations to non-equilibrium data of real, vibrating molecules for which, up to 1958, only the ground-state effective r_o structures had been available. These structures possess characteristically sizeable isotopic mass dependence. For isotopic substitution (r_s structures), a portion of the vibrational effects cancels out due to the operative definition of the r_s coordinates, i.e., due to forming first differences of I_o^α or P_o^α effective principal moments, or principal planar moments of inertia. These are eigenvalues of the effective I and P tensors, respectively. For the simplest case of a linear molecule the substitution coordinate is given by

$$[z_s(i)]^2 = (I'_o - I_o)\mu_i^{-1} = \Delta I_o \, \mu_i^{-1} \tag{18}$$

where the prime denotes the isotopically substituted molecule, and μ_i is the so-called "reduced mass" of substitution:

$$\mu_i = \Delta m_i \, M(\Delta m_i + M)^{-1} \approx \Delta m_i, \text{ for } \Delta m_i \ll M, \tag{19}$$

(under such conditions μ_i and Δm_i differ by $(\Delta m_i)^2 M^{-1}$). When $\mu_i = \Delta m_i$ is substituted into Eq. (18) it is seen that the square of the substitution coordinate is approximated by the difference ratio $\Delta I_o/\Delta m_i$. In Sec. IV it will be seen that this approximation may be refined and further generalized [29].

Costain showed [19a] for linear molecules and, in particular, for diatomics that the substitution bond length is usually closer to the equilibrium value r_e than is the effective r_o value. Specifically,

$$r_s \approx (r_e + r_o)/2, \quad \text{or} \quad r_e \approx 2r_s - r_o. \tag{20}$$

For diatomic molecules, Eq. (20) and the corresponding inertial moment equation;

$$I_e \approx 2I_s - I_o \tag{21}$$

are simply related to each other. For polyatomic molecules this situation is not easily obtained. However, Costain did find Eq. (21) to hold empirically for triatomics, such as the molecules HCN and OCS.

In the use of the Kraitchman-Costain method, the basic question emerges as to how many isotopically substituted species of the parent molecule are needed, i.e., how many isotopomers are necessary for the complete determination of the geometry. This question is obviously related to the number of non-redundant structural parameters, mentioned at the beginning of this section, and was posed first by Kraitchman [18]. He showed that, for linear molecules and planar asymmetric tops constructed of n nuclei, one requires at least n - 2 isotopic substitutions (i.e., n - 1 isotopomers) for complete structures, whereas a minimum of only n - 3 substitutions are needed for the general (non-planar) asymmetric top. Kraitchman's original argument is that, when the substitution coordinates of n - 3 nuclei are determined, the nine remaining coordinates may be calculated from the three first moment constraints locating the center-of-mass:

$$\sum_i m_i \, r_i = 0, \tag{22}$$

the three second moment constraints originating from the principal axis transformation:

$$\sum_i m_i \, r_{i\alpha} \, r_{i\beta} = 0 \quad \text{where } \alpha,\beta = x,y; \; y,z; \; x,z \tag{23}$$

and from the three principal planar moments

$$P_\alpha = \sum_i m_i r_{i\alpha}^2 \qquad (\alpha = x, y \text{ and } z).$$

Recently van Eijck [30] has pointed out that the six constraints in Eqs. (22) and (23), in addition to the definition equations for P_α, are not independent since the unsubstituted three atoms are, of necessity, in a plane. Thus, even for the general asymmetric top, n - 1 isotopomers are required for deriving the complete geometry. This statement holds only when a rigid (equilibrium) model is used because vibrational effects normally necessitate the use of one or several more isotopomers.

There appear to be three basic problems with the determination of the r_s coordinates, and the efforts to cope with them are highly characteristic in the development of microwave spectroscopic structural work.

(i) For chemical elements that possess only a single stable nuclide, isotopic substitution is clearly impossible. There are twenty-one such elements: Be, F, Na, Al, P, Sc, Mn, Co, As, Y, Nb, Rh, I, Cs, Pr, Tb, Ho, Tm, Au, Bi and the semi-stable Th. Most of the chemical elements important for structural work have two or more stable isotopes. It may sometimes be important to have access to more than two isotopes for a given nucleus in order to assess the vibrational effects in the r_s structures.

When isotopic substitution is impossible, first or second moments, of the type Eq. (22) and Eq. (23) represent, become indispensable for the location of the atom in question [31].

(ii) Another typical difficulty is encountered for nuclei that fall close to one, two or all three principal axes of inertia. When the magnitude of the absolute value of the substitution coordinate $|z_s|$ is 0.15 Å (15 pm), or smaller, vibrational effects may lead to a complex number in the calculation of z_s, as Coles and Hughes [32] noted in 1949. In even moderately complicated asymmetric top structures more than one nucleus may lie at one of these difficult positions.

In order to reduce vibrational effects and thus free small z_s coordinates from the load of vibrational contributions, Pierce [33] extended the original, first difference Costain expression, Eq. (18), by formulating second differences: $\Delta\Delta I_o^{jk,j} = \Delta I_o^{jk} - \Delta I_{o'}^{j}$, where ΔI_o^{j} is involved in Eq. (18), and ΔI_o^{jk} expresses the change upon substituting two different atoms, j and k. The method was originally elaborated for linear molecules and for the axial nucleus substitution of symmetric tops, i.e., when there is no rotation of the inertial axes upon substitution.

The essence of the Pierce method lies in the modification of Eq. (18) to include the higher-order isotopic mass dependence of ΔI_o^{j}:

$$\Delta I_o^{j} = (z_s(j))^2 \, \mu_j + \frac{\partial \varepsilon}{\partial m_j} \, \Delta m_j + \frac{\partial^2 \varepsilon}{\partial m_j \partial m_j} \, \Delta m_j \, \Delta m_j + \ldots$$

$$= (z_s(j))^2 \, \mu_j + \varepsilon_j^{(1)} \, \Delta m_j + \varepsilon_{jj}^{(2)} \, \Delta m_j \, \Delta m_j + \ldots \qquad (24)$$

where $\varepsilon = I_o - I_e$, the empirical vibrational correction for the parent molecule. Equation (24) contains a part of the general Taylor expansion of ε according to the substitution mass changes at any nucleus. When $z_s(j)$ is a small quantity, an appropriate substitution is made at another nucleus k that shifts the principal axis system by a conveniently large amount. For the determination of the substitution z coordinate of atom j one needs four iso-topomers; two parents with moments of inertia I_o and I_o^k, and two substituted molecules with I_o^j and I_o^{jk}. This increased requirement is considered worth-while because a better approximation to the r_e geometry is obtained, as was shown by Pierce [33] for the linear NNO molecule.

Pierce's method was later extended to C_s asymmetric tops by Krisher and Pierce [34] who gave explicit formulas for the transformation of the inertial axes upon isotopic substitution. This transformation consists of a rotation about a single axis. Penn and Buxton [35] further extended the Pierce method insofar as these authors eliminated the coefficients of transformation between the two principal axis systems and, by using the planar dyadic P formalism, they generalized the method for asymmetric tops with no plane of symmetry.

(iii) Vibrational effects in positional coordinates may not be compensated by the Kraitchman-Costain method. On the contrary, they may even be magnified when isotopic substitution leads to a gross rotation of the inertial axes relative to the parent isotopomer. This rotation may be so extensive in nearly symmetric oblate rotors that the a and b inertial axes are inter-changed, as was pointed out by Beaudet et al. [36-39] and by Kuczkowski et al. [40,41]. Such rotations may sometimes be avoided by symmetry-preserving multiple isotopic substitutions.

An important result of studies of large axis rotations is the development of a technique called "isotopic pulling" for the determination of relative signs of substitution coordinates [37-39]. The term is derived from the fact that when the substitution coordinate of a nucleus is calculated using a different reference, or parent molecules, the sign of this coordinate may be inferred from the change in the magnitude of the coordinate as the center-of-mass is pulled by the different substitutions.

The Pierce double substitution method is an example of the applicability of the "isotopic pulling" technique. A more general motivation for the use of this technique is that the isotopic transformation of inertial axes can only be correctly described by the Cartesian nuclear coordinates in the related axis systems when these coordinates possess internally consistent relative signs.

The problems listed under items (i), (ii) and (iii) bring us to the question of the uncertainty of the substitution coordinates, and also to the problem of what extent the r_s structures can be regarded as approximations to the equilibrium geometry. It is important to realize that these two aspects are different, as was pointed out quite early by Graner [42]. One could expect r_s coordinates to be fairly consistent internally when derived from a family of isotopically substituted molecules, and still significantly different from equilibrium values. Whereas the question of internal consistency can be decided purely on a statistical basis, the rigorous study of the $r_s \approx r_e$ approximation is very difficult mainly because of the complexity of the effects of vibration-rotation interactions in molecular geometry.

On a simple statistical ground, Costain [19a] derived uncertainty estimates for the equilibrium coordinates from Kraitchman's equations of the Eq. (18) type, assuming that it is derived solely from the error of the ΔI_e isotopic differences:

$$\delta z_e(i) = \delta(\Delta I_e) \, (2\mu_i |z_e(i)|)^{-1}.$$ (25)

On this basis, Costain [19b] suggested the following estimate for the substitution coordinates:

$$|a_s(i)| \delta a_s(i) = 0.0012 \ \overset{\circ}{A}^2 \ (12 \ pm^2).$$ (26)

In order to include an estimate for the deviations from the corresponding equilibrium value, Schwendeman [31] suggested the so-called "pseudo inertial defect". This quantity was introduced by Herschbach and Laurie [7] through relationships among the eigenvalues of the tensors P and I (see Eq. (14)). Thus, e.g.,

$$2P_e^z = I_e^x + I_e^y - I_e^z.$$ (27)

An analogous relationship may be constructed for the ground-state effective values:

$$2P_o^z = I_o^x + I_o^y - I_o^z.$$ (28)

The difference between Eq. (27) and Eq. (28) is called the "pseudo inertial defect", Δ_z:

$$\Delta_z = 2(P_e^z - P_o^z) = (I_e^x - I_o^x) + (I_e^y - I_o^y) - (I_e^z - I_o^z)$$

$$= \varepsilon_z - \varepsilon_x - \varepsilon_y.$$ (29)

For molecules with a plane of symmetry $\sigma(xy)$, Costain's Eq. (25) may be written using ground-state effective values of the principal moments P_z:

$$\delta z_s(i) = \delta(P_o^{z'} - P_o^z)(2\mu|z_s(i)|)^{-1}. \tag{30}$$

As for isotopic substitution in the $\sigma(xy)$ plane, the difference $P_e^{z'} - P_e^z$ is zero; in place of $\delta(P_o^{z'} - P_o^z)$ one can write $(1/2)|(\Delta_z' - \Delta_z)|$ from Eq. (29). Schwendeman [31,46] assumed that the isotopic variation of Δ_z is approximately equal to 0.006 $u \cdot \overset{\circ}{A}{}^2$, and thus obtained a modified estimate for cases when $\mu \underset{\sim}{\sim} \Delta m_i = 1u$;

$$|z_s(i)| \, \delta z_s(i) = 0.0015 \, \overset{\circ}{A}{}^2(15 \text{ pm}^2). \tag{31}$$

This estimate is now accepted as an independent measure of the uncertainty of substitution geometrical parameters [43]. In a recent paper van Eijck [44] has reported an extensive statistical test of the original Costain estimate in Eq. (25). For single isotopic substitution, small inertial axis rotation and harmonic vibrational effects, van Eijck [44] has done statistics on a large body of carefully chosen microwave spectroscopic data [45]. For $\delta(P_o^{z'} - P_o^z)$ to be used, the square root of weighted second moments $\langle \Delta P^2 \rangle$ is calculated as:

$$\delta(P_o^{z'} - P_o^z) = \langle P^2 \rangle^{1/2} = \{\sum_i w_i (\Delta P_o^z)_i^2 / \sum_i w_i\}^{1/2}, \tag{32}$$

where the weight factors w_i are given in a parametrically variable form which also contains σ_i^{-2} formed from the experimental errors. From his results, van Eijck [44] has concluded that, for the internal consistency of r_s coordinates, Costain's original estimate (12 pm^2 in Eq. (26)) holds well and is even a little overestimated. He emphasized that this high degree of consistency is compatible with significant deviations from the equilibrium structures.

Both Eqs. (25) and (31) reflect the well known difficulty with small substitution coordinates. For such small coordinates special computation schemes may become necessary to determine both the r_s structure coordinate and its uncertainty. One customary procedure is to use the principal axis condition, Eq. (22) and/or Eq. (23), in which case the uncertainty $\delta a_s(i)$ should be assessed by variance propagation formulas [46]. In a recent review and compilation at the National Bureau of Standards [43] such problems are discussed and, additionally, a rating scheme for various levels of uncertainties of spectroscopic structures is proposed.

---

†The author is grateful to Dr. Georges Graner, Laboratoire d'Infrarouge, Orsay, France, for calling his attention to the systematic differences between $a_s(i)$ and $a_e(i)$ coordinates[42].

B. Modern Extensions of the Kraitchman-Costain Method

In the past few years there have been attempts to broaden the scope of the isotopic substitution method. Essentially, two routes have been followed which include the extension of the basic algebra of the Kraitchman method and the application of the method of least-squares to calculate r_s structures. Advances along the former route are due to Wilson and Smith [47] and to Rudolph [48]. Both of these works aim at the evaluation and use of the transformation of inertial axes upon isotopic substitution. Although Kraitchman's equilibrium structure equations are independent of such axis transformations, in the previous subsection A it has been noted that such transformations may lead to magnified vibrational effects. The development of explicit algebraic tools to calculate such axis transformations is then potentially important for dealing with large vibrational effects.

For the case of planar molecules or for molecules that preserve their symmetry plane upon isotopic substitution, Wilson and Smith [47] have sketched a formalism to take inertial frame rotation and translation explicitly into account. When the distinguished plane is $\sigma(xy)$, formulas like the one below can be derived for multiple isotopic substitution:

$$\sum_i \Delta m_i x_i^2 = (P^{x'} - P^x) + (P^{y'} - P^{x'}) \sin^2\alpha + M' \delta_x^2, \tag{33}$$

where all substituted atoms have $\Delta m_i \neq 0$, the prime denotes the substituted isotopomer, α and δ_x are the rotation angle and the translational shift of axes in the $\sigma(xy)$ plane, respectively, and M' is the molecular mass. The rotation α and the translations (δ_x, δ_y) can be evaluated from Kraitchman coordinates and experimental principal planar moments for both isotopomers. Iterative schemes may be used when the r_s structure coordinates of a single nucleus are unknown.

In Rudolph's paper [48] the algebra of the above-mentioned transformation of axes is given in a detailed fashion. Some typical uses of the newly derived relationships are also given.

For the general case, the position vectors of an arbitrary atom in the parent axis frame ($\underset{\sim}{r}$), and in the frame of the singly substituted isotopomer ($\underset{\sim}{r}'$) are given by:

$$\underset{\sim}{r} = R \underset{\sim}{r}' + \underset{\sim}{t}, \tag{34}$$

where R represents the orthogonal transformation between the two reference systems, $(R^T)^{-1} = R$, and $\underset{\sim}{t}$ is the translation vector pointing from the center-of-mass (COM) of the parent isotopomer toward the COM' of the substituted one. The matrix elements of R are basically direction cosines describing the relative orientation of one system with respect to the other[†]. Rudolph [48], however, avoids the use of angular functions (e.g., direction cosines of Eulerian angles [25]). His method for the analytic construction of R is based on the similarity transformation diagonalizing P of the isotopically substituted molecule:

$$R^{-1} P^{(')} R = P' \text{ (diagonal)}, \tag{35}$$

where $P^{(')}$ is constructed using coordinates referring to the parent axis system, hence it is non-diagonal.

As a result of some algebraic manipulations, the matrix of R can be built from dimensionless combinations of the principal planar moments of both isotopomers.

When the r_s structure coordinate of a given nucleus is unmanageably small with respect to a common parent in a substitution sequence and therefore it is loaded with vibrational effects, Rudolph's method allows one to alleviate the problem by using, for instance, a singly substituted intermediate parent isotopomer. The determination of the Kraitchman coordinate in question is then made first with reference to the intermediate parent, and the appropriate R transformation is subsequently used to re-transform this conveniently sized r_s structure coordinate into the axis system of the common parent. This "trick" is suggested to remove much of the uncertainty of small $a_s(i)$ coordinates by operationally reducing the vibrational effects incorporated in them.

Rudolph [48] has used the 1-H position of skew-1-butene to illustrate his method. Other possible recommended uses are a general formulation of Pierce's double substitution method, the determination of the true orientation of the electric dipole moment p, and a generalized method for isotopic substitution sequences.

When the inertial ellipsoid of the molecule [25] has axial symmetry, and when isotopic substitution occurs in that axis of the inertial plane, the derivation of substitution structures is usually done through specific forms of the Kraitchman equations. These specific forms are obtained from the general Kraitchman equation expressed by isotopic differences of principal planar moments, e.g.,

---

† One might remark that the ideas of the R transformation are dormant in earlier works applying "isotopic pulling" [37-39].

$$|z| = (\frac{\Delta P^z}{\mu})^{1/2} \{1 + (\frac{\Delta P^x}{P^x-P^z})\}^{1/2} \{1 + (\frac{\Delta P^y}{P^y-P^z})\}^{1/2} \tag{36}$$

by requiring the isotopic difference of the appropriate pseudo inertial defect to vanish. When, e.g., a nucleus on the z-axis is isotopically substituted, ΔP^x and/or ΔP^y may be taken to be zero. This is obviously a neglect of vibrational effects (because rigorously only $\Delta P^x_e = \Delta P^y_e = 0$ in such a sub-stitution), and eventually leads to a simplification of Eq. (36) to a type of Eq. (18):

$$|z| = (\frac{\Delta P^z}{\mu})^{1/2}. \tag{37}$$

Rudolph has recently shown [49] that, at least for the equilibrium geometry, Eq. (37) is rigorous when solely the symmetry of the problem is considered, and is in fact a Chutjian-Nygaard-type simplification [27,28] of Eq. (36). Similar simplification is achieved for $|x|$ and $|y|$ (when isotopic substitution occurs in the $\sigma(xy)$ plane) by using the Chutjian-Nygaard modi-fications of the Kraitchman equations. In this case only one of the three parentheses vanishes from the original equations of the type of Eq. (36).

When the simplification of equations like Eq. (36) is achieved through a requirement of a vanishing pseudo inertial defect, a number of possibilities arise. These are obtained by first replacing ΔP^x and/or ΔP^y and/or ΔP^z by the appropriate combination(s) of ΔI^x, ΔI^y and ΔI^z differences (on the basis of Eq. (28) and similar ones), then eliminating some of the ΔI^x, ΔI^y or ΔI^z terms through vanishing pseudo inertial effect relations (for z-axis substitu-tion see Eqs. (2a) and (2b) in Ref. [49]). It has been shown that sub-stitution structures, which are well balanced with respect to the various forms of neglect of vibrational effects, can only be obtained when all possi-bilities for simplifying Eq. (36) (and the other equations for $|x|$ and $|y|$) through the "no pseudo inertial defect" conditions are considered.

Rudolph [49] has argued that it is better to use the Chutjian-Nygaard symmetrical forms than an incomplete set of simplified Kraitchman equations obtained through the use of the vanishing pseudo inertial defects. The symmetry argument is rigorously valid only for the equilibrium structure, or for vibrational mass-distribution averages symmetrical to the z-axis or to the $\sigma(xy)$ plane.

The application of least squares methods for r_s structure calculations is also a comparatively new development and constitutes the alternative route to extensions of the Kraitchman method. The application of regression methods opens up new possibilities for the estimation of uncertainties and for correlations among the structural parameters. By the least squares method, the principal moments of inertia and their isotopic differences can be utilized for single or multiple isotopic substitutions in any set of isotopomers.

Vibrational effects incorporated in r_s structures might and usually do cause deviations from the principal axis conditions represented by Eqs. (22) and (23). In regression methods, substitution structures may be obtained which are balanced with respect to such constraints. Because the second moments of inertia are not linear functions of nuclear Cartesian coordinates or combinations of them, iterative regression steps are necessary for the optimalization of structures. Schwendeman [31] reviewed both the mathematics and the main characteristics of such least squares calculations.

There are a number of approaches available for regression structure calculations. In addition to Schwendeman's method [31], Nösberger et al. [50,51], and Typke [52,53] have also published computational schemes based on least squares routines. All of these are based on the solution of the normal equations:

$$D^T WD \; \delta \underset{\sim}{X} = D^T W \; \delta \underset{\sim}{B}, \tag{38}$$

where $\delta \underset{\sim}{X}$ and $\delta \underset{\sim}{B}$ are vectors of the $\underset{\sim}{X}$ parameter corrections and of the errors of the estimated principal second moments (or their isotopic differences), respectively; D is the Jacobian matrix containing $\partial B / \partial X$ derivatives, and W is usually a diagonal statistical weight matrix.

When Eq. (38) is solved for $\delta \underset{\sim}{X}$, problems typical of least squares calculations may be encountered when $D^T WD$ is singular or nearly so. In these cases the former matrix can not be simply inverted. Singularities may arise from small D elements, or from the fact that some components of the $\underset{\sim}{X}$ parameter vector, which we wish to obtain individually, are in fact strongly correlated, so that only their linear combinations can be determined in a statistically meaningful way. Singularities are manifested in small eigenvalues of $D^T WD$. These singularities are best eliminated by enlarging the basis of input data, i.e., a larger and more varied set of experimental moments of inertia should be applied. Alternatively, independent structural assumptions may be used which are available from other experimental sources, such as gas-phase electron diffraction [2]. Except when such assumptions refer to the position of the H nuclei, structural hypotheses are dangerous for the final uncertainties in the calculated structures [43].

An interesting mathematical way to handle the singularity problem was proposed by Nösberger et al. [50]. The essentials are the following: the

Jacobian matrix D of size n × k (n is the number of components in B, while k is the number of components in X) is of rank r. The matrix $D^T WD$, or simply $D^T D$, is singular when r < k. In such cases, following the algorithm of Golub et al. [55,56], $[D^T D]^{-1}$ is replaced by d^{-1}, the pseudo inverse. The pseudo inverse matrix is obtained from the principles of singular value decomposition, first described by Lánczos [54], as:

$$d^{-1} = V \wedge V^T, \qquad (39)$$

where V is the matrix of the orthonormal eigenvectors of $D^T D$, and the diagonal $\wedge$ matrix contains the elements σ_i^{-2} and zero, depending on whether σ_i is larger or smaller, respectively, than $\varepsilon\sigma_{max}$. The quantity σ_{max} is the largest singular value of D, while ε is a threshold parameter set by considerations of computational accuracy (the σ_i elements are the positive square root of the eigenvalues of $D^T D$). The estimated variance of the elements of vector $\underset{\sim}{X}$ (the vector of structural parameters) in a singular system is obtained as:

$$\sigma^2 \{X_i\} = \sigma^2 \{d^{-1}\}_{ii} (n - p)^{-1}, \qquad (40)$$

where p is the number of non-zero singular values.

Nösberger et al. have tested their method on furan, ethyl fluoride and acetaldehyde [50], and also on the r_s structure of nitro ethylene [51]. It has been shown by experience that when complete r_s structures are available from Kraitchman's equations (such as Eq. (36)) the least squares method yields almost congruent structures, and so it is rightly called a "pseudo-Kraitchman" structure [31], or denoted simply by r_s, as in Ref. [43].

The least squares variant developed by Typke [52,53] is different from that of Nösberger et al. in several aspects. Typke's program utilizes the P formalism and, in contrast to the Nösberger method, refines the nuclear Cartesian coordinates. Therefore, the calculations need not involve the position coordinates of the unsubstituted nuclei.

When single isotopic substitution occurs at nucleus k, with a reduced mass of substitution μ, the tensor $P^{(')}$ of the substituted molecule is constructed from the diagonal P tensor of the parent molecule, as follows

$$P^{(')} = P + \mu \underset{\sim}{r}_k \underset{\sim}{r}_k \qquad (41)$$

using the dyadic notation of Sec. III. In Eq. (41) $\underset{\sim}{r}_k$ is composed of the substitution coordinates of nucleus k (of course, in the parent axis system). The general, non-diagonal $P^{(')}$ tensor may then be transformed as in Eq. (35) to yield the principal planar moments of the substituted isotopomer.

For multiple substitution, Eq. (41) is extended over the substitution coordinates of all nuclei involved (see Eq. (8) in Ref. [52]). The normal equation, Eq. (38), is then solved for $\delta\underset{\sim}{X}$, and now contains the corrections of Cartesian substitution coordinates.

As in the case of the Nösberger method [50] it is possible to include the principal axis conditions in Eqs. (22) and (23) as constraints. Such constraints were found by Typke to create correlations among the $\underset{\sim}{X}$ elements, and these correlations help in the determination of the relative signs of the substitution coordinates. For small Kraitchman coordinates, the steps of coordinate refinement oscillate about zero, and so it is recommended to constrain such coordinates to zero. The method was tried out on ethylene epoxide; the r_s structure of this molecule was published by Hirose [57] prior to Typke's work. Apart from differences in error estimates of the parameters, the agreement was found to be satisfactory [52].

Typke tested his regression method on vibrational ground-state average structures (r_z) as well [53]. We shall return to these calculations in Sec. IV.

IV. ESTIMATION OF EQUILIBRIUM GEOMETRIES

The best available representation of molecular geometries is the equilibrium geometry, r_e. The final aim of any analysis of structural vibrational effects is to estimate the r_e structure. Apart from 'ab initio' and other quantum-chemical estimates of r_e, in experimental molecular spectroscopy there are basically three different approaches for the estimation of r_e parameters from experimental rotational constants.

When it is possible to perform rotational analyses of vibrationally excited states (in the ground electronic state), the following expansion in vibrational quantum numbers v can be used [58]:

$$B_v^\alpha = B_e^\alpha - \sum_s \alpha_s^B \left(v_s + \frac{d_s}{2}\right) + \sum_{r<s} \sum \gamma_{rs}^B \left(v_r + \frac{d_r}{2}\right)\left(v_s + \frac{d_s}{2}\right) + \ldots \qquad (42)$$

where d_s and d_r are vibrational degeneracies and α and γ are the empirical expansion coefficients to be determined. For the ground vibrational state, and with the neglect of the quadratic term in Eq. (42),

$$B_e^\alpha = B_o^\alpha + (1/2) \sum_s \alpha_s d_s. \qquad (43)$$

The rotational constants B^α are reciprocally related to the principal moments of inertia I^α, so that $B^\alpha I^\alpha$ is a constant. When B^α is measured in MHz units, while I^α in $u \cdot \overset{\circ}{A}{}^2$ units ($1 \ u \cdot \overset{\circ}{A}{}^2 = 10^{-2} \ u \cdot nm^2 = 10^4 \ u \cdot pm^2$), the product $B^\alpha I^\alpha$ has a numerical value 505379.08 (for values of fundamental constants see Ref. [4]). Equation (43) applies only when the effects of strong resonances (Coriolis and Fermi-type ones) among vibrationally excited states have been removed, i.e., when the constants B_o^α are nonperturbed with respect to the effect of such resonances. Most of the r_e structures in the spectroscopic literature have been obtained by this method (for r_e structures see Refs. [21,43,45] and also Ref. [59]).

In the present review the other two methods for estimation of equilibrium structures shall be discussed, viz, the mass dependence (r_m) method, and the method of anharmonic correction of the vibrational ground-state average structures (r_z).

A. Mass Dependence Estimates for r_e

In Sec. III.A, Eqs. (20) and (21) were found by Costain [19a] to apply for diatomic molecules, and approximately for some triatomic molecules as well. Watson [29] provided foundations for Costain's proposition through a mathematical analysis of the isotopic mass dependence of molecular quantities involved in the derivation of Eq. (21). Thus Watson has introduced the mass dependence estimate of the equilibrium moments of inertia:

$$I_m^\alpha = 2I_s^\alpha - I_o^\alpha \sim I_e^\alpha \tag{44}$$

where the substitution moment of inertia I_s^α was defined by Costain [19a] for linear molecules as

$$I_s \equiv \sum_i m_i \ (z_s(i))^2. \tag{45}$$

Watson [29] uses I_s^α as a generalization of I_s in Eq. (45) for polyatomic molecules. For the r_e estimates in Eq. (44), it is crucial that all I_s^α values be obtained either by the original Kraitchman formulas [18] or by their symmetrized versions [27,28]. Occasional negative values for $(z_s(i))^2$ must also be used in Eq. (45) because of the mathematical principles used in the derivation of Eq. (44). Due to this requirement the r_m method can not be applied when the molecule has a nucleus or several nuclei that possess just a single isotopic form (see Sec. III.A). In order to obtain I_m^α values for all isotopomers, extensive isotopic sequences may be necessary, involving in the general case not only all singly substituted molecules, but a number of doubly substituted ones as well. Watson [29] has tested the r_m method on the diatomic model, and on the CO molecule specifically. Furthermore, he tested it on triatomics such as NNO, OCS, SO_2 and HCN. The conclusions were that, in addition to the limitations stemming from the proper derivation and use of the I_s^α values, the r_m procedure in its existing form is not applicable for molecules with H atoms. Also the calculation of geometrical parameters from I_m^α isotopic values by a least squares fit has a tendency for magnifying the experimental errors in the input I_o^α isotopic moments. Therefore only the most precise microwave data may be used for r_m calculations with sufficiently narrow error margins.

On the other hand, Watson emphasized that the r_m method is insensitive to perturbations and resonances among excited vibrational-rotational levels. This is, of course, due to the fact that one is using effective values for the vibrational ground state rotational constants. Watson [29] expressed hopes that, in general, the absolute error $|r_e - r_m|$ in the structural estimates shall be less than 0.1 pm. Difficulties were, however, encountered [29] for the presumably well-behaved OCS molecule. In this case the $|r_e - r_m|$ deviations were 0.44 and 0.35 pm for the C=O and C=S bonds, respectively.

This finding has led to a re-investigation of the OCS case and the r_m method itself by Smith and Watson [60]. The systematic deviation of Kraitchman's substitution coordinates from the equilibrium is given by

$$(z_s(i))^2 - (z_e(i))^2 \approx \frac{\partial \varepsilon^\alpha}{\partial m_i} + \frac{1}{2M} \Delta m_i \frac{\partial^2 (M\varepsilon^\alpha)}{\partial m_i^2} \tag{46}$$

where higher derivatives of ε^α with respect to the isotopic mass are neglected. The vibrational correction ε^α is defined as:

$$\varepsilon^\alpha = I_o^\alpha - I_e^\alpha = (1/2) \sum_s \varepsilon_s^\alpha \tag{47}$$

where the sum is over all normal modes Q_s, and α = a, b or c. As a first order approximation to Δm_i effects, Watson proposed to neglect the second and higher terms in Eq. (46) so that the systematic error is simply $\partial \varepsilon^\alpha / \partial m_i$. From Eqs. (44), (46) and (47) and from the consequence of Euler's theorem for homogeneous functions (see Ref. [29]),

$$\sum_i m_i \left(\frac{\partial \varepsilon^\alpha}{\partial m_i} \right) = \varepsilon^\alpha / 2 \tag{48}$$

one obtains

$$I_m^\alpha \underset{\sim}{\sim} I_e^\alpha + \frac{1}{M} \sum_i m_i \, \Delta m_i \, \frac{\partial^2 (M\varepsilon^\alpha)}{\partial m_i^2}. \tag{49}$$

For single isotopic substitution, the sum in Eq. (49) simplifies to one term corresponding to the substituted nucleus. Upon defining the r_m structure coordinates as

$$(z_m(i))^2 = (I_m^{\alpha'} - I_m^\alpha) \, \mu_i^{-1} \underset{\sim}{\sim} (I_m^{\alpha'} - I_m^\alpha) \Delta m_i^{-1}, \tag{50}$$

one obtains from Eq. (50), through isotopic difference formation, a systematic error estimate for the r_m structure coordinates, bond lengths, and bond angles [60]:

$$(z_m(i))^2 - (z_e(i))^2 \underset{\sim}{\sim} - \frac{2m_i}{M} \, \frac{\partial^2 (M\varepsilon^\alpha)}{\partial m_i^2}, \tag{51}$$

where the factor 2 on the right hand side arises from the two Δm_i terms of opposite sign adding up instead of compensating, in the isotopic difference formation. By the same Euler's theorem that leads to Eq. (48), it is possible to simplify Eq. (51) to an order-of-magnitude estimate that comes out to be essentially the same as for $(z_s(i))^2 - (z_e(i))^2$ [60]. Thus the systematic errors of the r_s and r_m structural coordinates are commensurate. It could happen that both r_s and r_m geometries approximate the equilibrium geometry when the normal coordinate components of ε^α in Eq. (47) cancel fortuitously in the summation. This, of course, can not be always expected.

There are two ways to avoid significant systematic errors in I_m values, as shown by Watson [29]. One possibility is to apply a procedure analogous to that of Laurie [61] for diatomic molecules. This consists of extrapolating $(z_s(i))^2$ quantities to $\Delta m_i = 0$ (infinite mass of the substituted nucleus). Thereafter such $(z_s(i))^2$ values can be used to calculate I_s^α and thus I_m^α through Eq. (44). This procedure assures that $I_m^\alpha = I_e^\alpha$ so that a least squares fit to such I_m^α values automatically yields r_e geometrical parameters. This method requires at least two different Δm_i values for each substituted nucleus and so the necessary isotopic sequence is an extensive one and is seldom available.

Instead of the long extrapolation to $\Delta m_i = 0$, vibrational effects can be averaged out by using a purposefully chosen, limited isotopic set of I_m^α moments. When the set of isotopomers is "symmetric" with regard to the changes Δm_i at each nucleus, the second term in Eq. (49) averages out in least squares fits to such a set of I_m^α values:

$$(I_m^\alpha)_{av} \approx (I_e^\alpha)_{av}. \tag{52}$$

In least squares calculations of r_m structural parameters, such as bond lengths and bond angles, it is possible to obtain statistically uncorrelated parameter combinations. These are formed using the eigenvector elements of the matrix $D^T W D$ in Eq. (38) as coefficients of combination. Smith and Watson [60] have shown that there will be n combinations that provide good approximations to the corresponding equilibrium combinations. The value of n is the number of independent moments of inertia of one isotopomer and it can be decided on the basis of symmetry ahead of the analysis. The rest of the parameter combinations (their number is given by the number of totally symmetric vibrational modes less the number of independent rotational constants) shall carry distortions to the r_e geometry. When, for this number of combinations, independent equilibrium estimates are available, the full r_e geometry can be calculated. The following example illustrates the approach.

Nakata et al. [62] have recently extended the r_m method to four-atomic molecules and, in particular, to the planar phosgene molecule ($COCl_2$) of C_{2v} symmetry. The available isotopic molecules offered an excellent opportunity for examining the validity of the approximation in Eq. (52). It is an important property of the r_m geometries that in this representation the inertial defect should vanish for planar molecules;

$$\Delta_m \equiv 2\Delta_s - \Delta_o \approx \Delta_e = 0, \tag{53}$$

where Δ_o is given essentially by Eq. (28), while Δ_s is defined by Watson [29] as

$$\Delta_s = \sum_i m_i \left(\frac{\partial \Delta_o}{\partial m_i} \right) \approx \sum_i m_i (\Delta_o' - \Delta_o) \Delta m_i^{-1}, \tag{54}$$

where in parentheses stands the isotopic difference of Δ_o.

Nakata et al. [62] used the r_s structures of eight $COCl_2$ isotopomers to test the validity of Eq. (52). Using in shorthand the number of nucleons in the constituent atoms C, O and Cl of phosgene, the isotopic sequence may be represented as $\{(12,16,35),(13,18,37)\}$; $\{(12,16,37),(13,18,35)\}$; $\{(13,16,37),(12,18,35)\}$ and $\{(13,16,35),(12,18,37)\}$. This sequence contains an ^{16}O and an ^{18}O group; a member of each enters the four "complementary" pairs indicated in curly braces. Within each such pair the Δm_i values in Eq. (49) are opposite.

The averaging procedure for the phosgene structure was carried out for two geometrical parameters h and g, which are related to the bond lengths and angles as follows (for isotopomers of C_{2v} symmetry):

$$h = r(C\text{-}Cl)\ \sin(\alpha/2)$$
$$g = r(C\text{=}O) + r(C\text{-}Cl)\ \cos(\alpha/2) \tag{55}$$

where α denotes the angle $\sphericalangle(Cl\text{-}C\text{-}Cl)$. The quantity h is simply related to the moment of inertia I^b:

$$h = (I^b/2m_{Cl})^{1/2} \tag{56}$$

where m_{Cl} is the mass of chlorine. An average of h_m values over the set of eight isotopomers should yield $h_m \sim h_e = r_e(C\text{-}Cl)\ \sin(\alpha_e/2)$. If one knew $r_e(C\text{-}Cl)$ from an independent source, this relation would determine α_e.

The quantity g_m is not so simply related to I_m^a isotopic values as was the case for h. Its complementary average is determined from fitting the isotopic I_m^a values to the sum and the difference of the two components of g, viz, $r(C\text{=}O)$ and $x = r(C\text{-}Cl)\ \cos(\alpha/2)$. For each I_m^a, a curve is thus obtained and the intersection area of these isotopic curves defines the complementary average value for g_m that is expected to yield g_e. Again, if we knew $r_e(C\text{=}O)$ and $r_e(C\text{-}Cl)$ from an independent source, another α_e estimate would be obtained. As the equilibrium estimates are available from a recent combined electron diffraction and spectroscopic study [63], the complete r_e geometry was derived in Ref. [62]: $r_e(C\text{=}O) = 117.56(23)$ pm, $r_e(C\text{-}Cl) = 173.81(19)$pm, and $\sphericalangle_e(Cl\text{-}C\text{-}Cl) = 111.79(24)°$. Additionally, it was possible to check the condition in Eq. (53), and it was found that, for the eight isotopomers of phosgene, Δ_m ranged from 0.00(6) to 0.03(8) $u{\cdot}Å^{2}$, and $(\Delta_m)_{av} = 0$, indeed.

Nakata et al. [64] used a practical extension to the r_m method for planar molecules. The extension consisted of the utilization of the inertial defect expressed by a linear relationship among ε^{α} values (see Eq. (29)) in the isotopic averaging procedure. From the B_0^{α} rotational constants of four isotopomers of OCl_2 of C_{2v} symmetry (by the nucleon number: {(16,35),(18,37)} and {(16,37),(18,35)}), the four isotopic r_m structures were calculated. There are two r_m parameters; $r_m(O\text{-}Cl)$ and $\sphericalangle_m(Cl\text{-}O\text{-}Cl)$. The isotopic variation of the former is 0.041 pm, whereas it is 0.02° for the latter. Both ranges are about ten times the individual isotopic dispersions and are due to the second term in Eq. (49). The r_e geometry was then estimated by the previously described averaging procedure. The isotopic I_m^α values were fit to the parameters x and h which are related to the principal moments of inertia as follows (for C_{2v} geometry):

$$x = \{(1/2m_{Cl} + 1/m_O) \ I^a\}^{1/2}$$

$$h = (I^b/2m_{Cl})^{1/2}.$$

(57)

Using the symbol r_c instead of r_m to refer to the averaging over complementary sets, the agreement between r_c and r_e parameters [65] is the following: $r_c(O-Cl) = 169.587(7)$ pm, $r_e(O-Cl) = 169.591(13)$ pm, and $\}_c(Cl-O-Cl) = 110.886(6)°$, while $\}_e(Cl-O-Cl) = 110.876(15)°$.

In the Appendix to Ref. [64] the r_c structure is derived for SO_2 using Watson's original data [29], and the results were compared to an independent r_e structure [66,67]: $r_c(S=O) = 143.071(4)$ pm and $\}_c(O=S=O) = 119.332(6)°$ as opposed to $r_e(S=O) = 143.077(4)$ pm and $\}_e(O=S=O) = 119.329(6)°$. Watson's original estimates [29], which were also based on the mean of estimates from various inertial defect relationships, are: $r_m(S=O) = 143.074(7)$ pm and $\}_m(O=S=O) = 119.333(33)°$.

B.   Zero-point Vibrational Average Structures

As already mentioned in Sec. II, the basic theory for zero-point or ground-state vibrational average geometry was provided by Oka [9], Laurie and Herschbach [6] and Toyama et al. [11]. General reviews on molecular geometry by Kuchitsu [12,13], and by Laurie [73] contain sections dealing with r_z structures. Various bond angle averages were defined by Kuchitsu [70]. For molecules with one or two internal symmetric rotors, the r_z formalism was elaborated by Iijima [71], and by Iijima and Tsuchiya [72]. The practical significance of the r_z structure is understood when one considers that

(i) The zero-point vibrational averaging correction to the effective r_0 geometries is easily calculated, and

(ii) The spectroscopic r_z parameters may be directly compared to the r_α^0 parameters of gas-phase electron diffraction (see Sec. II).

The α_s coefficients in Eq. (43) are split into three components:

$$\alpha_s^B = \alpha_s^B(\text{harm}) + \alpha_s^B(\text{Coriolis}) + \alpha_s^B(\text{anharm}), \tag{58}$$

where the superscript B refers to the rotational constants A, B or C, and the first two components are obtainable from the harmonic part of the vibrational potential function (see e.g. Eq. (13) in Ref. [58]).

The rotational constants in the r_z geometry are calculated using the corrections:

$$B_z^\alpha - B_o^\alpha = (1/2) \sum_s \{\alpha_s^B(\text{harm}) + \alpha_s^B(\text{Coriolis})\}d_s. \tag{59}$$

With the corresponding ε_s^α components in Eq. (47):

$$I_z^\alpha - I_o^\alpha = -(1/2) \sum_s \{\varepsilon_s^\alpha(\text{harm}) + \varepsilon_s^\alpha(\text{Coriolis})\} \, d_s. \tag{60}$$

The harmonic vibrational parts of α_s^B and ε_s are related by

$$\varepsilon_s^\alpha(\text{harm}) = (\frac{2hc}{\hbar^2}) (I_o^\alpha)^2 \, \alpha_s^B(\text{harm}) \, \{(\frac{\hbar^2}{2hc}) + I_o^\alpha \sum_s \alpha_s^B \, d_s\}^{-1}. \tag{61}$$

In Eq. (61) $\hbar^2/2hc$ has the value $2.77932009 \times 10^{-39}$ g·cm, and applies when ε_s^α is given in g·cm^2 units, while α_s^B is in cm^{-1} units.

For more exact calculations $\varepsilon_s^\alpha(\text{harm})$ should contain also quartic centrifugal distortion contributions [69], and an electronic mass contribution of π-electrons and lone-pair electrons [9,68].

Geometric parameters derived from isotopic I_z^α moments of inertia possess isotopic mass dependence. Thus, when more than three independent structural parameters are to be determined, these isotopic effects must be considered. The best solution for this task is to obtain equilibrium structure estimates by

$$I_e^\alpha - I_z^\alpha = -(1/2) \sum_s \varepsilon_s^\alpha(\text{anharm}) \, d_s \tag{62}$$

where, for the evaluation of the sum, a number of cubic anharmonic force constants, normal mode wavenumbers, and derivatives of the moments of inertia with respect to normal coordinates are used. Equation (8) in Sec. II expresses a correction analogous to Eq. (62).

The isotopic mass variation of the r_z geometrical parameters can be minimized through anharmonicity corrections for the bond stretching motion, and by the use of vibrational amplitudes calculated from Wilson's GF matrix formalism [16]. A discussion of this procedure was given by Kuchitsu and Cyvin [12], and by Kuchitsu [13]. The r_z structures calculated through mass dependence corrections are expected to be closer to the r_e structure and therefore these should be distinguished from r_z structures which do not contain such corrections. This requirement is not always considered; for instance, in Ref. [43] no such distinction is made. In this review emphasis shall be placed on r_z geometries obtained with isotopic mass dependence corrections. Before such calculations are reviewed, a few less sophisticated, modern r_z estimates are mentioned as representative examples.

Robiette [74] derived r_z parameters for arsenic tribromide $AsBr_3$; $r_z(As-Br)$ = 232.4(3) pm, $\vartheta_z(Br-As-Br)$ = 99.8(2)°. Cox et al. [75] determined the ground-state average geometry of CF_3Br and CF_3I. For CF_3Br their results are: $r_z(C-F)$ = 132.65(23) pm, $r_z(C-Br)$ = 192.34(31) pm and $\vartheta_m(F-C-F)$ = 108.81(25)°. For trifluoromethyl iodide, $r_z(C-F)$ = 132.85(23) pm, $r_z(C-I)$ = 214.38(27) pm and $\vartheta_z(F-C-F)$ = 108.42(23)°.

As a matter of interest the work of Chidichimo et al. [76] may be mentioned. These authors have used proton magnetic resonance spectroscopy for the derivation of the r_z geometry of 1,2,5-selenadiazole. The planar structure of this molecule is given in the form of the Cartesian nuclear coordinates in the $\sigma(xy)$ plane.

C. The Isotopic Mass Dependence of r_z Geometries

The fact that r_z geometrical parameters are not isotope-mass invariant has been known since the birth of the ground-state vibrational average structure concept. Laurie and Herschbach [6] pointed out that the D → H substitution leads to a shortening of 0.3 - 0.5 pm in the $r_z(C-H)$ bond lengths, whereas for heavy isotope substitution the shortening is only about 0.01 pm. Lafferty et al. [77] determined the r_e structure for cyanogen chloride, ClCN from the rotational data of four isotopomers. They found significant isotopic variation in the $r_z(C\equiv N)$ bond length, and also a deviation of about 0.3 - 0.6 pm from $r_e(C\equiv N)$ = 116.02(70) pm. The isotopic variation reached 1 pm. Neglect of these mass effects (vibrational effects) may lead to larger errors in regression calculations of structural parameters [12].

Kuchitsu et al. [1,79] analyzed carefully the difference between r_α^o and r_z geometries for acrolein, $H_2C=CH-CHO$, by using data from an earlier microwave study [80]: $r_z(=C-C)$ = 147.43(53) pm, while $r_\alpha^o(=C-C)$ =

148.21(40) pm [1]. This difference, and similar ones in other bonds and angles, were attributed to the isotopic sensitivity of the heavy nuclear skeleton of acrolein [79].

Analytical expressions for the isotopic variation of r_z structural parameters were given by Kuchitsu and coworkers [78,79,81]. A recent summary of these relationships is found in the work of Nakata et al. [63]. The isotopic r_z geometry differences are derivable from the equilibrium correction of r_z geometries [12]:

$$r_z - r_e \cong \langle \Delta z \rangle_0 \tag{63}$$

(see also Eqs. (8) and (6) in Sec. II).

The $\langle \Delta z \rangle$ linear vibrational average contains the projection of the instantaneous displacement Δr (which is in general not bond-directed) upon the equilibrium internuclear axis. This is an important point as $\langle \Delta r \rangle_0 = r_g^0 - r_e \neq \langle \Delta z \rangle_0$ [81].

The general evaluation of $\langle \Delta z \rangle_0$ requires the ground-state linear averages of all normal coordinates which involve a number of cubic force constants (see Eq. (8) in Sec. II). When, however, the geometric parameter involved in the correction that Eq. (63) represents corresponds to a chemical bond, an important simplification is achieved through the diatomic-bond approximation [13]. This approximation is a natural one for electron diffraction experiments and can be transferred to spectroscopy. Accordingly $\langle \Delta z \rangle_0$ in Eq. (63) may be approximated by:

$$\langle \Delta z \rangle_0 = \langle \Delta r \rangle_0 - K_0 \tag{64}$$

where K_0 contains the mean square amplitudes perpendicular to the direction of the diatomic equilibrium bond, whose length is r_e:

$$K_0 = \{ \langle \Delta x^2 \rangle_0 + \langle \Delta y^2 \rangle_0 \}/2r_e. \tag{65}$$

Using Bartell's estimate [82]:

$$\langle \Delta r \rangle_0 \approx (3/2)\, a_3\, \langle \Delta z^2 \rangle_0, \tag{66}$$

where Δz stands for the bond-directed, parallel vibrational amplitude, and a_3 is the cubic anharmonicity constant in the empirical Morse function [83-85]. For diatomic molecules, a_3 (sometimes simply denoted as a) is calculated from the Herzberg formula [86]:

$$a_3 = 10^{-10} \left(\frac{hc}{2N_A \hbar^2}\right)^{1/2} \{\mu(u) \, \omega_e^2/D_e\}^{1/2} \, pm^{-1}. \tag{67}$$

When ω_e (harmonic wavenumber) and D_e (dissociation energy) are measured in cm^{-1} units, and μ (reduced mass of the two nuclei) is in units u ($u = 1/N_A = 1.6605655(86) \times 10^{-27}$ kg [4]), the proportionality constant in Eq. (67) is 1.21779×10^{-3}.

Although it is not evident from Eq. (67), a_3 is invariant to isotopic mass, and therefore the isotopic dependence of r_z chemical bond lengths is given by:

$$\delta r_z = (3/2) \, a_3 \, \delta<\Delta z^2>_0 - \delta K_0, \tag{68}$$

where δ is the operational symbol for isotopic difference formation. Once the three types of mean square Cartesian amplitudes are obtained for both isotopomers from a harmonic normal coordinate calculation (see e.g. Ref. [87]), and a_3 for the given bond is estimated through Herzberg's relationship above, estimates for δr_z are available. These δr_z quantities can then be used to correct the r_z geometrical parameters of isotopically substituted molecules, relative to those of the parent.

Equation (68) represents primary isotope effects. Secondary mass effects are manifested in the changes of the ground-state average positions of nuclei not involved in the diatomic bond in question. In principle the isotopic substitution of a given nucleus generally has secondary effects over the whole of the molecule.

Kuchitsu and Oyanagi [81] have dealt extensively with primary and higher order δr_z corrections for polyatomics. With respect to the magnitude of δr_z effects, their findings agree with those of Laurie and Herschbach [6].

For multiple isotopic substitution, δr_z values for given bonds were found [81] to be synthesizable from single substitution components. For instance, when an $ABC \rightarrow A^{\cdot}B^{\cdot}C^{\cdot}$ triple substitution is made, the δr_z of say the A-B bond is given approximately by the sum:

$$\delta r_z^{AB} (ABC \rightarrow A^{\cdot}B^{\cdot}C^{\cdot}) \approx \delta r_z^{AB}(ABC \rightarrow A^{\cdot}BC) +$$

$$\delta r_z^{AB}(ABC \rightarrow AB^{\cdot}C) + \delta r_z^{AB}(ABC \rightarrow ABC^{\cdot}). \tag{69}$$

The first two terms of Eq. (69) represent primary isotopic effects, while the third corresponds to a secondary effect. It has empirically been found that secondary isotopic mass effects may be commensurate to, or even larger than, first order ones. These secondary effects arise mainly from the δK_0 term in Eq. (68) and can overcompensate the anharmonic bond shortening because of the first term, thus leading to an effective lengthening of the bond in question.

Several recent papers have dealt with the estimation of δr_z isotope effects in order to calculate dependable and internally consistent r_z geometrical parameters from a set of isotopic I_z^α moments of inertia. A few modern, polyatomic r_z estimates shall now be reviewed.

Duncan [88] estimated the δr_z effects in the geometry of formaldehyde, H_2CO, and ethylene, $H_2C=CH_2$, and provided the three independent equilibrium geometrical parameters for both molecules. For H_2CO he used the rotational constants of seven isotopomers [89-92]. For the ethylene geometry, rotational data from the vibration-rotation spectra of four isotopomers were used [93-96]. In the correction represented by Eq. (60), the normal coordinates from Refs. [97,98] were utilized. For both molecules in question, which possess planar equilibrium geometries, the r_z structure moments of inertia satisfied the requirement: $\Delta_z \simeq 0$ (similar to Eq. (53) in Sec. IV.A).

Duncan [88] assumed that the substitution $^{12}C \rightarrow {}^{13}C$ or $^{16}O \rightarrow {}^{18}O$ does not induce secondary effects on the C-H (C-D) bond lengths or ⟩(H-C-H) and ⟩(D-C-D) angles. The primary δr_z effects involved in his r_z calculations were:

$$\delta r_z(\text{C-H}) \ (\text{H} \rightarrow \text{D}) = 0.41(5) \ \text{pm},$$

$$\delta r_z(\text{C=O}) \ (^{12}\text{C} \rightarrow {}^{13}\text{C}) = 0.003(5) \ \text{pm},$$

$$\delta r_z(\text{C=O}) \ (^{16}\text{O} \rightarrow {}^{18}\text{O}) = 0.010(5) \ \text{pm}, \qquad\qquad (70)$$

$$\delta r_z(\text{C=O}) \ (\text{H}_2 \rightarrow \text{D}_2) = 0.05(2) \ \text{pm},$$

$$\delta\text{⟩}_z(\text{H-C-H}) \ (\text{H}_2 \rightarrow \text{D}_2) = -5'(2').$$

In equilibrium structure calculations the following Morse-type anharmonicity constants were applied: $a_3(\text{C-H})(\text{H}_2\text{CO}) = 0.027 \ \text{pm}^{-1}$, $a_3(\text{C-H})(\text{H}_2\text{CCH}_2) = 0.022 \ \text{pm}^{-1}$, $a_3(\text{C=O}) = a_3(\text{C=C}) = 0.02 \ \text{pm}^{-1}$, and $a_3(\text{O}\cdots\text{H}) = a_3(\text{C}\cdots\text{H}) = 0.01 \ \text{pm}^{-1}$. The last a_3 constant reproduces the observed δr_z bond shortening in the non-bonded distances upon deuteration. The r_e geometry of formaldehyde is then: $r_e(\text{C-H}) = 110.05(20) \ \text{pm}$, $r_e(\text{C=O}) = 120.33(10) \ \text{pm}$ and ⟩$_e$(H-C-H) $= 116°18'(15')$, while r_e of ethylene is: $r_e(\text{C-H}) = 108.1(2) \ \text{pm}$, $r_e(\text{C=C}) = 133.4(2) \ \text{pm}$, and ⟩$_e$(H-C-H) $= 117°22'(20')$ [88].

Hirota et al. [99] have recently obtained the rotational constants of the ethylene isotopomers, $H_2C=CD_2$, $H_2C=CHD$ and cis-HDC=CHD, through isotopic substitution induced polarity (see Sec. V for a general discussion). As the rotational data of these authors are much more precise than those used by Duncan [88], the Japanese authors have repeated the determination of the r_z geometry of ethylene. From the I_z^α moments of the three isotopomers mentioned above, they determined only four structural parameters: $r_z(C-H)$, $r_z(C=C)$, $\vartheta_z(C=C-H)$ and $\delta r_z(C-H)(H \rightarrow D)$, for which two different estimates were obtained according to two different sets of statistical weights. Their results are identical to those of Duncan [88] within the estimated error limits in Refs. [88] and [99]. This agreement proves the dependability of Duncan's estimates.

Mallinson and Nemes [100] have derived the r_z structure of ketene, H_2CCO, from the rotational constants of five isotopomers [101-103], and used a harmonic force-field [100] based on some newly derived vibration-rotation data [104] to calculate the $I_z^\alpha - I_0^\alpha$ corrections. The inertial defect formed of I_z^α moments practically vanishes ($\Delta_z < 7$ u·pm^2) which strongly supports the planarity of the molecule. Only the primary δr_z shortening of the C-H bond could be derived from the rotational data for D $\rightarrow$ H substitution: $\delta r_z(C-H)(D \rightarrow H) = 0.10(3)$ pm. Values for δr_z of the bonds C=C and C=O for isotopic substitutions $^{13}C \rightarrow {}^{12}C$, $^{18}C \rightarrow {}^{16}O$ and D $\rightarrow$ H were taken from Duncan's work [88] (see Eq. (70)). The empirically found $\delta r_z(C-H)(D \rightarrow H)$ value yields $a_3 = 0.023$ pm^{-1} from Eq. (68), which is in good agreement with the case of ethylene [88].

Kuchitsu and Takabayashi [105] used gas-phase electron diffraction data to improve the r_z structure of ketene. Table 1 contains some refined δr_z estimates.

TABLE 1

Isotopic Mass Effects for Ketene [105] (in 10^{-3}pm and 10^{-2} degrees)

Substitution	$\delta r_z(C-H)$	$\delta r_z(C=C)$	$\delta r_z(C=O)$	$\delta\vartheta_z(C=C-H)$
$D_2 \rightarrow H_2$	-108(15)	-3.0(6)	-19.0(20)	0(3)
$^{13}CCO \rightarrow {}^{12}CCO$	-4.2(6)	-4.6(8)	2.7(2)	0(3)
$CC^{18}O \rightarrow CC^{16}O$	-4.5(5)	-7.9(5)	0.0(1)	0(3)

A comparison of the ground-state average structural data from Refs. [100] and [105] is given in Table 2.

Hegelund et al. [106] derived the r_z geometry of allene, H_2CCCH_2. Their work was based on rotational constants from various sources (infrared, Raman and microwave spectroscopy) [107-110], and on a harmonic force-field more dependable than any reported previously (e.g., Ref. [111]). The r_z

TABLE 2

The Ground-state Vibrational Average Structure of H_2CCO (pm or degrees)

Parameter	Ref. [100]	Ref. [105]
r(C-H)	107.97(10)	107.99(20)
r(C=C)	131.71(20)	131.52(14)
r(C=O)	116.08(20)	116.27(17)
ꝫ(H-C=C)	119.02(10)	119.04(16)

parameters and their isotopic corrections were estimated as: r_z(C-H) = 108.62 pm, r_z(C=C) = 130.91 pm, $ꝫ_z$(H-C-H) = 118.28° and, with associated isotopic mass effects, $δr_z$(C-H)(H → D) = 0.13 pm, $δr_z$(C=C)(H_2 → D_2) = 0.01 pm and $δꝫ$(H-C-H) (H_2 → D_2) = 0.03°.

Demaison et al. [112] have made a very thorough spectroscopic analysis of the geometry of acetonitrile, H_3C-CN. The available isotopomers (twelve different ones) which are indicated by the nucleon number of the constituent atoms H, C, C, and N from one end of the molecule to the other, were: (1,12,12,14); (1,13,12,14); (1,12,13,14); (1,12,12,15); (2,13,12,14); (2,12,13,14); (2,12,12,15); (1,12,13,15) and (2,12,12,14) for the C_{3v} geometry, and the two partially deuterated isotopomers $^{12}CH_2D^{12}C^{14}N$ and $^{12}CHD_2^{12}C^{14}N$. This extensive isotopic sequence and a good-quality force-field [113] made it feasible to calculate a dependable r_s and r_z geometry. These structures were then compared to previously reported r_o, r_z and $r_α^o$ geometries [113-115]. The $δr_z$ isotopic effects, both primary and secondary ones, were also taken into account (see Table X in Ref. [112]). The r_z geometry was obtained by different types of regression calculations, such as

(i) fitting $B_z^α$ isotopic rotational constants to r_z bond lengths and bond angles,

(ii) fitting isotopic $ΔI_z^α$ differences to r_z bond lengths and bond angles, and finally

(iii) fitting $ΔB_z^α$ differences to the Cartesian coordinates of the nuclei by taking $δr_z$ bond length effects into account (see Ref. [53] and a discussion at the end of this section).

The mean values of r_z parameters as obtained from these approaches are: r_z(C-C) = 146.17(6) pm, r_z(C≡N) = 115.67(6) pm, r_z(C-H) = 109.47(24) pm and $\mathcal{3}_z(C-C-H) = 109.85(10)°$. The comparison among electron diffraction r_α^o and spectroscopic r_z parameters revealed no significant differences.

In an earlier paper, Halonen and Mills [116] reported the microwave rotational spectra of twice deuterated acetonitrile (HD_2C-CN) and twice deuterated methyl isocyanide (HD_2C-NC). They have derived a harmonic force-field also for these isomeric molecules, and obtained the I_z^α moments of inertia for five isotopomers of acetonitrile and methyl isocyanide, respectively. The corrections for isotopic mass effects were somewhat larger than those applied by Demaison et al. [112], which supports the view that these corrections have a systematic uncertainty of about 10% due to acceptable differences in various force-field estimates.

As a result of infrared spectroscopic studies on partially deuterated ethanes, Duncan et al. [117] determined the r_z geometry within narrow limits, and gave an estimate of the equilibrium geometry for this molecule. The determinable r_z parameters were found by Duncan et al. [117] to be $r_z(C-H)$ = 109.42(4) pm, $r_z(C-C)$ = 153.51(2) pm and $\mathcal{3}_z(H-C-C)$ = 111.13(3)°, with $\delta r_z(C-H)(H \rightarrow D)$ = 0.15(3) pm, $\delta r_z(C-C)(^{12}CC \rightarrow ^{13}CC)$ = 0.014(10) pm and $\mathcal{3}_z(H-C-C)(H \rightarrow D)$ = -0.010(5)°. These r_z parameters reproduce thirteen inertial moments of ten isotopomers of ethane to within 0.2% for the A_o, and 0.01% for the B_o rotational constant. It is interesting that the r_z value of the bond length $r(C-C)$ is significantly longer than a previous electron-diffraction ground-state average estimate [78], and in this sense deviates significantly from the electron diffraction $r_g^o(C-C)$ bond length for H_3C-CH_3 and D_3C-CD_3 [118] as well. In a similar way, former spectroscopic $r_s(C-C)$ values [119-121] are definitely shorter than Duncan et al.'s $r_z(C-C)$ estimate [117]. These authors explain the difference by a neglect for contraction in the heavy nuclear framework upon $^{13}C \rightarrow ^{12}C$ substitution (see Ref. [79] for analogous arguments).

In a recent paper Hirota et al. [122] have addressed once again the problem of the r_z structure of ethane on the basis of new, precise microwave data of four deuterated isotopomers. Their new results are: $r_z(C-H)$ = 109.40(2) pm, $r_z(C-C)$ = 153.51(1) pm and $\mathcal{3}_z(H-C-C)$ = 111.17(1)°, with isotopic mass effects $\delta r_z(C-H)(H \rightarrow D)$ = 0.12(2) pm, $\delta r_z(C-C)(^{12}C \rightarrow ^{13}C)$ = 0.015(10) pm, and $\delta\mathcal{3}_z(H-C-C)(H \rightarrow D)$ = 0.040(10)° together with the assumption: $\delta r_z(C-C)/D$ atom = -5 x 10^{-3} pm. Hirota et al.'s new study [122] corroborated the finding that the $r_z(C-C)$ bond length is longer than electron diffractionists had thought earlier.

Davies et al. [123] have calculated for the first time the r_z geometry of formic acid HCOOH from spectroscopic data of three isotopomers: (1,13,16,16,1); (1,12,18,16,1) and (1,12,16,18,1). In regression calculations of nine r_z parameters they have also used the harmonically corrected rota-

TABLE 3

Molecular Structure Representations for Formic Acid [123] (pm or degrees)

Parameter	r_s	r_z	r_e
r(C-H)	109.7(5)	109.7(5)	109.1(5)
r(C=O)	120.4(5)	120.5(5)	120.1(5)
r(C-O)	134.2(5)	134.7(5)	134.0(5)
r(O-H)	97.2(5)	96.6(5)	96.9(5)
r(O··O)	225.7(5)	226.2(5)	225.3(5)
∢(O=C=O)	124.82(40)	124.80(40)	124.80(40)
∢(H-C=O)	123.21(150)	123.26(150)	123.26(150)
∢(H-C-O)	111.97(150)	111.94(150)	111.94(150)
∢(C-O-H)	106.34(40)	106.61(40)	106.61(40)

tional constants of the H/D isotopomers: (1,12,16,16,1); (1,12,16,16,2); (2,12,16,16,1) and (2,12,16,16,2). Their own refined harmonic force-field was put to use in that work where, in addition, r_e estimates are also given. Table 3 contains Davies et al.'s r_z and r_e estimates which are closely similar to previous r_s structures [124,126], and to electron diffraction r_α^0 structure estimates [125]. The close agreement among the r_s and r_e parameters is surprising in light of what has been said in Secs. III and IV.

Fusina and Mills [127] calculated the vibrational coordinates for isocyanic acid, HNCO, using a variety of vibration-rotation data. This analysis was then utilized in the calculation of the r_z geometry. The results support an earlier estimate of the substitution geometry [128] and are fully consistent with an 'ab initio' theoretical geometry [129]. Thus HNCO possesses a slightly bent N=C=O backbone with the H nucleus in the plane of the heavy nuclei in a trans-position. From the I_z^α moments of inertia for the iso-topomers: (1,14,12,16); (1,15,12,16); (1,14,13,16); (1,14,12,18) and (2,14,12,16), the r_z parameters were derived as shown in Table 4.

It is interesting to note that the harmonic corrections for the I_0^a moments are three orders of magnitude larger than the I_z^b - I_0^b and I_z^c - I_0^c correc-tions. This has to do with the strong a-type Coriolis interactions in this molecule. The isotopic mass-effects used by Fusina and Mills are δr_z(H-N)(H $\rightarrow$ D) = 0.17 pm and $\delta\vartheta_z$(H-N=C) (H $\rightarrow$ D) = 0.6°. The r_z geometry satisfies the planarity constraint, i.e., for most isotopomers $|\Delta_z| < 60$ u·pm^2.

TABLE 4

A Comparison of Geometrical Parameters for Isocyanic Acid (pm or degrees)

Parameter	r_z[127]	r_s[128]	r_e[129]
r(H-N)	101.27(10)	99.46(64)	100.1(5)
r(N=C)	121.75(35)	121.40(24)	120.8(5)
r(C=O)	116.54(34)	116.64(8)	116.6(5)
⩗(H-N=C)	124.0(1)	123.9(17)	124.0(10)
⩗(N=C=O)	172.1(1)	172.6(27)	174.0(5)

Nakata et al. [63] used the phosgene molecule, $COCl_2$, as a model for joint spectroscopic and gas-phase electron diffraction analysis. Applying the geometric parameter h (see Eq. (56) in Sec. IV.A) which is simply related to the rotational constant B, h_z values were calculated for each isotopomer from the harmonic correction in Eq. (60). The δh_z's calculated from the h_z values of the seven isotopomers relative to h_z of the common parent (12,16,35) were found to be additive, according to Eq. (69). The δr_z(C=O) and δr_z(C-Cl) effects could not be derived that simply because of the planarity interdependence between the A_z and C_z rotational constants, and therefore these were estimated by Eq. (68). The a_3 values are customarily transferred from diatomics, as has formerly been noted here. The authors [63], however, were able to fit the electron diffraction intensity data and eight rotational constants to the four parameters r_z(C=O), r_z(C-Cl), a_3(C=O) and a_3(C-Cl). Among these, the r_z bond lengths are those of the common parent $^{12}C^{16}O^{35}Cl$. As the Morse parameters are invariant to isotopic mass, the a_3 values were inferred from a plot showing the interdependence of a_3(C=O) and a_3(C-Cl) for each isotopomer less the common parent. For these plots the r_z bond lengths were fixed at their converged values. The determined values were a_3(C=O) = 0.029(9) pm^{-1} and a_3(C-Cl) = 0.016(4) pm^{-1} which can be compared to the diatomic values of 0.0239 pm^{-1} and 0.0192 pm^{-1}, respectively. From Eq. (68) the δr_z(C=O) and δr_z(C-Cl) quantities could then be obtained. These quantities also satisfied the approximation in Eq. (69). The r_z geometry is thus the following: r_z(C=O) = 117.85(26) pm, r_z(C-Cl) = 174.24(13) pm and $⩗_z$(Cl-C-Cl) = 111.83(11)$^{\circ}$. From these data and the anharmonicity coefficients, r_e estimates for the bond lengths were obtained (see Sec. IV.A, the discussion of Ref. [63] therein).

Finally in this section we deal with the influence of δr_z isotopic effects upon the regression calculation of geometrical parameters. It is well known that δr_z corrections may become magnified in the calculated structures. To take these effects explicitly into account, algebraic methods were given by Laurie and Herschbach [6] for linear molecules, by Hilderbrandt and Wieser [130] for symmetric tops, and by Kuchitsu et al. [79] for planar asymmetric tops. These principles were generalized by Typke [53] for non-planar asymmetric tops. The method consists essentially of incorporating δr_z bond length effects into least-squares refinements of nuclear Cartesian coordinates [52].

Upon defining the $\delta \underset{\sim}{r}_i$ isotopic mass effect vector pointing from nucleus i to nucleus j:

$$\delta \underset{\sim}{r}_i = \underset{\sim}{e}_{ij} \, \delta r_{i,z},$$
(71)

where $\underset{\sim}{e}_{ij}$ is the bond-directed unit vector:

$$\underset{\sim}{e}_{ij} = (\underset{\sim}{r}_i - \underset{\sim}{r}_j)(|\underset{\sim}{r}_i - \underset{\sim}{r}_j|)^{-1},$$
(72)

and estimating $\delta r_{i,z}$ from Eq. (68), Eq. (41) may be modified [53]:

$$p^{(')} =$$

$$P + \sum_i \Delta m_i \, \underset{\sim}{r}_i \underset{\sim}{r}_i - \frac{\underset{\sim}{a}\,\underset{\sim}{a}}{M'} + \sum_i m_i'(\underset{\sim}{r}_i \delta \underset{\sim}{r}_i + \delta \underset{\sim}{r}_i \underset{\sim}{r}_i + \delta \underset{\sim}{r}_i \delta \underset{\sim}{r}_i)$$
(73)

where $m_i' = m_i + \Delta m_i$, $M' = M + \sum_i \Delta m_i$.

The vector $\underset{\sim}{a}$ is:

$$\underset{\sim}{a} = \sum_i \Delta m_i \, \underset{\sim}{r}_i + \sum_i m_i' \, \delta \underset{\sim}{r}_i.$$
(74)

The main difference with respect to an r_s regression calculation [52] is that, even in the case of single isotopic substitution, all nuclear Cartesian coordinates should be refined. In Eqs. (73) and (74) the i sum extends over all nuclei, but for unsubstituted ones $\Delta m_i = 0$, of course. For good convergence the $\delta \underset{\sim}{r}_i$ vectors should be kept constant during iterations.

When in Eqs. (73) and (74) only one $\delta \underset{\sim}{r}_i$ quantity is different from zero (the assumption of a single end-atom (k) primary δr_z effect) the solution may analytically be given as

$$p^{(')} = P + \mu \underset{\sim}{r}_k \underset{\sim}{r}_k + \kappa(\underset{\sim}{r}_k \, \delta \underset{\sim}{r}_k + \delta \underset{\sim}{r}_k \, \underset{\sim}{r}_k) + \lambda \delta \underset{\sim}{r}_k \delta \underset{\sim}{r}_k,$$
(75)

where μ is the usual reduced mass (M/M') Δm_k, while κ and λ are also mass factors:

$$\kappa = (M/M')\, m_k'; \qquad \lambda = \kappa - (m_k\, m_k'/M'). \tag{76}$$

Using the mathematical substitution:

$$\underset{\sim}{R} = \underset{\sim}{r}_k + (m_k'/\Delta m_k)\, \delta\underset{\sim}{r}_k, \tag{77}$$

an approximate, alternative form of Eq. (75) is:

$$P^{(')} \underset{\sim}{\sim} P + \mu\, \underset{\sim}{R}\, \underset{\sim}{R}, \tag{78}$$

where a term containing $\delta\underset{\sim}{r}_k \delta\underset{\sim}{r}_k$ has been neglected. As Eq. (78) has the same form as Eq. (41), $\underset{\sim}{R}$ may be identified with the usual substitution coordinate vector $\underset{\sim}{r}_{k,s}$ (an extra index s is used to emphasize this fact), so that from Eq. (77):

$$\underset{\sim}{r}_k = \underset{\sim}{r}_{k,s} - (m_k'/\Delta m_k)\, \delta\underset{\sim}{r}_k. \tag{79}$$

The difference $\underset{\sim}{r}_k - \underset{\sim}{r}_{k,s}$ is thus proportional to $\delta\underset{\sim}{r}_k$, and the proportionality constant is usually much larger than unity (e.g., for the substitution $^{13}C \rightarrow$ ^{12}C the constant is 13, and for $^{18}O \rightarrow {}^{16}O$ the constant is 9). Equation (79) demonstrates the structural magnification of δr_z isotopic mass effects that had formerly been noted by Nygaard [131].

V. MOLECULAR SHAPE INFORMATION FROM "FORBIDDEN" SPECTRA

For the determination of vibrational effects in molecular geometry, one normally needs the very high precision of microwave spectroscopy. Pure rotational spectra are, however, obtainable only for polar molecules possessing a permanent electric dipole moment in the equilibrium configuration. This has been the general attitude towards rotational spectroscopy up to relatively recent times, and so geometrical problems of non-polar molecules were not regarded as the territory of the microwave spectroscopist. Highly symmetric molecules were rather thought of as the perfect models for the gas-phase electron diffractionist.

High resolution infrared and Raman spectra allow one to overcome this difficulty since vibration-rotation transitions have more relaxed rotational selection rules than pure rotational spectra. Even though lack of polarity does not pose a problem for the vibration-rotation spectroscopist, characteristic difficulties are encountered in the determination of the axial rotational constants of symmetric top molecules. These shall be covered in this section.

It is a relatively new development in spectroscopy that through the application of sensitive modern experiments for the detection of subtle inter-actions between molecules and electromagnetic waves it has become possible to "see" rotational transitions that had formerly been regarded as forbidden ones. "Forbiddenness" is thus a term that keeps losing its constraining implications as the level of spectroscopic techniques is continuously being elevated. Such advances allow the spectroscopist to determine rotational constants for his or her "difficult" molecules with the much desired high precision of microwave spectroscopy.

A. Vibrationally Induced Rotational Transitions of Non-polar Molecules

For molecules in which the equilibrium geometry can uniquely be classi-fied by their point group symmetry, the criterion of polarity is that at least one or more translations T_α in the direction of the fixed molecular axes x, y or z must belong to $\Gamma^{(s)}$, the totally symmetric representation. Thus, when the equilibrium geometry of the molecule is such that none of the T_α's belong to $\Gamma^{(s)}$, the ground-state vibrational average values of the components of p, the electric dipole moment, vanish. Thus the molecule is non-polar and has no easily detectable pure rotational spectrum. The polar point groups are C_1, C_s, C_n and C_{nv} (n = 2, 3, ..., ∞). The remaining point groups are non-polar.

It is, however, possible for non-polar molecules to develop a small effec-tive electric dipole moment in vibrationally or rotationally excited states, provided the symmetry of these states satisfies certain criteria. The rota-tional spectra of non-polar molecules and their geometry may be studied via basically two intramolecular mechanisms.

A very important new branch of molecular spectroscopy is the study of "forbidden" spectra arising through centrifugal distortion effects. Oka [132] has provided an excellent review of centrifugally induced rotation spectra covering research up to 1976. Although no such induced spectra shall be systematically reviewed here, it may be noted that their study started in earnest with the theoretical papers by Watson [133] and Fox [134], and experimentally with the observation of the vibrational ground-state electric dipole moment of methane, CH_4, by Ozier [135]. These three works all appeared in 1971.

Only a fraction of the papers published in the field of induced rotational spectra report rotational constants or geometrical data derived from "forbidden" transitions. This is especially true for centrifugally-induced spectra where most of the authors concentrated on the determination of the induced electric dipole moment.

Vibrationally induced rotational spectra may basically be divided into two groups. To the first group belong those emerging in certain degenerate, vibrationally excited states; their appearance is mostly due to Coriolis vibration-rotation interactions.

Zero-point vibrational motions may also lead to small, induced electric moments in non-polar molecules that are unsymmetrically isotopically substituted. Such spectra form the second group covered here.

Mizushima and Venkateswarlu [136] were the first authors to suggest that, in some non-polar point groups, small electric dipole moments may be vibrationally induced. This is the case for the doubly degenerate vibrational states of point group D_{2d}, or the F_2 triply degenerate states of T_d molecules. These authors have given transition dipole matrix elements for these "forbidden" rotational transitions in groups D_{2d} and T_d, and found them to be very similar to those of the first overtone vibrational transitions. Estimates for these weak rotational transition frequencies in allene, H_2CCCH_2, and its D_4-isotopomer were also given.

Mills, Watson and Smith [137] have further refined the theory of the Mizushima-Venkateswarlu spectra and have extended it to all relevant point groups: T, T_d, S_{2n} and D_{nd} (n = even); D_{nh}, D_n and C_{nh} (n = odd). They have shown that the group-theoretical symmetry condition for the induced rotational spectra in an excited vibrational state v is:

$$(\Gamma_v)^2 \supset \Gamma_{T_\alpha} \tag{80}$$

where Γ_v is the representation of the vibrational state, while Γ_{T_α} is the species of the corresponding translation. For non-polar molecules Γ_{T_α} can not be $\Gamma^{(s)}$, as discussed previously, so $(\Gamma_v)^2$ must contain a non-totally symmetric species. Therefore Γ_v can only be a degenerate species.

The authors have also shown that linear molecules and those with a center of symmetry can not possess Mizushima-Venkateswarlu spectra. The intensity of these transitions derives from terms of the electric dipole moment expansion quadratic in the normal coordinates. Mills et al. [137] have estimated that the intensity of such vibrationally induced rotational spectra are almost four orders of magnitude lower than the intensity of the usual rotational transitions. Induced rotational spectra may be put to use in the determination of axial rotational constants of symmetric tops, but only for molecules that contain C_3 and S_4 rotational symmetry axes.

Mills [138] attempted to observe the $\Delta K = \pm 1$ Mizushima-Venkateswarlu transitions in the microwave spectrum of trifluoromethyl-acetylene, $F_3C-C\equiv C-H$, of C_{3v} symmetry, but has failed to find the expected features due to instrumental sensitivity problems. Such weak spectra are indeed difficult to detect in the usual way.

To improve detectability, non-linear spectroscopic techniques, such as infrared-microwave double resonance and precision laser-Stark spectroscopy, have been introduced by Brewer [139,140], Luntz [141], Luntz and Brewer [142], Curl et al. [143-145], and Takami et al. [146].

Venkateswarlu wrote a review [147] on double resonance experiments on methane, and indicated possible microwave transitions and double resonances in allene and allene-D_4. He noted that modern spectroscopy has proved the correctness of the Mizushima-Venkateswarlu theory. Jagannadham and Venkateswarlu [148] indicated coincidences with laser frequencies among vibrational-rotational transitions in silane, SiH_4, germane, GeH_4, and the deuterated isotopomers. Possible double-resonances and Lamb-dip experiments were suggested for the detection of "forbidden" lines, and for the measurement of the vibrationally induced electric dipole. Ozier and Rosenberg [149] provided in 1978 a complete confirmation of the theoretical predictions in Refs. [136] and [137] by measuring in the 80-180 cm^{-1} infrared range the vibrationally induced rotational spectrum of methane. The authors have reported the value of the electric dipole moment in the $v_4 = 1$ state ($p_v = 1.38(20) \times 10^{-2}D = 4.60(67) \times 10^{-32}C \cdot m = 5.43(79) \times 10^{-3}au$), but no rotational constant estimate.

An interesting extension of the theory of the Mizushima-Venkateswarlu spectra was reported by Rosenberg and Susskind [150]. It has been shown that an effective electric dipole arises through torsion-vibration-rotation interactions in ethane, H_3C-CH_3, and similar molecules. The interactions are the same as those activating "forbidden" pure torsional transitions in ethane. The emergence of these, which are essentially Coriolis-activated transitions, would be forbidden in the rigid geometrical model of ethane. In point group D_{3d} there is a center of symmetry. The flexibility of ethane, however, requires a non-rigid symmetry description, and so its proper symmetry group is $G_{36}^{\dagger}$ as was shown by Hougen [151]. In this group no center of symmetry element exists so that, according to the symmetry arguments for the Mizushima-Venkateswarlu spectra [136,137], "forbidden" rotational spectra may still appear.

The simplest non-polar molecule that exhibits an isotope-substitution induced rotational spectrum (this is the second main type of "forbidden" transitions reviewed here) is HD. The first experimental investigations of the "forbidden" $\Delta J = +1$ transitions in the far infrared spectrum of HD were made by Trefler and Gush [152], who deduced the value of the induced dipole moment: $p_o = 5.85(17)$ x $10^{-4}D = 1.95(6)$ x $10^{-33}C \cdot m = 2.30(7)$ x 10^{-4}au.

Blinder [153], Kolos and Wolniewicz [154], and Wolniewicz and Kowalski [155] computed the value of the induced electric dipole moment by quantum-theoretical methods. Their results are (in $C \cdot m$ units): 1.89 x 10^{-33}, 5.14 x 10^{-33} and 3.05 x 10^{-33}, respectively. According to Ref. [154] the sense of the dipole is $H^{+}D^{-}$. Wolniewicz and Kowalski [155] showed that the corresponding spectra would be forbidden within the Born-Oppenheimer approximation and, therefore, they are due to non-adiabatic effects.

Bunker [156] has re-investigated the HD problem to draw general conclusions with respect to "forbidden" spectra of homopolar isotopically unsymmetrical diatomics. He has shown the various contributions to the effective vibronic ground-state dipole moment. Whereas the centrifugal distortion contribution is negligibly small, the rotational contribution is not, and it has the sense $H^{-}D^{+}$. Bunker has pointed out that the agreement between Blinder's estimate [153] and the experimental dipole moment [152] is fortuitous. The rotational dependence of the induced dipole moment in the vibronic ground-state of HD was derived in Ref. [156].

Polyatomic "forbidden" spectra induced by unsymmetrical isotope substitution would emerge even in the absence of non-adiabatic effects, as was stressed by Gangemi [157]. These weak rotational spectra have proved to be very useful in the determination of the geometry of non-polar molecules. Some such studies overlap with the next subsection for the discussion of axial rotational constant determinations, for which unsymmetrical isotope substitution is applied widely for that purpose as well.

Gangemi [157] calculated the zero-point dipole moments of $^{18}O^{12}C^{16}O$ and mono-deutero-methane CH_3D, and concluded that they are on the order of 1 x $10^{-3}D \underset{\sim}{} 3$ x $10^{-33}C \cdot m$, and 4 x $10^{-3}D \underset{\sim}{} 1$ x $10^{-32}C \cdot m$, respectively. These predictions were later put to experimental tests. Ozier et al. [158] observed the isotopic substitution induced pure rotational spectrum of CH_3D between 40 and 120 cm^{-1}. The experimental value is $p_o = 1.89(10)$ x $10^{-32}C \cdot m$ The zero-point effective rotational constant is $B_o = 3.882(2)$ cm^{-1}. Wofsy et al. [159] have refined the estimate for the induced dipole by molecular beam electric resonance studies on CH_3D, and reported $p_o = 1.882$ x $10^{-32}C \cdot m$, and 1.894 x $10^{-32}C \cdot m$ for the rotational states $J = 1$, $K = 1$, and $J = 1$, $K = 2$, respectively. These values are in reasonable agreement with Gangemi's theory.

Endo et al. [160] have measured the "forbidden" microwave spectrum of the $^{18}O^{12}C^{16}O$ molecule. From five rotational transitions the effective rotational constant was determined, $B_o = 11037.892(14)$ MHz = 0.368184(15) cm^{-1}, which is compatible with recent infrared estimates: 11037.930(39) MHz [161], and 11037.46(37) MHz [162]. The isotope substitution induced dipole moment is p_0 = 2.34(50) x 10^{-33} C·m, which is in fair agreement with Gangemi's former computations.

Muenter and Laurie [163] were the first to report isotope substitution allowed rotational spectra of polyatomics. In 1964 they published the rotational constant of mono-deutero-acetylene, DC≡CH, B_0 = 29725.3(1) MHz = 0.991527(3) cm^{-1}. This molecule was later revisited by Matsumura et al. [166].

Extensive polyatomic studies on the type of "forbidden" spectra discussed here have been pursued by Hirota and his coworkers since 1971. Their results on the r_z structure of ethylene [99], allene [110] and ethane [120,121] have already been quoted in Sec. IV. In addition, Hirota et al. have discussed the isotope substitution induced spectra of dideutero-methane, CH_2D_2 [164], dideutero-ethane, CH_3-CHD_2 [165], and H-C≡CD [166]. From the "forbidden" spectra of CH_2D_2 Hirota and Imachi [164] deduced the combination A_0 - C_0 = 37555.758 MHz, and B_0 - C_0 = 13664.280 MHz. The induced ground-state dipole moment is 4.7(1.7) x 10^{-32} C·m.

Before leaving the subject of vibrationally induced rotational spectra it is instructive to return to Eq. (80), the symmetry condition for the Mizushima-Venkateswarlu spectra. In order to decide on the observability of "forbidden" rotational transitions, it pays to investigate the symmetry condition for the first-order (linear) Stark effect. This is justified because, while for polar molecules the symmetry restriction for the existence of a permanent electric dipole moment is not simply identical to the corresponding restriction for a linear Stark effect, for non-polar molecules the knowledge of the latter provides a practical answer to the question of the observability of forbidden spectra.

Watson [167] has given the strict symmetry condition for first-order Stark effect for the case of negligible nuclear quadrupole hyperfine splittings and no accidental degeneracies:

$$(\Gamma^2)_{antisym} \supset \Gamma(p_f), \tag{81}$$

where on the left-hand side is the antisymmetric product of the ro-vibrational (ro-vibronic) species, and f denotes the space-fixed component of the electric dipole moment operator. In order for Eq. (81) to hold, Γ should be a degenerate species. For instance, for the polar C_{3v} symmetric tops, when Γ = E, $(\Gamma^2)_{antisym}$ = A_2 and $\Gamma(p_z)$ = A_2, so that such levels possess linear Stark effects, as is well known. On the other hand, for a D_{2d} molecule such as allene, H_2CCCH_2, when Γ = E, $(\Gamma^2)_{antisym}$ = A_2 while $\Gamma(p_z)$ = B_1. These levels are, therefore, not characterized by linear Stark effects [167].

For spherical top molecules Eq. (81) is satisfied only for T and T_d symmetry, in the case of doubly degenerate E levels. The corresponding first-order Stark effects have been observed, e.g., by Luntz [141] and Luntz and Brewer [142]. Oka in a preceeding paper [168] had already shown that the Stark effect is first-order only for levels of degenerate (double) parity and, because molecules with a center of symmetry do not have such levels, no linear Stark effect is observable for them.

B.   The Determination of Axial Rotational Constants of Symmetric Tops

The purely spectroscopic determination of the geometry of axially symmetric molecules encounters difficulties. The rotational constant A for prolate symmetric tops, while the constant C for oblate symmetric tops, poses characteristic problems. Similar problems arise in the case of asymmetric tops of axial symmetry that are inertially close to the symmetric top limit (usually prolate tops). The underlying reason for these difficulties is found in the rotational selection rules for pure rotation, and vibration-rotation spectra. The determination of the axial rotational constant is not directly possible from the microwave spectra, nor from the parallel vibration-rotation bands in the infrared spectra. The reason is that for such transitions the quantum numbers k and ℓ do not change during the transitions. Neither is it possible to derive axial rotational constants from a single perpendicular band in the infrared spectrum, because in such transitions the values of Δk and $\Delta \ell$ are proportional to each other.

Since the rotational constant B of symmetric tops can be precisely determined from pure rotation or vibration-rotation spectra, the difficulty in the derivation of axial constants is a very bothersome one. It impedes the spectroscopic determination of the geometry of axially symmetric molecules at a precision level that can easily be reached for linear molecules and for asymmetric tops without axial symmetry. For this reason many authors have tried to develop methods for circumventing this difficulty.

The traditional method for resolving this dilemma is described by Herzberg (see pp. 404-406 in Ref. [169]), and by Allen and Cross (see p. 153 in Ref. [170]). Albeit one can derive only the product $A\zeta_t$ from perpendicular infrared vibration-rotation bands (ζ_t is the Coriolis constant of the coupling of the components of a degenerate mode), such products may be derived for all degenerate modes of a symmetric top. Therefore the effective A rotational constant can be estimated via the use of harmonic sum rules for Coriolis constants [171]. The axial constant obtained in this way is not, however, the quantity A_o or C_o.

A further method was suggested by Edwards et al. [172-174] for the determination of A of prolate symmetric tops, and was mainly used for C_{3v} molecules. Basically, the method consists of the simultaneous analysis of the rotational structure of a degenerate infrared band v_t and that of the perpendicular component of the corresponding first overtone band $2v_t$. From the subband origins of the fundamental v_t, the combination $(A - (A\zeta_t) - B)_{v_t=1}$ whereas from the first overtone $2v_t$, the combination $(A + 2(A\zeta_t) - B)_{v_t=2}$ may be derived [169,170].

As Mills has outlined [58] it is possible to obtain the following relationships:

$$A_o - (A\zeta_t)_{v_t=1} = A_o - (A\zeta_t)_{v_t=0} + \alpha_t^{A\zeta_t} \tag{82}$$

and

$$A_o + 2(A\zeta_t)_{v_t=2} = A_o + 2(A\zeta_t)_{v_t=0} - 4\alpha_t^{A\zeta_t} \tag{83}$$

where $\alpha_t^{A\zeta_t}$ is an expansion coefficient in the series:

$$(A\zeta_t)_v \approx (A\zeta_t)_e - \sum_s \alpha_s^{A\zeta_t}(v_s + \frac{d_s}{2}). \tag{84}$$

From Eqs. (82) and (83) the combinations

$$\{(A\zeta_t)_o - (5/3)\,\alpha_t^{A\zeta_t}\} \quad \text{and} \quad \{A_o - (2/3)\,\alpha_t^{A\zeta_t}\}$$

can be derived. The method elaborated by Edwards et al. yields actually

$$A \equiv A_o - (2/3)\alpha_t^{A\zeta_t}$$

as was shown by Mills (see p. 138 in Ref. [58]), and by Watson (see p. 487 in Ref. [29]). The advantage of the Edwards method, as was emphasized by Watson [29], is that the vibrational dependence of the combination denoted by A above is comparatively simple.

Barnett and Edwards [172] used the above method for the v_4 and $2v_4$ infrared bands of CH_3Br and CH_3I and, for methyl iodide, they determined the combination A as $5.134(3)$ cm^{-1}.

Sarka and Rao [175] have shown that because the application of the Edwards method usually involves the neglect of vibrational dependence, the accuracy relative to A_o is only about 1%. This is a smaller uncertainty than that characterizing the traditional zeta sum rule method [169,170]. The latter has an uncertainty around 5% [58]. Sarka and Rao [175] worked out another method, similar to the Edwards procedure, that reduces the uncertainty to about 0.01% of A_o.

In the foregoing (Sec. V.A), it has been indicated how useful certain vibrational and centrifugal distortion effects may prove in the determination of the geometry of non-polar molecules. The "forbidden" rotational transitions discussed there are also important for the determination of the axial rotational constants of polar molecules. Additionally, certain intramolecular dynamic couplings and perturbations can also be utilized to furnish information relevant to axial constants. Such effects are the various types of Coriolis coupling, k-, and ℓ-type doubling, etc. As a result of these intramolecular effects $\Delta k \neq 0$ transitions become allowed. Through the use of combination-differences [169,170] these perturbation-allowed or -induced transitions can be analyzed either separately from, or together with, normal dipole-allowed transitions. Such "forbidden" transitions are usually observed in high resolution infrared spectra. However, for axial constants it is normally necessary to have access to microwave rotational constants as well. Thus the determination of axial rotational constants constitutes a good example for the combination of different branches of molecular spectroscopy.

Maki and Hexter [176] were pioneers in the determination of symmetric top A_o constants avoiding the use of Coriolis sum rules. In the infrared spectrum of CH_3I a Coriolis perturbation is observable between v_5 and $v_3 + v_6$, both of which are degenerate modes. The effect of this resonance upon the infrared band contours may be simulated by a computer, and thus allows the determination of the constant A_o, which in this case is $5.158(20)$ cm^{-1}.

Matsuura et al. [177] were the first authors to analyze Coriolis-induced rotational transitions in the infrared spectrum of methyl iodide. In this molecule the x-y axis Coriolis interactions were observed to take place among the rovibrational states $v_5 = 1$, $K = 3$ and $v_3 = 1$, $v_6 = 1$, $K = 4$. From the analysis of the Coriolis-doublets the combination $A_o - 29D_K^o$ was determined. A_o was then approximated to lie between the limits 5.1723 cm^{-1} and 5.1766 cm^{-1}.

Lovejoy and Olson [178] analyzed the weak, perturbation-allowed transitions in the v_1 and v_4 bands of mono-deutero-silane, $^{28}SiH_3D$, and deduced a value for $A_o - B_o - 5D_K^o = 0.76406(45)$ cm^{-1}.

Olson wrote a theoretical paper [179] dealing with perturbation-induced vibration-rotation transitions and applied his theory to the determination of the geometry of C_{3v} tops. He has given a combination-difference procedure utilizing "forbidden" and allowed transitions involving the same perturbed upper vibrational state thus yielding ground-state effective A and D_K values.

By the symmetry selection rules of Amat [180] and Hougen [181], Olson derived the requirement for C_{3v} molecules that the k value of the two vibrational ground-state levels must differ by multiples of three:

$$\Delta k = k_2'' - k_1'' = 3n, \tag{85}$$

where n is an integer, negative or positive.

For CH_3D Olson [179] obtained A_o = 5.250703(56) cm^{-1} and, by using the B_o constant of methane [182] (B_o = 5.24059(6) cm^{-1}), he was able to obtain r_o(C-H) = 109.4030(7) pm that leads to $\sphericalangle$(H-C-D) = 109°37'36"(6") in CH_3D, and r_o(C-D) = 109.1722(22) pm.

Maki et al. [183] determined C_o of the oblate top phosphine molecule (PH_3). They showed that in this molecule Coriolis interactions arise among rotational levels in the $3\nu_2$ state with $\Delta k = 3$, and these interactions lead to the emergence of "forbidden" $\Delta|k - \ell| = \pm 3$ transitions. A value of 3.91894(40) cm^{-1} for C_o was obtained, and using B_o = 4.4524183(46) cm^{-1} the authors derived the structure r_o(H-P) = 142.002(6) pm and $\sphericalangle_o$(H-P-H) = 93.3454(43)°.

Olson et al. [184] studied the perturbation-allowed transitions in the infrared spectrum of arsine, AsH_3, that contain many intensive "forbidden" transitions. They thus derived C_o = 3.498579(21) cm^{-1}, whereas B_o = 3.75161435(120) cm^{-1}. The r_o geometry is then: r_o(As-H) = 152.0143(4) pm and $\sphericalangle_o$(H-As-H) = 91.9758(1)°. From the results of earlier high resolution infrared analyses [185] it proved possible to give an r_e structure estimate of r_e(As-H) = 151.08(4) pm, $\sphericalangle_e$(H-As-H) = 92.083(43)°.

Olson and Lovejoy [186] observed a number of perturbations in the ν_1 and ν_4 infrared bands of $^{28}SiH_3D$. The emergence of "forbidden" transitions is due in this case to k-, and ℓ-type doubling effects and to various forms of Coriolis resonances. The A_o value deduced therefrom is 2.86334(45) cm^{-1}. The assumption of tetrahedral geometry results in unrealistic bond lengths, such as 148.008 pm for r_o(Si-H), which leads to r_o(Si-D) = 148.092 pm > r_o(Si-H) when combined with B_o = 2.099477 cm^{-1}. When the assumption of tetrahedral geometry is dropped but the hypothesis that r_o(Si-H) is identical in the r_o structures of silane and SiH_3D is used, an r_o(Si-H) value of 148.11893 pm is obtained. Therefrom $\sphericalangle_o$(H-Si-D) = 109°35.54' is calculated. The authors used their B_o constant (see above) and calculated, from the former bond angle, r_o(Si-D) = 147.9041(95) pm < r_o(Si-H).

Graner [187] made use of "forbidden" vibration-rotation transitions in the infrared spectrum of CH_3F to obtain A_o via a novel combination-difference technique. Accordingly A_o = 5.182009(12) cm^{-1}. We shall return to Graner's work on C_{3v} tops later in connection with the CH_3Br molecule.

In an elegant theoretical paper, Sarka [188] emphasized that the axial rotational constants of symmetric tops are most precisely determined with the help of perturbation-induced transitions. On the basis of Hougen's group theory [181], the author derived the symmetry restrictions for the existence of perturbation-allowed transitions in vibration-rotation spectra. A systematic discussion is given in Sarka's paper [188] for molecules containing the C_3 axis for perturbations of various order that may lead to the emergence of transitions forbidden in the zeroth order.

In contrast to the case of "forbidden" pure rotational spectra, centrifugal distortion-induced vibration-rotation transitions are important in geometrical work. Oka in his review [132] discussed that centrifugal distortion leads to a selection rule $\Delta k = \pm 3$ for C_{3v} tops. This is the consequence of an induced electric dipole perpendicular to the C_3 axis. Such transitions were theoretically forecast by Hanson [189] and by Shimizu et al. [190]. These selection rules follow from the fact that a C_3 rotation spoils the "goodness" of the quantum number k, and so $\Delta k = 3n(n = \pm 1, \pm 3, \ldots)$ becomes valid (see also Ref. [179]). The theory provided by Watson [133] for centrifugally-induced rotational spectra is thus applicable to the determination of axial rotation constants of polar molecules as well.

Ammonia, phosphine (PH_3) and arsine (AsH_3) were among the first, light C_{3v} tops for which centrifugal distortion-allowed spectra were utilized to refine existing axial rotational constant estimates. For NH_3 the rotational lines with $\Delta k = \pm 3$ appear in the far infrared region and therefore their analysis becomes difficult due to overlapping with strong dipole-allowed transitions. For phosphine and arsine, B_0 and C_0 are nearly the same and so the $\Delta k = \pm 3$ transitions appear in the microwave region. As PH_3 and AsH_3 do not possess allowed rotational spectra in this region, the detection of centrifugal distortion-induced lines is much easier than for ammonia.

For AsH_3, Chu and Oka [191] derived, from "forbidden" $\Delta k = \pm 3$ Q-branch ($\Delta J = 0$) frequencies, not only the constant $C_0 = 3.4985064(670)$ cm^{-1}, but also the r_z and the equilibrium geometry. Helms and Gordy [192] later refined the rotational constants of AsH_3 and reported $C_0 = 3.4985758(14)$ cm^{-1} and $B_0 = 3.7516153(10)$ cm^{-1}, but gave no geometrical estimates.

Olson et al.'s estimate [184] for the axial rotational constant of phosphine was refined by Chu and Oka [191,193], Helms and Gordy [194], and by Belov et al. [195]. Chu and Oka's first estimate [193] was $C_0 = 3.919011(70)$ cm^{-1}, whereas in their subsequent paper [191] they gave 3.918981(73) cm^{-1} for PH_3, and 1.96710(10) cm^{-1} for PD_3. In this latter work r_z structures for the two isotopomers were derived, and the equilibrium geometry of PH_3 was reported. Helms and Gordy [194] measured also the "forbidden" mm-wave $\Delta k = \pm 3$ transitions and obtained a C_0 value for PH_3 of 3.9190062(53) cm^{-1}, and

TABLE 5

Geometrical Representations for the Phosphine Molecule (pm or degrees)

Molecule	r_0	r_z	r_e	Ref.
PH$_3$		r(P-H) = 142.731(10)		
		∢(HPH) = 93.280(20)	r(P-H) = 141.154(50)	[191]
PD$_3$		r(P-D) = 142.287(10)	∢(HPH) = 93.36(8)	
		∢(DPD) = 93.301(10)		
PH$_3$	r(P-H) = 142.00	r(P-H) = 142.699(20)		
	∢(HPH) = 93.345	∢(HPH) = 93.2287(50)	r(P-H) = 141.159(60)	[194]
PD$_3$	r(P-D) = 141.76	r(P-D) = 142.265(10)	∢(HPH) = 93.328(20)	
	∢(DPD) = 93.359	∢(DPD) = 93.2567(40)		

for PD$_3$, $1.967173(2)$ cm^{-1}. It is interesting to compare their revised geometrical data to those of Chu and Oka [191]. The comparison is given in Table 5.

Belov et al. [195] have recently measured the sub-mm wave, centrifugally-induced rotational spectrum of PH$_3$ along with dipole-allowed transitions. There are two slightly different estimates for C_0 in this paper, $3.9190058(7)$ cm^{-1} and $3.9190010(33)$ cm^{-1}.

Laughton et al. [196] have determined the ground-state effective rotational constants for NH$_3$ from "forbidden" $\Delta k = \pm 3$ transitions in the ν_2 infrared band. Because of the weakness of these transitions, laser-Stark spectroscopy and infrared-microwave double resonance methods were used. The rotational constants of ammonia were then estimated as $B_0 = 9.94413(9)$ cm^{-1} and $C_0 = 6.2280(8)$ cm^{-1}.

For the axial rotational constant problem, Raman spectroscopy is a very important tool. Mills [197] showed that in symmetric tops of D_{2d} symmetry the selection rules for Raman-scattering allow one to determine the axial constant A_0 through a combination of infrared and Raman-active vibration-rotation transitions without recourse to the Coriolis zeta sum rules. In this way Mills determined an A_0 value for the allene molecule of $4.816(5)$ cm^{-1} from which, through the use of Stoicheff's B_0 value of $0.2965(1)$ cm^{-1} [198], he obtained an r_0 structure: r_0(C-H) = 108.4(5) pm, r_0(C=C) = 130.9 pm and $∢_0$(H-C-H) = 118.2(5)°. These results comply very well with subsequent studies by Hirota and Matsumura [110] who deduced the following structure from the pure rotational spectrum of 1,1-dideutero-allene: $r_o(C-H) =$ 108.72(13) pm, $r_o(C=C) = 130.84(3)$ pm and $\vartheta_o(H-C-H) = 118°10'(10')$.

The fact that A_o may be directly calculated for degenerate vibrational modes from the rotational structure in Raman spectra was pointed out already by Herzberg (see p. 444 in Ref. [169]). Richardson et al. [199] derived the A_o constant for CH_3D from its v_4 Raman band and obtained 5.243(2) cm^{-1}.

The determination of A_o is rendered possible by the fact that the Raman bands of degenerate vibrations of C_{3v} tops are built from two components whose K-selection rules are $\Delta K = +1$, and $\Delta K = -2$, respectively. As Edwards and Brodersen [200] described, the rotational structure of such Raman bands allows the derivation of $A_o - B_o$ by combination-differences. The knowledge of B_o then leads to an estimate for A_o. Edwards and Brodersen [201] thus deduced a value of 5.1800(10) cm^{-1} for the A_o constant of CH_3Br, and vibrational corrections yield $A_e = 5.2442(15)$ cm^{-1}. This value then allows one to calculate the equilibrium distance of the H nuclei from the C_3 axis which turns out to be 32.077(5) pm.

For the oblate top trideutero-methane, CD_3H, Kattenberg and Brodersen [202] determined C_o from the Raman-active fundamental v_4. Using the infrared B_o value they obtained $C_o = 2.6297(3)$ cm^{-1}.

Deroche et al. [203] analyzed the rotational structure of the v_3 and v_5 fundamentals in Raman scattering for the dideutero-methane molecule (CH_2D_2), and from the latter band could determine A_o as 4.3027777 cm^{-1}.

Hegelund [204] carried out analyses for a number of vibrational-rotational Raman bands of dideutero-allene, D_2CCCH_2, and by the use of its B_o constant [110] derived $A_o = 3.2137(19)$ cm^{-1}.

Brodersen [205] calculated A_o for CD_3Cl through the rotational analysis of its v_4 Raman band and found it to be 2.6136(5) cm^{-1}, which is accidentally close to the zeta sum rule estimate of 2.614 cm^{-1}. Jensen et al. [206] have combined in their work the analyses of the v_4 infrared and Raman fundamentals of $CH_3{}^{35}Cl$ and reported its A_o value as 5.20530(10) cm^{-1}. For the isotopomer with the ^{37}Cl nucleus only the Raman data were used to give a corresponding approximate value for A_o of 5.2182(10) cm^{-1}. These authors have given equilibrium structure parameters for methyl chloride: $r_e(C-H) =$ 108.54(5) pm, $r_e(C-Cl) = 177.60(3)$ pm, and $\vartheta_e(H-C-H) = 110.35(5)°$.

A review paper rather relevant to the general problem of the determination of A_o for C_{3v} tops has recently been published by Graner [207]. In this systematic work all fundamental problems are emphasized for the determination of axial rotational constants for the particular case of the methyl bromide molecule. Graner has derived equilibrium rotational constants for a number of isotopomers, and gave the following equilibrium geometry for

TABLE 6

Equilibrium Geometry Estimates for CH_3X Molecules [209] (pm or degrees)

Parameter	CH_3F	CH_3Cl	CH_3Br	CH_3I
r(C-H)	109.5	108.6(4)	108.6(3)	108.5(3)
r(C-X)	138.2	177.8(2)	193.3(2)	213.3(2)
∢(HCH)	110°27'	110°40'(40')	111°10'(25')	111°17'(25')

CH_3Br: r_e(C-H) = r_e(C-D) = 108.23 pm, r_e(C-Br) = 193.40 pm, and ∢$_e$(H-C-H) = 111.157°.

The importance of perturbation-induced transitions in the axial rotational constant problem is stressed in Graner's work [207]. Tarrago and Dupre-Maquaire [208] have recently described a new method for determining vibronic ground-state constants for C_{3v} tops. This method allows one to utilize perturbation-allowed transitions and involves the use of combination-differences. The authors demonstrate their procedure on the K-dependent rotational constants, e.g., C_0 for the oblate top $^{12}CD_3H$. Thus C_0 = 2.62896(4) cm^{-1}, comparable to the Raman spectroscopic estimate of 2.6297(3) cm^{-1} in Ref. [202].

In Sec. V.A isotope substitution induced pure rotational spectra have been reviewed. It is a fairly logical step then to introduce axially unsymmetric substitution to circumvent the problems caused by axial symmetry. In fact, many such studies have been reported in recent times.

Before a short resume of these studies is given, Duncan's estimates for the r_e geometry of the CH_3X methyl halide molecules will be noted. He has given a detailed discussion of the effects of Coriolis-, and Fermi-perturbations upon the effective rotational constants for which corrections have to be made prior to the calculation of the molecular geometry. Duncan [209] used the then most dependable A_0 axial constants, plus microwave estimates for B_0, and the α_s^A and α_s^B coefficients in Eq. (42) for vibrational corrections. His results for the four CH_3X molecules are shown in Table 6.

Duncan and Mallinson [210] determined A_0 for dideutero-methyl-iodide. In a molecule like CD_2HI the separation of Q-branches does not depend on the product $A\zeta_t$ as in the case for symmetric tops, and so when B_0 is known from microwave spectroscopy, A_0 may precisely be determined. From the vibration-rotation analysis of five infrared bands (v_1, v_5, v_7, v_8 and v_9) an average A_0 value of 3.154(5) cm^{-1} was derived. For methyl iodide an r_0 structure was calculated: r_0(C-H) = 108.8(2) pm, r_0(C-D) = 108.6(2) pm,

r_0(C-I) = 214.23(15) pm and $\sphericalangle_0$(H-C-H) = 111°31(15)'.

Duncan et al. [211] used the above method also for A_0 of the twice deuterated methyl chloride and methyl acetylene. They stressed the point that, while for C_{3v} tops the effects of Coriolis resonances is of first order upon the effective A value, partial deuteration of the methyl group reduces it to second order. Duncan et al. [211] stated that at least one directly determined isotopic A_0 value is necessary to arrive at a realistic r_0 structure. Their results for the r_0 geometry are then: for CH_3Cl, r_0(C-H) = 109.0(2) pm, r_0(C-D) = 108.8(2) pm, r_0(C-Cl) = 178.54(10) pm, $\sphericalangle_0$(H-C-H) = 110°45'(15'); for CH_3CCH, r_0(C-H) = 109.6(2) pm, r_0(C-D) = 109.4(2) pm, r_0(C-C) = 145.96(10) pm, r_0(C≡C) = 120.73(10) pm, r_0(≡C-H) = 106.0(2) pm, r_0(≡C-D) = 105.8(2) pm and $\sphericalangle_0$(H-C-H) = 108°17'(15').

For the molecule CD_2HCN Duncan et al. [114] derived A_0 and a complete r_0 geometry. McKean et al. [212] determined A_0 and the ground-state effective geometry likewise for methyl isocyanide (CH_3NC): A_0 = 3.201(5) cm^{-1}, and the r_0 geometry of r_0(C-H) = 109.34(15) pm, r_0(C-D) = 109.14(15) pm, r_0(C-N) = 142.66(15) pm, r_0(N≡C) = 116.65(15) pm and $\sphericalangle_0$(H-C-H) = 109°32'(12').

Mallinson [213] determined A_0 very precisely for $^{12}CH_2DI$ from its dipole-allowed microwave spectrum and obtained 3.966266(1) cm^{-1}. The r_s geometry was also derived for methyl iodide in this paper: r_s(C-H) = 108.4 pm, r_s(C-I) = 213.6 pm, and $\sphericalangle_s$(H-C-H) = 107.47°. In Mallinson's other paper [214] on the axial constant of $CHD_2{}^{35}Cl$ and $CHD_2{}^{37}Cl$ he gave estimates for A_0 as 3.183071(2) cm^{-1} and 3.183043(4) cm^{-1}, respectively. An empirical relationship between infrared and microwave estimates for molecules of the type CHD_2Cl was also given:

$$A_0(\text{microwave}) - A_0(\text{infrared}) \underset{\sim}{\sim} 0.0033 \text{ cm}^{-1} \tag{86}$$

Eggers [215] has discussed in detail the r_0 geometry of methyl fluoride, and the role of the A_0 value for CH_2DF in fixing the r(C-H) bond length. For the ground-state geometry the rotational constants of five methyl fluoride isotopomers were used, the results of which are r_0(C-H) = 109.47(11) pm, r_0(C-D) = 109.27(11) pm, r_0(C-F) = 138.90(8) pm and $\sphericalangle_0$(H-C-H) = 110°19'(7').

A paper by Duncan [216] should be mentioned in connection with axial rotational constants. The author devoted himself to the problem of the methyl group substitution structure. In the absence of experimental A_0 values the r(C-H) bond length estimates are usually too long so that the calculated geometry does not satisfy the Costain-Watson relationship (see Eqs. (20) and (44)). Duncan [216] pointed out that for the equilibrium geometry one could use Kraitchman's relationship (see Eq. (15) in Ref. [18]):

$$I_e^c = I_e^{c'} + I_e^{b'} - I_e^{a'},$$
(87)

where the prime denotes the isotopically substituted molecule. By Eq. (87) it is possible to eliminate the axial moment of inertia for the parent (symmetric top) isotopomer. When one uses effective ground-state moments of inertia, Eq. (87) yields erroneous r_s geometries. This is clearly a consequence of vibrational effects.

To avoid this systematic error one has to use those forms of the Kraitchman equations (e.g., Eq. (36)) that explicitly contain the axial principal moments (or planar moments) of inertia, and not the modified forms obtained through Eq. (87). In this case the r_s geometry shall conform to the Costain-Watson relationships, within their corresponding limits of course, and the methyl group geometries shall be more acceptable for chemical thinking.

VI.   CONCLUSIONS

The study of vibrational effects in spectroscopic geometries is a subject that still attracts attention notwithstanding its state of theoretical maturity. It is the reviewer's hope that some reasons for this sustained interest have become clear from the present summary. One of the main reasons is that the physical content of the most useful structural representation of the substitution geometry is still in need of theoretical clarification. The relationship between substitution and equilibrium geometries is not always easy to establish. It is clearly important to have independent theoretical estimates of r_e, preferably from quantum-chemical calculations.

This is indeed a current tendency in theoretical chemistry, a few examples are listed in the References [217-225]. The interplay between accurate quantum-chemical calculations and high precision spectroscopic studies is expected to shed light on basic questions so far unanswered. It shall probably provide a boost also for the development of structure-oriented molecular spectroscopy.

REFERENCES

1. K. Kuchitsu, T. Fukuyama and Y. Morino, J. Mol. Struct., 1, 463 (1967-68).

2. K. Kuchitsu, MTP International Review of Science, Molecular Structure and Properties, Phys. Chem. Series One, Vol. 2 (G. Allen, ed.), Butterworths, London, 1972.

3. M. Nakata, T. Fukuyama and K. Kuchitsu, J. Mol. Struct., 81, 121 (1982).

4. E. R. Cohen and B. N. Taylor, J. Phys. Chem. Ref. Data, 2, 663 (1973).

5. D. R. Herschbach and V. W. Laurie, J. Chem. Phys., 37, 1668 (1962).

6. V. W. Laurie and D. R. Herschbach, J. Chem. Phys., 37, 1687 (1962).

7. D. R. Herschbach and V. W. Laurie, J. Chem. Phys., 40, 3142 (1964).

8. E. B. Wilson, Jr. and J. B. Howard, J. Chem. Phys., 4, 260 (1936); H. H. Nielsen, Rev. Mod. Phys., 23, 90 (1951); Encyclopedia of Physics, Vol. 37/1 (S. Flügge, ed.), Springer, Berlin, 1959.

9. T. Oka, J. Phys. Soc. Jpn., 15, 2274 (1960).

10. Y. Morino, K. Kuchitsu and T. Oka, J. Chem. Phys., 36, 1108 (1962).

11. M. Toyama, T. Oka and Y. Morino, J. Mol. Spectrosc., 13, 193 (1964).

12. K. Kuchitsu and S. J. Cyvin, in Molecular Structures and Vibrations, Chapter 12 (S. J. Cyvin, ed.), Elsevier, Amsterdam, 1972.

13. K. Kuchitsu, in Diffraction Studies on Non-crystalline Substances, (I. Hargittai and W. J. Orville-Thomas, eds.), Akademiai Kiado, Budapest, 1981.

14. I. M. Mills, J. Phys. Chem., 80, 1187 (1976).

15. A. R. Hoy, I. M. Mills and G. Strey, Mol. Phys., 24, 1265 (1972).

16. E. B. Wilson, Jr., J. C. Decius and P. C. Cross, Molecular Vibrations, McGraw-Hill, New York, 1955.

17. A. Ya. Tsaune, N. T. Storchai, L. V. Belyavskaya and V. P. Morozov, Optics and Spectrosc., 26, 502 (1969).

18. J. Kraitchman, Am. J. Phys., 21, 17 (1953).

19a. C. C. Costain, J. Chem. Phys., 29, 864 (1958).

19b. C. C. Costain, Trans. Am. Cryst. Assoc., 2, 157 (1966).

20. J. E. Wollrab, Rotational Spectra and Molecular Structure, Chapter 4, Academic Press, New York, 1967.

21. W. Gordy and R. L. Cook, Microwave Molecular Spectra, Chapter 13, Interscience, New York, 1970.

22. H. W. Kroto, Molecular Rotation Spectra, Chapter 6.11 and Appendix A6, Wiley and Sons, London, 1975.

23. J. K. G. Watson, J. Mol. Spectrosc., 41, 229 (1972).

24. Gy. Varsányi and L. Nemes, Kémiai Közlemények, 51, 277 (1979).

25. H. Goldstein, Classical Mechanics, 2nd ed., Addison-Wesley Series in Physics, Addison-Wesley, Reading, Massachusetts, 1980, Chapter 5.

26. J. K. G. Watson, Mol. Phys., 15, 479 (1968).

27. A. Chutjian, J. Mol. Spectrosc., 14, 361 (1964).

28. L. Nygaard, J. Mol. Spectrosc., 62, 292 (1976).

29. J. K. G. Watson, J. Mol. Spectrosc., 48, 479 (1973).

30. B. P. van Eijck, J. Mol. Spectrosc., 63, 152 (1976).

31. R. H. Schwendeman, Critical Evaluation of Chemical and Physical Structural Information, (D. R. Lide, Jr. and M. A. Paul, eds.), Chapter 2, National Academy of Sciences, Washington, 1974.

32. D. K. Coles and R. H. Hughes, Phys. Rev., 76, 178 (1949).

33. L. Pierce, J. Mol. Spectrosc., 3, 575 (1959).

34. L. C. Krisher and L. Pierce, J. Chem. Phys., 32, 1619 (1960).

35. R. E. Penn and L. W. Buxton, J. Chem. Phys., 67, 831 (1977).

36. R. A. Beaudet and R. L. Poynter, J. Chem. Phys., 53, 1899 (1970).

37. J. P. Pasinski and R. A. Beaudet, J. Chem. Phys., 61, 683 (1974).

38. H. N. Rogers, Kar-Kuen Lau and R. A. Beaudet, Inorg. Chem., 15, 1775 (1976).

39. D. Schwoch, A. B. Burg and R. A. Beaudet, Inorg. Chem., 16, 3219 (1977).

40. R. L. Kuczkowski, C. W. Gillies and K. L. Gallaher, J. Mol. Spectrosc., 60, 361 (1976).

41. U. Mazur and R. L. Kuczkowski, J. Mol. Spectrosc., 65, 84 (1977).

42. G. Graner, Spectrochim. Acta, 19, 2113 (1963); Deuxieme These, Faculté des Sciences, Orsay, 1965.

43. M. D. Harmony, V. W. Laurie, R. L. Kuczkowski, R. H. Schwendeman, D. A. Ramsay, F. J. Lovas, W. J. Lafferty, A. G. Maki, J. Phys. Chem. Ref. Data, 8, 619 (1979).

44. B. P. van Eijck, J. Mol. Spectrosc., 91, 348 (1982).

45. B. Starck, R. Mutter, C. Spreter, K. Kettemann, A. Boggs, M. Botskor and M. Jones, Bibliography of Microwave Spectroscopy 1945-1975, Zentralstelle für Atomkernenergie-Dokumentation, Physik Daten, Sektion für Strukturdokumentation der Universität Ulm, Ulm, West Germany, 1977.

46. F. L. Tobiason and R. H. Schwendeman, J. Chem. Phys., 40, 1014 (1964).

47. E. B. Wilson and Z. Smith, J. Mol. Spectrosc., 87, 569 (1981).

48. H. D. Rudolph, J. Mol. Spectrosc., 89, 430 (1981).

49. H. D. Rudolph, J. Mol. Spectrosc., 89, 460 (1981).

50. P. Nösberger, A. Bauder and Hs. H. Günthard, Chem. Phys., 1, 418 (1973).

51. P. Nösberger, A. Bauder and Hs. H. Günthard, Chem. Phys., 8, 245 (1975).

52. V. Typke, J. Mol. Spectrosc., 69, 173 (1978).

53. V. Typke, J. Mol. Spectrosc., 72, 293 (1978).

54. C. Lánczos, Am. Math. Monthly, 65, 665 (1958); C. Lánczos, Linear Differential Operators, Van Nostrand, London, 1964, Chapter 3.

55. G. H. Golub and W. Kahan, J. SIAM Numer. Anal., Series B, 2, 205 (1965).

56. G. H. Golub and C. Reinsch, Num. Math., 14, 403 (1970).

57. C. Hirose, Bull. Chem. Soc. Jpn., 47, 976, 1311 (1974).

58. I. M. Mills, Molecular Spectroscopy: Modern Research (K. Narahari Rao and C. W. Mathews, eds.), Academic Press, New York, 1972, Chapter 3.2.

59. Landolt-Börnstein, Numerical Data and Functional Relationships in Science and Technology, New Series, Group II, Atomic and Molecular Physics (K. H. Hellwege and A. M. Hellwege, eds.), Vol. 4 (1967); Vol. 6 (1974), and Vol. 7 (1976), Springer, Berlin.

60. J. G. Smith and J. K. G. Watson, J. Mol. Spectrosc., 69, 47 (1978).

61. V. W. Laurie, J. Chem. Phys., 28, 704 (1958).

62. M. Nakata, T. Fukuyama and K. Kuchitsu, J. Mol. Spectrosc., 83, 118 (1980).

63. M. Nakata, K. Kohata, T. Fukuyama and K. Kuchitsu, J. Mol. Spectrosc., 83, 105 (1980).

64. M. Nakata, M. Sugie, H. Takeo, C. Matsumura, T. Fukuyama and K. Kuchitsu, J. Mol. Spectrosc., 86, 241 (1981).

65. M. Sugie, H. Takeo and C. Matsumura, to be published.

66. Y. Morino, Y. Kikuchi, S. Saito and E. Hirota, J. Mol. Spectrosc., 13, 95 (1964).

67. S. Saito, J. Mol. Spectrosc., 30, 1 (1969).

68. T. Oka and Y. Morino, J. Mol. Spectrosc., 6, 472 (1961).

69. K. Kondo, H. Hirakawa, A. Miyahara ad K. Shimoda, J. Phys. Soc. Jpn., 15, 303 (1960).

70. K. Kuchitsu, Bull. Chem. Soc. Jpn., 44, 96 (1971).

71. T. Iijima, Bull. Chem. Soc. Jpn., 45, 3526 (1972).

72. T. Iijima and S. Tsuchiya, J. Mol. Spectrosc., 44, 88 (1972).

73. V. W. Laurie, Critical Evaluation of Chemical and Physical Structural Information, (D. R. Lide, Jr. and M. A. Paul, eds.), Chapter 2, National Academy of Sciences, Washington, 1974.

74. A. G. Robiette, J. Mol. Struct., 35, 81 (1976).

75. A. P. Cox, G. Duxbury, J. A. Hardy and Y. Kawashima, J. Chem. Soc. Faraday Disc. II, 76, 339 (1980).

76. G. Chidichimo, F. Lelj, P. L. Barili and C. A. Veracini, Chem. Phys. Lett., 55, 519 (1978).

77. W. J. Lafferty, D. R. Lide, and R. A. Toth, J. Chem. Phys., 43, 2063 (1965).

78. K. Kuchitsu, J. Chem. Phys., 49, 4456 (1968).

79. K. Kuchitsu, T. Fukuyama and Y. Morino, J. Mol. Struct., 4, 41 (1969).

80. E. A. Cherniak and C. C. Costain, J. Chem. Phys., 45, 104 (1966).

81. K. Kuchitsu and K. Oyanagi, Faraday Discuss. Chem. Soc., 62, 20 (1977).

82. L. S. Bartell, J. Chem. Phys., 38, 1827 (1963).

83. P. M. Morse, Phys. Rev., 34, 57 (1929).

84. K. Kuchitsu and L. S. Bartell, J. Chem. Phys., 36, 2460, 2470 (1962).

85. K. Kuchitsu and Y. Morino, Bull. Chem. Soc. Jpn., 38, 805, 814 (1965).

86. G. Herzberg, Spectra of Diatomic Molecules, Van Nostrand, Princeton, New Jersey, 1950, p. 100.

87. R. Stølevik, H. M. Seip and S. J. Cyvin, Chem. Phys. Lett., 15, 263 (1972).

88. J. L. Duncan, Mol. Phys., 28, 1177 (1974).

89. T. Oka, H. Hirakawa and K. Shimoda, J. Phys. Soc. Jpn., 15, 2265, 2274 (1960).

90. K. Takagi and T. Oka, J. Phys. Soc. Jpn., 18, 1174 (1963).

91. D. R. Johnson, F. R. Lovas and W. Kirchhoff, J. Phys. Chem. Ref. Data, 1, 1011 (1972).

92. F. Y. Chu, S. M. Freund, J. W. C. Johns and T. Oka, J. Mol. Spectrosc., 48, 328 (1973).

93. H. C. Allen and E. K. Plyler, J. Am. Chem. Soc., 80, 2673 (1958).

94. J. C. Dowling and B. P. Stoicheff, Can. J. Phys., 37, 703 (1959).

95. D. Van Lerberghe, I. J. Wright and J. L. Duncan, J. Mol. Spectrosc., 42, 251 (1972).

96. G. K. Speirs, J. L. Duncan and D. Van Lerberghe, J. Mol. Spectrosc., 51, 524 (1974).

97. J. L. Duncan and P. D. Mallinson, Chem. Phys. Lett., 23, 597 (1973).

98. J. L. Duncan, D. C. McKean and P. D. Mallinson, J. Mol. Spectrosc., 45, 221 (1973).

99. E. Hirota, Y. Endo, S. Saito, K. Yoshida, I. Yamaguchi and K. Machida, J. Mol. Spectrosc., 89, 223 (1981).

100. P. D. Mallinson and L. Nemes, J. Mol. Spectrosc., 59, 470 (1976).

101. A. P. Cox, L. F. Thomas and J. Sheridan, Spectrochim. Acta, 15A, 542 (1959).

102. J. W. C. Johns, J. M. R. Stone and G. Winnewisser, J. Mol. Spectrosc., 42, 523 (1972).

103. L. Nemes and M. Winnewisser, Z. Naturforsch., 31A, 272 (1976).

104. L. Nemes, J. Mol. Spectrosc., 72, 102 (1978).

105. K. Kuchitsu and F. Takabayashi, Collected Abstracts for the IXth Hungarian Diffraction Conference, Pécs, Hungary, 14-19 August, 1978, p. 25.

106. F. Hegelund, J. L. Duncan and D. C. McKean, J. Mol. Spectrosc., 65, 366 (1977).

107. R. J. Butcher and W. J. Jones, J. Raman Spectrosc., 1, 393 (1973).

108. F. Hegelund and H. B. Andersen, J. Raman Spectrosc., 3, 73 (1975).

109. A. G. Maki and R. A. Toth, J. Mol. Spectrosc., 17, 136 (1965).

110. E. Hirota and C. Matsumura, J. Chem. Phys., 59, 3038 (1973).

111. L. Nemes, J. L. Duncan and I. M. Mills, Spectrochim. Acta, 23A, 1803 (1967).

112. J. Demaison, A. Dubrulle, D. Boucher and J. Burie, J. Mol. Spectrosc., 76, 1 (1979).

113. J. L. Duncan, D. C. McKean, F. Tullini, G. D. Nivellini and J. Perez-Peña, J. Mol. Spectrosc., 69, 123 (1978).

114. J. L. Duncan, D. C. McKean and N. D. Michie, J. Mol. Struct., 21, 405 (1974).

115. K. Karakida, T. Fukuyama and K. Kuchitsu, Bull. Chem. Soc. Jpn., 47, 299 (1974).

116. L. Halonen and I. M. Mills, J. Mol. Spectrosc., 73, 494 (1978).

117. J. L. Duncan, D. C. McKean and A. J. Bruce, J. Mol. Spectrosc., 74, 361 (1978).

118. L. S. Bartell and H. K. Higginbotham, J. Chem. Phys., 42, 851 (1965).

119. W. J. Lafferty and E. K. Plyler, J. Res. Nat. Bur. Stand., A67, 225 (1963).

120. E. Hirota and C. Matsumura, J. Chem. Phys., 55, 981 (1971).

121. E. Hirota, K. Matsumura, M. Imachi, M. Fujio, Y. Tsuno and C. Matsumura, J. Chem. Phys., 66, 2660 (1977).

122. E. Hirota, Y. Endo, S. Saito and J. L. Duncan, J. Mol. Spectrosc., 89, 285 (1981).

123. R. W. Davies, A. G. Robiette, M. C. L. Gerry, E. Bjarnov and G. Winnewisser, J. Mol. Spectrosc., 81, 93 (1980).

124. G. H. Kwei and R. F. Curl, Jr., J. Chem. Phys., 32, 1592 (1960).

125. A. Almenningen, O. Bastiansen and T. Motzfeld, Acta Chem. Scand., 23, 2848 (1969).

126. J. Bellet, A. Deldalle, C. Samson, G. Steenbeckeliers and R. Wertheimer, J. Mol. Struct., 9, 65 (1971).

127. L. Fusina and I. M. Mills, J. Mol. Spectrosc., 86, 488 (1981).

128. K. Yamada, J. Mol. Spectrosc., 79, 323 (1980).

129. A. D. McLean, G. H. Loew and D. S. Berkowitz, J. Mol. Spectrosc., 72, 430 (1978).

130. R. L. Hilderbrandt and J. D. Wieser, J. Chem. Phys., 56, 1143 (1972).

131. L. Nygaard, private communication to Dr. V. Typke.

132. T. Oka, in Molecular Spectroscopy: Modern Research, Vol. 2, (K. Narahari Rao, ed.), Academic Press, New York, 1976, Chapter 5.

133. J. K. G. Watson, J. Mol. Spectrosc., 40, 536 (1971).

134. K. Fox, Phys. Rev. Lett., 27, 233 (1971).

135. I. Ozier, Phys. Rev. Lett., 27, 1329 (1971).

136. M. Mizushima and P. Venkateswarlu, J. Chem. Phys., 21, 705 (1953).

137. I. M. Mills, J. K. G. Watson and W. L. Smith, Mol. Phys., 16, 329 (1969).

138. I. M. Mills, Mol. Phys., 16, 345 (1969).

139. R. G. Brewer, Phys. Rev. Lett., 25, 1639 (1970).

140. R. G. Brewer, Science, 178, 247 (1972).

141. A. C. Luntz, Chem. Phys. Lett., 11, 186 (1971).

142. A. C. Luntz and R. G. Brewer, J. Chem. Phys., 54, 3641 (1971).

143. R. F. Curl, Jr., J. Mol. Spectrosc., 48, 165 (1973).

144. R. F. Curl, Jr., and T. Oka, J. Chem. Phys., 58, 4908 (1973).

145. R. F. Curl, Jr., T. Oka and D. S. Smith, J. Mol. Spectrosc., 46, 518 (1973).

146. M. Takami, K. Uehara and K. Shimoda, Jpn. J. Appl. Phys., 12, 924 (1973).

147. P. Venkateswarlu, Ind. J. Phys., 50, 100 (1976).

148. A. V. Jagannadham and P. Venkateswarlu, Ind. J. Phys., 50, 214 (1976).

149. I. Ozier and A. Rosenberg, J. Chem. Phys., 69, 5203 (1978).

150. A. Rosenberg and J. Susskind, Phys. Rev. Lett., 42, 1613 (1979).

151. J. T. Hougen, Can. J. Phys., 42, 1920 (1964).

152. M. Trefler and H. P. Gush, Phys. Rev. Lett., 20, 703 (1968).

153. S. M. Blinder, J. Chem. Phys., 32, 105, 582 (1960); 35, 974 (1961).

154. W. Kolos and L. Wolniewicz, J. Chem. Phys., 45, 944 (1966).

155. L. Wolniewicz and T. Kowalski, Chem. Phys. Lett., 18, 55 (1973).

156. P. R. Bunker, J. Mol. Spectrosc., 46, 119 (1973).

157. F. A. Gangemi, J. Chem. Phys., 39, 3490 (1963).

158. I. Ozier, W. Ho and G. Birnbaum, J. Chem. Phys., 51, 4873 (1969).

159. S. C. Wofsy, J. S. Muenter and W. Klemperer, J. Chem. Phys., 53, 4005 (1970).

160. Y. Endo, K. Yoshida, S. Saito and E. Hirota, J. Chem. Phys., 73, 3511 (1980).

161. G. Guelachvili, J. Mol. Spectrosc., 79, 72 (1980).

162. R. Paso, J. Kauppinen and R. Anttila, J. Mol. Spectrosc., 79, 236 (1980).

163. J. S. Muenter and V. W. Laurie, J. Am. Chem. Soc., 86, 3901 (1964).

164. E. Hirota and M. Imachi, Can. J. Phys., 53, 2023 (1975).

165. E. Hirota, S. Saito and Y. Endo, J. Chem. Phys., 71, 1183 (1979).

166. K. Matsumura, T. Tanaka, Y. Endo, S. Saito and E. Hirota, J. Phys., Chem., 84, 1793 (1980).

167. J. K. G. Watson, J. Mol. Spectrosc., 50, 281 (1974).

168. T. Oka, J. Mol. Spectrosc., 48, 503 (1973).

169. G. Herzberg, Molecular Spectra and Molecular Structure II, Infrared and Raman Spectra of Polyatomic Molecules, D. Van Nostrand, Toronto, 1949, pp. 404-406.

170. H. C. Allen, Jr., and P. C. Cross, Molecular Vib-rotors, Wiley, New York, 1963.

171. L. Nemes, in Vibrational Spectra and Structure, (J. R. Durig, ed.), Vol. 10, Elsevier, Amsterdam, 1981, Chapter 6.

172. T. L. Barnett and T. H. Edwards, J. Mol. Spectrosc., 20, 347, 352 (1966); 23, 302 (1967).

173. W. E. Blass and T. H. Edwards, J. Mol. Spectrosc., 24, 111, 116 (1967).

174. R. W. Peterson and T. H. Edwards, J. Mol. Spectrosc., 38, 1 (1971).

175. K. Sarka and K. Narahari Rao, J. Mol. Spectrosc., 46, 223 (1973).

176. A. G. Maki and R. M. Hexter, J. Chem. Phys., 53, 453 (1970).

177. H. Matsuura, T. Nakagawa and J. Overend, J. Chem. Phys., 53, 2540 (1970).

178. R. W. Lovejoy and W. B. Olson, J. Chem. Phys., 57, 2224 (1972).

179. W. B. Olson, J. Mol. Spectrosc., 43, 190 (1972).

180. G. Amat, Compt. Rend. Acad. Sci., 250, 1439 (1960).

181. J. T. Hougen, J. Chem. Phys., 37, 1433 (1962); 39, 358 (1963).

182. W. L. Barnes, J. Susskind, R. H. Hunt and E. K. Plyler, J. Chem. Phys., 56, 5160 (1972).

183. A. G. Maki, R. L. Sams and W. B. Olson, J. Chem. Phys., 58, 4502 (1973).

184. W. B. Olson, A. G. Maki and R. L. Sams, J. Mol. Spectrosc., 55, 252 (1975).

185. K. Sarka, D. Papoušek and K. Narahari Rao, J. Mol. Spectrosc., 37, 1 (1971).

186. W. B. Olson and R. W. Lovejoy, J. Mol. Spectrosc., 66, 314 (1977).

187. G. Graner, Mol. Phys., 31, 1833 (1976).

188. K. Sarka, Coll. Czech. Chem. Comm., 41, 2817 (1976).

189. H. M. Hanson, J. Mol. Spectrosc., 23, 287 (1967).

190. F. O. Shimizu, T. Shimizu and T. Oka, Symp. Mol. Struct. Spectrosc., Columbus, Ohio, Sept., 1969.

191. F. Y. Chu and T. Oka, J. Chem. Phys., 60, 4612 (1974).

192. D. A. Helms and W. Gordy, J. Mol. Spectrosc., 69, 473 (1978).

193. F. Y. Chu and T. Oka, J. Mol. Spectrosc., 48, 612 (1973).

194. D. A. Helms and W. Gordy, J. Mol. Spectrosc., 66, 206 (1977).

195. S. P. Belov, A. V. Burenin, L. I. Gershtein, A. F. Krupnov, V. N. Markov, A. V. Maslowsky and S. M. Shapin, J. Mol. Spectrosc., 86, 184 (1981).

196. D. Laughton, S. M. Freund and T. Oka, J. Mol. Spectrosc., 62, 263 (1976).

197. I. M. Mills, Mol. Phys., 8, 363 (1964).

198. B. P. Stoicheff, Can. J. Phys., 33, 811 (1955).

199. E. H. Richardson, S. Brodersen, L. Krause and H. L. Welsh, J. Mol. Spectrosc., 8, 406 (1962).

200. T. H. Edwards and S. Brodersen, J. Mol. Spectrosc., 54, 121 (1975).

201. T. H. Edwards and S. Brodersen, J. Mol. Spectrosc., 56, 376 (1975).

202. H. W. Kattenberg and S. Brodersen, J. Mol. Spectrosc., 59, 126 (1976).

203. J. C. Deroche, G. Graner, J. Bendtsen and S. Brodersen, J. Mol. Spectrosc., 62, 68 (1976).

204. F. Hegelund, J. Raman Spectrosc., 6, 42 (1977).

205. S. Brodersen, J. Mol. Spectrosc., 71, 312 (1978).

206. P. Jensen, S. Brodersen and G. Guelachvili, J. Mol. Spectrosc., 88, 378 (1981).

207. G. Graner, J. Mol. Spectrosc., 90, 394 (1981).

208. G. Tarrago and J. Dupre-Maquaire, J. Mol. Spectrosc., 96, 170 (1982).

209. J. L. Duncan, J. Mol. Struct., 6, 447 (1970).

210. J. L. Duncan and P. D. Mallinson, J. Mol. Spectrosc., 39, 471 (1971).

211. J. L. Duncan, D. C. McKean, P. D. Mallinson and R. D. McCulloch, J. Mol. Spectrosc., 46, 232 (1973).

212. D. C. McKean, J. L. Duncan and M. W. Mackenzie, J. Mol. Struct., 42, 77 (1977).

213. P. D. Mallinson, J. Mol. Spectrosc., 55, 94 (1975).

214. P. D. Mallinson, J. Mol. Spectrosc., 68, 68 (1977).

215. D. F. Eggers, Jr., J. Mol. Struct., 31, 367 (1976).

216. J. L. Duncan, J. Mol. Struct., 22, 225 (1974).

217. L. S. Bartell, S. Fitzwater and W. J. Hehre, J. Chem. Phys., 63, 3042 (1975).

218. L. Schäfer, H. L. Sellers, F. J. Lovas and R. H. Schwendeman, J. Am. Chem. Soc., 102, 6566 (1980).

219. R. C. Woods, R. J. Saykally, T. G. Anderson, T. A. Dixon and P. G. Szanto, J. Chem. Phys., 75, 4256 (1981).

220. C. Van Alsenoy, J. N. Scarsdale and L. Schäfer, J. Chem. Phys., 74, 6278 (1981).

221. P. G. Szanto, T. G. Anderson, R. J. Saykally, N. D. Piltch, T. A. Dixon and R. C. Woods, J. Chem. Phys., 75, 4261 (1981).

222. J. E. Boggs, M. von Karlowitz and S. von Karlowitz, J. Phys. Chem., 86, 157 (1982).

223. J. E. Boggs, F. Pang and P. Pulay, J. Comput. Chem., 3, 344 (1982).

224. J. E. Boggs, J. Mol. Struct., 97, 1 (1983).

225. L. Hedberg, K. Hedberg and J. E. Boggs, J. Am. Chem. Soc., to be published.

Chapter 5
APPLICATIONS OF DAVYDOV SPLITTING FOR STUDIES
OF CRYSTAL PROPERTIES

G. N. Zhizhin and A. F. Goncharov

Institute for Spectroscopy
USSR Academy of Science
Troitsk, Moscow, USSR

I. INTRODUCTION

On the Nature of Davydov Splitting in the
Vibrational Spectra of Crystals

The theoretical analysis of vibrational exciton states in molecular crystals as well as crystals with complex ions is very similar to that of electronic exciton states. Each nondegenerate energy level of a free molecule in a crystal splits into Z components, where Z is the number of molecules in the primitive unit cell of a crystal. From these Z components, only some number may be active in infrared absorption and Raman scattering spectra, because they must have proper symmetry with respect to the factor group of the space group of a crystal. The polarization properties of the corresponding vibrational modes are determined by the same selection rules.

Such splitting was analyzed for the first time by Davydov [1] and Agranovich [2] and is known in the literature as Davydov splitting (or correlation field splitting) which is unlike the Bethe one [3], i.e., splitting of the degenerate energy levels of atoms and molecules in crystals due to the symmetry of the crystalline field that removes the degeneracy.

For the qualitative study of Davydov splitting it is worthwhile to use the group theoretical analysis suggested by Hornig [4], and Winston and Halford [5]. This approach allows one to observe the transformation of an energy level with given symmetry of a molecule or a complex ion to a group of levels in a crystal as well as to derive the vibrational selection rules [6-8].

In order to have qualitatively in mind a picture of Davydov splitting we shall consider the simplest vibrational model of a molecular crystal [9] as illustrated in Fig. 1. This figure represents three one-dimensional spring models for diatomic molecules and their corresponding phonon dispersion curves, i.e., the wave vector $\bar{q}$ dependence on the frequency ν. Figure 1(a) shows the energy levels of an isolated diatomic molecule with atoms of mass m coupled by a spring of force constant K_o. The corresponding vibrational spectrum consists of two frequencies: $\nu = \nu_o = (2K_o/m)^{\frac{1}{2}}$, where ν_o is a frequency of the stretching mode; and $\nu = 0$, a frequency of the rigid-molecule free translation. If a one-dimensional crystal with one molecule in the unit cell is formed (Fig. 1(b)), the vibrational spectrum of one molecule transforms into two bands $\nu(\bar{q})$; the branches contained in the right half of the Brillouin zone ($0 < q < \frac{\pi}{a}$) are shown in Fig. 1(b). The stretching vibration ν_o of a free molecule gives rise to a flat optical branch in the frequency range from ν_o (at the zone center) to $\nu_o + \Delta\nu$ (at the zone boundary). The $\nu = 0$ translational mode transforms into an acoustic branch covering the low

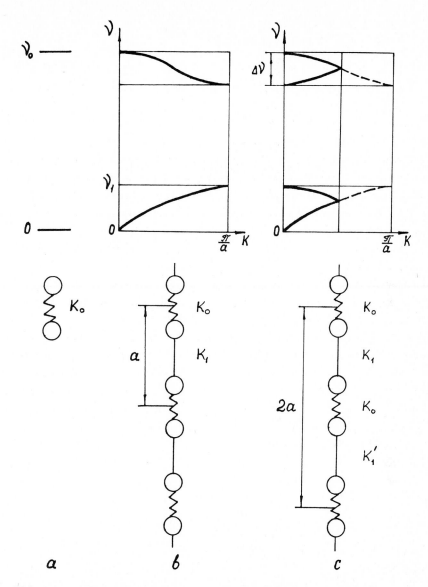

FIG. 1. Vibrational spectrum transformation (by column): diatomic molecule, linear chain, linear chain with the doubled lattice constant.

frequency range from 0 (at q = 0) to ν_1 (at q = $\frac{\pi}{a}$). If the intermolecular force constant K_1 is much smaller compared to the intramolecular constant K_o, then the ratios of the characteristic frequencies in this one-dimensional molecular lattice are given by

$$\frac{\nu_1}{\nu_0} = (\frac{K_1}{K_0})^{\frac{1}{2}} \tag{1}$$

$$\frac{\Delta\nu}{\nu_0} = \frac{1}{2} (\frac{K_1}{K_0}). \tag{2}$$

These relations illustrate the fact that the frequency range, in which phonon branches exist, depends upon the weakness of the intermolecular bonding as compared to the intramolecular one ($K_1 \ll K_0$). Let us thus consider $\beta = (K_1/K_0)$ as a parameter characterizing the value of dispersion of phonon branches in molecular crystals.

The next stage of complication of this one-dimensional molecular crystal model is introduced in Fig. 1(c) by doubling the unit-cell size to include two molecules in the unit cell. This effect may be achieved by alternating intermolecular force constants with the value K_1' slightly different from K_1. In this way phonon dispersion curves fold in two so that the second part of the curves is reflected from the vertical line $q = (\pi/2a)$ back to the zone center due to halving of the Brillouin zone. Thus, in the zone center (at $q = 0$) there appears a low frequency rigid-molecule mode (ν_1) and a Davydov splitting fine structure ($\Delta\nu_D$). In order to estimate the values of ν_1 and $\Delta\nu_D$ it is possible to use the relations (1) and (2). When the frequencies ν_1 and ν_0 are known from infrared or Raman spectra we should evaluate the Davydov splitting magnitude by the equation

$$\Delta\nu_D = \frac{1}{2} (\frac{\nu_1^2}{\nu_0}) , \tag{3}$$

which is a consequence of relations (1) and (2).

Usually an estimate of the magnitude of Davydov splitting in molecular crystals, depending on the properties of constituent molecules, is based on the assumption that the main contribution is determined by the electric dipole-dipole interaction, through which excitation exchange between neighboring molecules occurs. Thus, in the crystal with two molecules in the unit cell, the value of Davydov splitting is given by [2]:

$$|\Delta E_D(\bar{q})| = 2|\tilde{L}_{12}(\bar{q})|, \tag{4}$$

where $\tilde{L}_{12}(\bar{q}) = \tilde{L}_{21}(\bar{q})$ are the nondiagonal resonant interaction matrix elements. Generally we have:

$$L_{\alpha\beta}^{f}(\bar{q}) = \sum_{n}' M_{\bar{o}\beta,\bar{n}\alpha}^{f} \exp(i\bar{q}\bar{r}_n), \tag{5}$$

where

$$M_{\bar{o}\beta,\bar{n}\alpha}^{f} = \int \psi_{\bar{n}\alpha}^{*o} \psi_{\bar{o}\beta}^{*f} V_{\bar{n}\alpha,\bar{o}\beta} \psi_{\bar{o}\beta}^{o} \psi_{\bar{n}\alpha}^{f} d\tau \tag{6}$$

and the matrix elements characterizing the excitation exchange between the molecule $\bar{n}\alpha$ and the molecule $\bar{o}\beta$, $\bar{r}_n = \bar{n} + \bar{\rho}_\alpha - \bar{\rho}_\beta$, is the vector connecting the $\bar{o}\beta$ and $\bar{n}\alpha$ molecules.

In order to calculate these quantities, we should expand the interaction energy $V_{\bar{n}\alpha,\bar{o}\beta}$ of two molecules by including the Coulomb interaction between the electrons and nuclei of both molecules in a power series of reciprocal distances between the centers of gravity of molecules.

If the molecules are neutral, the first term of the series corresponds to the dipole-dipole interaction. In that approximation we have

$$M_{\bar{o}\beta,\bar{n}\alpha}^{f} =$$

$$\frac{d_\alpha^2}{|\bar{r}_n|} \{\cos \theta_{\bar{o}\beta}^{x} \cos \theta_{\bar{n}\alpha}^{x} + \cos \theta_{\bar{o}\beta}^{y} \cos \theta_{\bar{n}\alpha}^{y} - 2\cos \theta_{\bar{o}\beta}^{z} \cos \theta_{\bar{n}\alpha}^{z}\}, \tag{7}$$

where $\theta_{\bar{o}\beta}^{x} \ldots \theta_{\bar{n}\alpha}^{z}$ are the angles between the axes x, y, z and the directions of the molecular dipole moments.

To explain the splitting of electronic exciton states it is usually enough to take into consideration the dipole-dipole interaction. But in the vibrational spectra of molecular crystals the situation is quite different. The reason is that the oscillator strength of the electronic transitions is about 5 to 6 orders higher than that of the vibrational transitions [10]. In the latter case the multipoles of higher order should play an exceptional role [11].

II.   SYMMETRY PROPERTIES OF DAVYDOV COMPONENTS

The details of the vibrational spectra of liquids or gases essentially depend on the symmetry of the constituent molecules. In this case 3N - 6 (N is the number of atoms in the molecule) vibrational modes are classified by the irreducible representations of the molecular symmetry group, and these molecular vibrational modes will be called "fundamentals". Intermolecular interactions do not influence significantly the vibrational spectra and lead only to their distortion.

When a molecular crystal grows, interactions between oriented molecules and their mutual arrangement become important. Though the molecular symmetry still underlies the vibrational spectral analysis, one must find another point of view in order to account for a number of important changes which appear due to the lattice symmetry properties and translational periodicity. First, $3Z - 3$ translational and $3Z$ librational zone-center optical modes arise which provide direct information about intermolecular force constants. Second, there is a great change in the shape and fine structure nature of the intermolecular bands. Such transformations in the spectra, which include Davydov and Bethe splitting and the appearance of "forbidden" fundamentals, need a complete analysis.

As the first step to solve this problem an "oriented gas model" [12] was used. A crystal is considered as an ensemble of definitively oriented molecules which do not interact with one another in any way. This model allows one to predict the dichroic ratio for different vibrational transitions, but the best result may be achieved for the crystals with one molecule per unit cell.

In order to account for such effects as vibrational mode frequency shifts, splitting of degenerate modes (Bethe splitting), and the appearance of "forbidden" vibrational transitions, one must take into consideration the static crystalline field [13] which is a measure of the influence of the surrounding lattice on the molecule. Static field perturbation depends on the local symmetry of the molecule position, i.e., a site symmetry. The site group is a subgroup of the molecular group of symmetry. Thus, a mode classification should be done by the irreducible representations of the site group. In going from molecular symmetry species into site symmetry species, it is convenient to use the correlation tables [14].

The last step of this procedure includes resonant interaction between identical fundamentals of different molecules, constituting the primitive unit cell. As a result, for the crystal with Z molecules per unit cell, each non-degenerate fundamental splits into Z components, i.e., Davydov or correlation field splitting [1,4,5]. Acoustic modes give rise to the translational lattice modes. Because of the translational invariance of the vibrational transition moments under consideration, Davydov component classification should be carried out by the irreducible representations of the crystal factor group, which may be treated as the symmetry of the unit cell.

That last one is not exactly a point group although the factor groups are isomorphous to the thirty-two point groups, which may occur in crystallography. Thus, in order to take into account resonance interaction between molecules one should ascertain the correlation between the symmetry species of the site group and crystal factor group.

The results obtained by this total correlation procedure coincide with those derived by the general Bhagavantam method [15]. But the latter, derived from the total space distribution of atoms in the unit cell, does not allow one to trace the transformation of the molecular symmetry species to the crystalline ones. At the same time, using the correlation method, one would obtain not only the vibrational energy level symmetry transformations, but also the nature of expected resonant splitting in a crystal, i.e., the number of components and their symmetry with respect to the factor group.

The number of Davydov components n_ℓ^α of the certain factor group symmetry α, corresponding to the site group symmetry species ℓ, is determined by the formula

$$n_\ell^\alpha = \frac{Z}{F} \sum_R \chi_\alpha(R)\ \chi_\ell(R), \tag{8}$$

where Z is the number of molecules in the primitive unit cell; F is the order of the factor group; R is the respective symmetry operation, which is the same for the factor group and the site group; $\chi_\alpha(R)$ is the character of the irreducible representation α for the symmetry operation R of the factor group; $\chi_\ell(R)$ is the same but for the site group.

The spectral activity of each component of a multiplet, corresponding to a single band in a liquid or gas, is determined by its symmetry with respect to the factor group. Corresponding transition moments are the total vibrational transition moments of the unit cell rather than the molecular ones.

III. DAVYDOV SPLITTING IN THE VIBRATIONAL SPECTRA
OF MOLECULAR CRYSTALS

Studies of the shape of a multiplet in crystalline solutions of different concentrations [16-26] serve as reliable evidence for the Davydov splitting of infrared and Raman bands. As the concentration of the principal substance in the crystalline solution decreases, the splitting must diminish, and practically disappears, at concentrations on the order of a few per cent. As a solvent one uses either an isotropic analog of the principal molecule [17-19,23-25] or a molecule of closely related composition that forms iso-morphous crystals [30]. Having established the resonant nature of the splitting, one can pass on to the solution of a more complicated problem, i.e., to the study of the concentration dependence of the magnitude of the Davydov

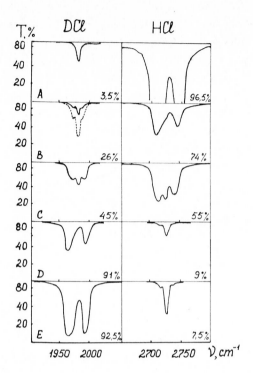

FIG. 2.   Infrared absorption spectra of $H_{1-x}D_xCl$ mixed crystals for different concentrations at 66K. Used by permission [17].

splitting in the simplest disordered media, which is the class to which isotopic solid solutions belong.

Such investigations have first been performed by Hornig and Hiebert [17] for the vibrational spectra of $H_{1-x}D_xBr$ and $H_{1-x}D_xCl$ mixed crystal systems. Figure 2 shows the change of the multiplet shape in the frequency region of the stretching mode of the $H_{1-x}D_xCl$ system at different concentrations. The single bands at small concentrations are transformed into doublets in the spectra of the pure crystals.

When the impurity molecules are distributed nearly uniformly in the host crystal, one would expect to observe a smooth decrease of the splitting as the impurity concentration is increased. However, the third component of splitting has been observed in the spectra at middle concentrations. This strange behavior may be accounted for by the existence of clusters, i.e., areas of the crystal with identical molecules. The experiment [17] proceeded from accidental distribution of molecules HX and DX between possible sites in the crystal, assuming hydrogen bonding between them, through dipole-dipole interaction (dipoles are located at the protons). In this way zigzag chains of the molecules are formed.

As a result of calculations it has been found that at middle concentrations the additional components may actually be observed because of the spectrum shape dependence of a chain length and length distribution of chains at the given concentration.

Mikhailov et al. [27] have observed the analogous concentration dependence of the shape of a multiplet corresponding to the O-H bond-stretching band in the spectra of the crystalline alcohols. The authors have suggested the method of construction of the ν_{OH} band envelope for different concentrations of deuterated molecules. The deuterated molecules lead to the effective subdivision of chains with the hydrogen bonds into shorter ones (it was supposed that the deuterated molecules are distributed randomly). The band shape of crystalline decanole calculated for different concentrations of the isotope impurity was in a good agreement with the results of the experiment [28]. The most characteristic feature of the spectra under consideration is the appearance of the third component in the middle of the two main components at the 30% concentration of deuterated molecules and then their transformation into the dominant one at the 70% concentration of deuterated molecules. These peculiarities are analogous to those in the spectra of crystalline solutions of DCl with HCl as in the previous case (Fig. 2).

Isotopic mixed crystals $CH_{3(1-x)}D_{3x}Cl$ are also very interesting compounds to study Davydov splitting. The crystal space group is C_{2v}^{12} (Cmc2$_1$); there are two molecules per primitive unit cell, both on sites C_s. The molecular symmetry group is C_{3v}. Thus one may expect that, in the vibrational spectra of the crystal, the A_1 fundamentals should split into two Davydov components. However, Jacox and Hexter [20,29] have observed three components corresponding to the $\nu_3(A_1)$ fundamental (Fig. 3). One of them they have assigned to the ^{37}Cl isotope effect. In mixed crystal spectra there is also a fourth additional component. For CH_3Cl-like modes, this extra band appears in the form of a shoulder at the 75% concentration of CH_3Cl and then splits out into the separate band at smaller concentrations. For CD_3Cl-like modes, the analogous band already exists as a shoulder in the pure CD_3Cl. The CH_3Cl-like Davydov components vanish at the 25% concentration while the CD_3Cl-like ones exist up to the 1% concentration. One must take into account that in the argon matrix (200:1 - Ar:$CH_3^{35}Cl$) at 4K only one band exists at 722 cm^{-1}. Probably, as in the previous case for the explanation of this complex picture, cluster effects should be taken into consideration.

Instead of using isotopic mixed crystal systems one may also study solid solutions in which the solvent is a closely related molecule. For the resonant splitting elucidation this method proves to be just as useful as the isotopic solution.

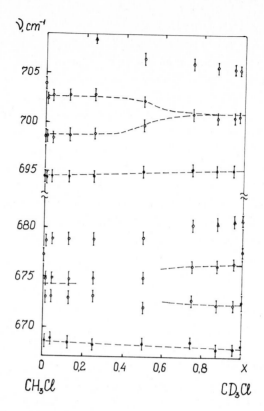

FIG. 3. Concentration dependence of maximum positions of Davydov components in the infrared absorption spectra of $CH_{3(1-x)}D_{3x}Cl_3$ mixed crystals: o - ^{35}Cl; ● - ^{37}Cl, Δ - shoulder. Used by permission [20,29].

Lisitsa and Kharchenko have performed the experiment [30] with chloroform which was dissolved in bromoform and carbon tetrachloride. It has been found that, at the 0.37 mole/liter concentration of chloroform, Davydov splitting on the v_4 mode of chloroform was not observed in the $CHCl_3$ + $CHBr_3$ system, whereas it was observed in the $CHCl_3$ + CS_2 and $CHCl_3$ + CH_2Cl_2 systems. The authors have concluded that the limit concentration at which Davydov splitting still exists is determined by the molecular size relation as well as the shape and the symmetry of molecules, which constitute the system under consideration.

Most experimental works [16-26] devoted to the study of the nature of a multiplet were performed without examination of polarization properties, since it is often impossible to grow single crystals of suitable sizes. Sometimes the splitting values reach 10-15 cm^{-1}, but for organic compounds they are usually found to be smaller than 3 cm^{-1}. In this case it is difficult to observe this vibrational band splitting because of the broad bandwidth of the components even at sufficiently low temperatures.

It is well known that the Davydov components usually have different polarization properties (this is not true only when the site group coincides with the factor group). This important property allows one to observe a splitting of rather small value (smaller than the own bandwidth) since in this case the relative shift of the spectra with different polarization is measured. The accuracy of these measurements is determined by the spectrometer reproducibility ($\pm$ 0.04 cm^{-1} in the best commercial models) and by the uncertainty in the determination of the band maximum position which is usually limited by its own shape only.

For the first time such dichroic properties in the infrared spectra of crystalline anthracene were observed by Fialkovskaya [31,32]. It has been found that the doublet observed at 742 cm^{-1} and 728 cm^{-1} has components with mutually perpendicular polarizations. In this way it has been shown for the first time that the conception of Davydov splitting is correct for the vibrational spectra of molecular crystals.

Since then there have been numerous studies of polarization properties of Davydov components [31-44]. In one of them [44] the resonant nature of the splitting was proved by the measurements of polarization properties as well as by the method of solid solutions. It is particularly important in the case of complex molecules when the multiplet may be constituted by non-resonant components due to accidental near degeneracy of their frequencies.

In Ref. [44] the spectra of 9,10-anthraquinone and its deuterated analog 9,10-anthraquinone-d_8 were studied. These crystals belong to the space group C_{2h}^2 with two molecules per unit cell. Davydov components A_u and B_u must have mutually perpendicular polarizations. Actually, Davydov pairs were observed in the spectra (four doublets for 9,10-anthraquinone and three for its deuterated analog). These doublets transformed into single bands at the 10% concentration of mutual solid solutions.

In order to interpret properly the vibrational spectra of molecular crystals it is very fruitful to use infrared and Raman spectroscopic methods simultaneously. Before lasers were discovered it was very difficult to obtain the Raman spectra of solid specimens at low temperatures with sufficiently high resolution. Since the laser has been discovered this problem became ordinary. Therefore, now there are many experimental works where details of the vibrational spectra of molecular crystals have been studied by the Raman spectroscopic method [45-50]. One of them is devoted to crystalline naphthalene [50]. It has been shown that the larger magnitude of splitting corresponds to the modes whose frequencies coincide or nearly coincide with the frequencies of strong polar optical modes. The conclusion may be drawn that the main part of Davydov splitting should be explained by the dipole-

dipole interaction between the molecules. In a more detailed way such problems will be discussed in the following sections.

IV. CONCENTRATION DEPENDENCE OF DAVYDOV SPLITTING MAGNITUDE IN THE ELECTRONIC EXCITON SPECTRA OF MOLECULAR CRYSTALS

The method of solid solutions (see Sec. III) may also be used for the investigations of Davydov splitting in the electronic exciton spectra. For the only two systems of mixed crystals, naphthalene+naphthalene-d_8 [51] and benzene+benzene-d_6 [52], the concentration dependence of the Davydov splitting magnitude of molecular electronic energy levels has been studied. Because of the almost identical physical and chemical properties of principal molecular crystals with their deuterated analogs, one may study perfect mixed crystals in the concentration range from 0 up to 100%.

For the naphthalene+naphthalene-d_8 mixed crystals in the spectral region of the pure electronic transition it has been found that each component of the alloy gives rise to the single nonpolarized band at a small concentration and then the doublet of strongly polarized bands at the concentration higher than 10%. The distances between the components of the doublets (the magnitude of splitting) increase with the increase of concentration of the corresponding substance.

Because of a small value of the isotopic shift, one may observe the effect of coupling between electronic energy levels, which was first considered by Rashba [53]. According to Rashba [53], if the isotopic (impurity) band and the corresponding band of the host crystal are placed at a distance shorter than the width of the excitonic energy gap, then the impurity absorption does not follow the oriented gas model and does not obey the Beer absorption law. This effect has been studied on six mixed crystals of different isotopic molecular constitutions [54]. Therefore, it was possible to vary the relative positions of the impurity and the host excitonic bands.

It has been observed that in the region of pure electronic transition a strong polarization ratio deviation from that predicted by the oriented gas model exists, and is dependent on the proximity of the impurity energy level to the exciton gap energy. For the nearest impurity band this value was 500 times as much. Simultaneously one could observe a great increase of the impurity band intensity, which becomes comparable with that of the host crystal bands [54].

The theoretical analysis of energy levels of such a complex system as disordered naphthalene+naphthalene-d_8 mixed crystals is very complicated. For the solution of this problem some approach is needed.

The first theoretical analysis of the concentration dependence of the Davydov splitting in crystals described above has been performed by Broude and Rashba [55]. They have assumed the same excitation amplitudes for all molecules of the same chemical composition occupying equivalent sites in the unit cell.

Using this approach, the simple relations for the energy absorption band positions and their intensity have been found. In this way the concentration dependence curves have been obtained which are in satisfactory agreement with the experiment. In this approximation the following relation is valid for the energy level $E_{\lambda\rho}$

$$\frac{1}{\varepsilon_\rho} = \sum_j \frac{C_j}{E_{\lambda\rho} - \varepsilon_j}, \tag{9}$$

where ρ is the polarization of radiation; λ characterizes the number of transitions corresponding to different types of molecules; j is the type of molecule; C_j is the concentration of the molecules which belong to j-th type; ε_j is the term position of the j-th isolated molecule in the crystalline alloy; and ε_ρ is the value which is inversely proportional to the interaction matrix elements and related to the excitation amplitudes of the different types of molecules.

Figure 4 shows the band positions of exciton multiplets in the spectrum of a two-component mixed crystal, where ε_1^{CR} and ε_2^{CR} are the unsplit term positions in pure crystals of the first and second types, respectively; ε_1^{b2} is the term position of the molecules of the first type which are a small impurity in the host crystal of the second one; ε_2^{b1} is the same for a small impurity of the second type in the host crystal of the first type; ε_1^o and ε_2^o are the term positions of noninteracting molecules of the first and second types (pairs); ε_I and ε_{II} are the distances between the exciton band and unsplit term in the pure crystal spectrum for the radiation polarizations ρ = I and ρ = II; E_{1II}, E_{1III}, E_{2I}, and E_{2II} are the absorption bands for two radiation polarizations ρ = I and ρ = II and two components of the alloy λ = 1 and λ = 2.

The theory of Broude and Rashba [55] does not allow one to analyze the absorption band shape and cannot be applied to the region of small concentrations, where the main assumption of the theory about the excitation amplitude equality for all molecules with the same chemical composition is no longer correct.

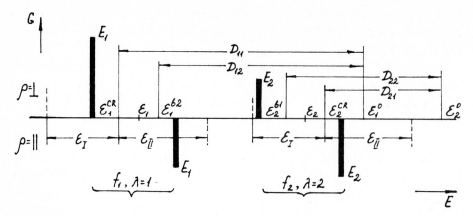

FIG. 4. The scheme of energy level positions in the spectrum of a two-component mixed crystal. Used by permission [53].

A more general consideration suitable for all concentration regions (0-100%) has been proposed in terms of Green's functions [56-60]. On the basis of the developed formalism [61], Onodera and Toyozawa [60] have found an interpolation expression for the exciton Green's function in the binary solid solution. This relation has permitted one to trace the development of the impurity band at an arbitrary concentration, and an arbitrary parameter of the theory $\frac{\Delta}{M}$, where Δ is the excitation energy difference between the main and impurity molecules, while M is the exciton gap energy bandwidth of the pure crystal. At the same time, the authors have restricted their consideration to the case of a crystal with one molecule per unit cell. Later Dubovskii and Konobeev [56,57] adopted this approach for the case of two molecules per unit cell. The interpolation expression for the exciton Green's function, which is able to describe the light absorption in mixed crystals at an arbitrary Δ/Δ_D parameter (Δ_D is the Davydov splitting in a pure crystal), was found.

The Hamiltonian describing Frenkel excitons in solid isotopic solution has the form

$$\hat{H} = \sum_{\bar{n}\alpha} \Delta_{\bar{n}\alpha} \hat{B}^+_{\bar{n}\alpha} \hat{B}_{\bar{n}\alpha} + \sum_{\bar{n}\alpha \neq \bar{m}\beta} V_{\bar{n}\alpha,\bar{m}\beta} \hat{B}^+_{\bar{n}\alpha} \hat{B}_{\bar{m}\beta}, \qquad (10)$$

where $\hat{B}^+_{\bar{n}\alpha}$ and $\hat{B}_{\bar{n}\alpha}$ are creation and annihilation Bose operators of the molecular excited state at the lattice site $\bar{n}\alpha$ ($\bar{n}$ is the primitive translation vector; α is the number of the molecule in the unit cell $\alpha = 1,2$). This expression does not take into account the molecular wavefunction coupling and is based on the framework of the two level scheme for an isolated molecule. The molecular excitation energy $\Delta_{\bar{n}\alpha}$ has either the value Δ_A or Δ_B depending on the type of molecule which occupies the lattice site $\bar{n}\alpha$. It is supposed that the molecules of different types are randomly distributed among the lattice sites and the resonant interaction matrix elements $V_{\bar{n}\alpha,\bar{m}\beta}$ do not depend on the isotope composition of the molecules occupying the lattice sites $\bar{n}\alpha$ and $\bar{m}\beta$.

The Hamiltonian $\hat{H}$ is divided into two parts $\hat{H}_0$ and $\hat{H}_1$ so that the averaging of $\hat{H}_1$ over the spatial configurations turns $\hat{H}_1$ to zero

$$\hat{H}_0 = \sum_{\bar{n}\alpha} (C_A \Delta_A + C_B \Delta_B) \, \hat{B}^+_{\bar{n}\alpha} \, \hat{B}_{\bar{n}\alpha} + \sum_{\bar{n}\alpha \neq \bar{m}\beta} V_{\bar{n}\alpha,\bar{m}\beta} \, \hat{B}^+_{\bar{n}\alpha} \, \hat{B}_{\bar{m}\beta} \tag{11}$$

$$\hat{H}_1 = \sum_{\bar{n}\alpha} (\Delta_{\bar{n}\alpha} - C_A \Delta_A - C_B \Delta_B) \, \hat{B}^+_{\bar{n}\alpha} \, \hat{B}_{\bar{n}\alpha}, \tag{12}$$

where C_A and $C_B = 1 - C_A$ are the concentrations of molecules of types A and B.

After several transformations with Bose operators, the Hamiltonian $\hat{H}_0$ may be diagonalized, while the eigenvalues $E_{\mu q}$ ($\mu = 1,2$) may be found as the roots of the secular equation for the system

$$(E - \bar{\Delta})u^\alpha = \sum_{\beta} V_{\alpha\beta}(\bar{q})u^\beta, \tag{13}$$

where

$$\bar{\Delta} = C_A \Delta_A + C_B \Delta_B, \tag{14}$$

$$V_{\alpha\beta}(\bar{q}) = \sum_{\bar{m}} V_{\bar{n}\alpha,\bar{m}\beta} \, e^{i\bar{q}(\bar{n} - \bar{m})}. \tag{15}$$

An explicit expression for the eigenvalues is given by

$$E_{(1,2)\bar{q}} \equiv \bar{\Delta} + \varepsilon_{(1,2)\bar{q}} = \bar{\Delta} + \frac{V_{11}(\bar{q}) + V_{22}(\bar{q})}{2} \pm$$

$$\sqrt{[\frac{V_{11}(\bar{q}) - V_{22}(\bar{q})}{2}]^2 + |V_{12}(\bar{q})|^2} \, . \tag{16}$$

The absorption band shape of the crystal is described by the expression [2]:

$$\chi_j(E) = \chi_0 \sum_{\alpha,\beta} (\bar{P}_\alpha\, e_{\bar{q}j})\, (\bar{P}_\beta^*\, e_{\bar{q}j})\, \text{Im} < G_{\bar{q}}^{\alpha\beta}\, (E)>_{Av}, \tag{17}$$

where χ_0 is the approximately constant value in the spectral region under consideration; $\bar{P}_\alpha$ ($\alpha = 1,2$) are the electric transition dipole moment vectors of the molecule α which are assumed to be the same for the molecules of types A and B; $e_{\bar{q}j}$ is the polarization vector of radiation with the wave vector $\bar{q}$. The brackets $< >_{Av}$ mean the Green's function averaging over the spatial configurations of impurity molecules in the crystal solution.

The Green's function is introduced in the form:

$$G_{\bar{q}}^{\alpha\beta}(E) = <0|\hat{B}_{\bar{q}\alpha} \frac{1}{E - \hat{H}} \hat{B}_{\bar{q}\beta}^+|0>, \tag{18}$$

where the $|0>$ function describes the ground state of the crystal.

The space averaging procedure was described in detail in the work of Dubovskii [57]. If we have $\Delta \gg V_{\alpha\beta}(\bar{q}) \sim \Delta_D$, then the averaged Green's function may be represented in the form:

$$<G_{\bar{q}}^{\alpha\beta}(E)>_{Av} = \sum_\mu \frac{u_{\mu\bar{q}}^\alpha\, u_{\mu\bar{q}}^{*\beta}}{(\frac{C_A}{E - \Delta_A} - \frac{C_B}{E - \Delta_B})^{-1} - \varepsilon_{\mu\bar{q}}}. \tag{19}$$

The poles of such an approximate Green's function coincide with the exciton energy values obtained by Broude and Rashba [55]. Thus, the last result is the particular case of the solution of the problem examined in Refs. [56] and [57].

The averaged Green's function, Eq. (19), allows one to analyze the concentration dependence of the spectrum shape, although for the precise calculation of these dependences the function of state density $\rho(E)$ is needed.

Assuming that for naphthalene the state density function has the form

$$\rho(E) = \begin{cases} \dfrac{8}{\pi\, \Delta_D^2}\, \sqrt{(\dfrac{\Delta_D}{2})^2 - (E - \Delta_A)^2} & \text{if} \quad |E - \Delta_A| < \dfrac{\Delta_D}{2} \\[4mm] 0 & \text{if} \quad |E - \Delta_A| \geq \dfrac{\Delta_D}{2} \end{cases} \tag{20}$$

(which is very close to the experimental density function $\rho(E)$), Dubovskii

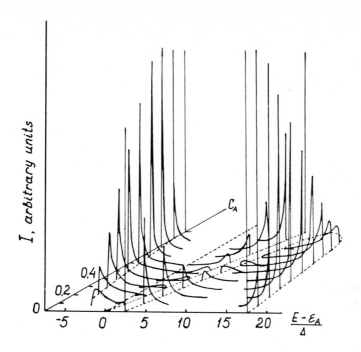

FIG. 5. Positions and shapes of absorption peaks due to the electronic transitions in mixed isotopic naphthalene+naphthalene-d_8 mixed crystals as a function of concentration. Used by permission [56].

and Konobeev calculated the concentration dependence of the Davydov splitting magnitude for the naphthalene-deuteronaphthalene mixed crystals (Fig. 5).

In Fig. 6 (taken from Ref. [62]), experimental exciton absorption band frequencies are presented and compared with the approximate Broude and Rashba theory results [55] and Green's function calculations [56]. Figure 6 also shows a calculated curve [56], corresponding to band centers of gravity, which is in better agreement with experimental results (also obtained from band centers of gravity) than the previous ones calculated for the peak absorption positions. The discrepancy between these theoretical curves is due to the absorption band asymmetry.

It has been found in the later measurements of the pure electronic transitions in mixed naphthalene+deuteronaphthalene crystals that, at the 4.2K temperature, the initial Davydov multiplet shows the fine structure which was unresolved before [64]. The electronic exciton absorption spectrum of the alloy is richer as compared with that of the pure crystal. Instead of the doublet predicted from the previous discussion one may observe at least six bands corresponding to the naphthalene impurity.

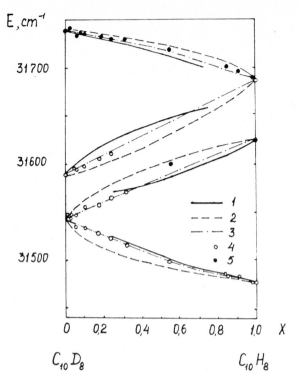

FIG. 6. Concentration dependence of absorption band positions in the electronic spectrum; 1,2 - maximum positions of absorption bands from Refs. [55,56]; 3 - curve of band centers of gravity calculated from Ref. [63]; 4 - experimental points for band centers of gravity; 5 - estimated positions of centers of gravity for which experimental data are absent.

When the impurity concentration varies, one may observe an appearance or disappearance of some of these bands and a variation of their relative and absolute intensities. The frequencies of these bands do not depend on the impurity content.

When the naphthalene impurity content rises, then all of the bands broaden, and the accompanying disappearance of the fine structure is observed. Broude and Leiderman [64] explained the fine structure by the existence in the crystal of not only single impurity molecules, but also of pairs or more complex aggregates of translationally nonequivalent molecules (clusters) coupled through the resonant intermolecular interaction. Thus, the naphthalene concentration increase leads to an increase of resonant pairs and then to a decrease of their relative number and an increased number of more complex aggregates.

In Ref. [62] the calculation of the number of single molecules, resonant pairs and other clusters for naphthalene concentrations of 2.5%, 6% and 10%

have been performed in the nearest-neighbor approximation.

The relative intensity of impurity bands should be proportional to the number of molecules in the cluster. The comparison with the results of the experiment confirms the reality of the cluster model.

In conclusion one should note, as shown by Agranovich [65], that the concentration effects discussed above may be studied and recognized in the framework of a simple extension of the Lorentz theory for anisotropic media.

V.   CONCENTRATION DEPENDENCE OF DAVYDOV SPLITTING IN THE INFRARED SPECTRA OF CRYSTALLINE NAPHTHALENE AND THIOPHTHENE

As far as the investigations of the features in the vibrational spectra of molecular crystals are concerned, one must properly choose objects in order to avoid some experimental difficulties. Among the most important ones are rigorous selection rules which often lead to the impossibility to use data from the same experimental method (Raman or infrared) in order to investigate a Davydov multiplet, small values of splitting, and sometimes large intensity differences between the components of the Davydov multiplet. The most suitable objects for this purpose are naphthalene and thiophthene (thieno-[3,2-b]-thiophene) [66-68].

There is one significant circumstance that differentiates the vibrational spectra of crystalline isotopic solid solutions from electronic ones. In the electronic spectra of naphthalene-d_8 the isotopic shift equals 115 cm^{-1}, i.e., about 0.3% of the 0-0 electronic transition frequency. That is why Broude and Rashba [55] interpreted the concentration experiment of the Davydov splitting as if the resonant interaction between molecules does not depend on the isotope molecular constitution. From the same point of view one may account for the strong polarization ratio effects as due to the coupling of electronic states. In all of these studies where such effects were examined, the ratio of the isotopic shift magnitude to the electronic transition frequency does not exceed 1%, and it is impossible to increase it.

In the case of vibrational spectra of molecular crystals the situation is completely different. An isotope shift leads to a set of frequency shifts depending on the vibrational eigenvector. The largest shifts, approximately 800 cm^{-1}, correspond to the C-H stretching vibrations when the modes from the 3000 cm^{-1} region shift into the 2200 cm^{-1} region. A smaller shift corresponds to the skeleton bending vibrations where the heavy atoms play the

main role. Therefore, in the vibrational spectra of isotope analogs one may
find the frequencies of corresponding modes which differ from 0.5% to 30%.
In this way the hypothesis about an isotope molecular constitution independent
of resonant interaction is, probably, not correct.

Note that the crystalline isotope solid solutions are very attractive
objects because of the large isotope shift variations and limited (~ 20 cm^{-1})
Davydov splitting values. Thus, it is possible to observe effects of shallow
and deep "traps" in the spectrum of the same crystal.

Mostly the concentration dependence of the Davydov splitting magnitude
was studied on polycrystalline samples without the use of polarized radiation.
The first such study of single crystals in polarized light was performed on
monocrystalline solid solutions of naphthalene+deuteronaphthalene [67]. The
structure of a naphthalene crystal is well known [69]. It forms a monoclinic
crystal with the C_{2h}^5 space group. The crystal unit cell contains two
molecules. The site symmetry of the molecule is C_i. The vibrational spectra
of naphthalene [70-71] and deuteronaphthalene [72] have been studied and
interpreted in detail.

The attenuated total reflection spectrum [73] of naphthalene single crystals from different cuts was obtained by Yamada and Susuki [74]. It has
been found that the splitting of the 480 cm^{-1} fundamental is approximately 10
cm^{-1}. This splitting is observed as a shift between the orthogonal polarization spectra. It is this mode (480 cm^{-1} in naphthalene which corresponds to
410 cm^{-1} in deuteronaphthalene) which was used by Yamada and Susuki [74].
The choice of these bands was also made because of the absence of other ones
in the same frequency range.

The Davydov splitting polarized infrared absorption spectra of nine
single crystals at different concentrations of naphthalene in deuteronaphthalene are shown in Fig. 7. For each concentration of the deuterated substance, the value of the Davydov splitting is determined as the distance
between peak positions for two different polarizations (solid and dotted lines).
Peak positions (Fig. 8) were measured by the chord method. By taking into
account the spectrometer reproducibility ($\lesssim 0.25$ cm^{-1}) and the accuracy of
the frequency measurements ($\lesssim 1$ cm^{-1}), one may conclude that the accuracy
of this procedure is determined by the band shape only.

The monocrystalline samples for each concentration were grown 3 to 4
times and for each time the infrared absorption was measured. Thus, the
points shown in Fig. 8 were averaged over several samples and several
measurements. Besides the splitting characteristics, the measurements of
band halfwidths and their integral intensities were performed. The peak
positions are practically within the limits shown in Fig. 8 which restrict the
confidence interval.

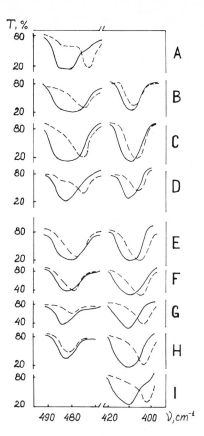

FIG. 7. Concentration dependence of the doublet shape (two polarizations - ‖ and ⊥) in the infrared absorption spectrum of naphthalene and
deuteronaphthalene mixed crystals (A - 100-0, B - 80-20, C - 70-30, D -
60-40, E - 50-50, F - 37-63, G - 23-77, H - 18-82, I - 0-100%).

As seen in Fig. 8, the high frequency components (which have identical
polarization - see the solid line in Fig. 7) of naphthalene as well as of
deuteronaphthalene remain practically at the same place, while the low frequency components shift linearly to lower frequency with the decreasing
corresponding substance content. At the 10% concentrations the resonance
splitting is almost unresolved and is found to be within the limits of measurement accuracy. The resonance splitting is a maximum for the pure crystals:
naphthalene - 10 cm^{-1}, deuteronaphthalene - 8 cm^{-1}. This difference may
probably be due to the oscillator strength difference between the naphthalene
crystal and its deuteroanalog. Therefore, the splitting magnitude sum
changes linearly with concentration, unlike the situation in the electronic
spectra where we have an almost constant sum [54].

The situation described may be examined as an example of the so-called

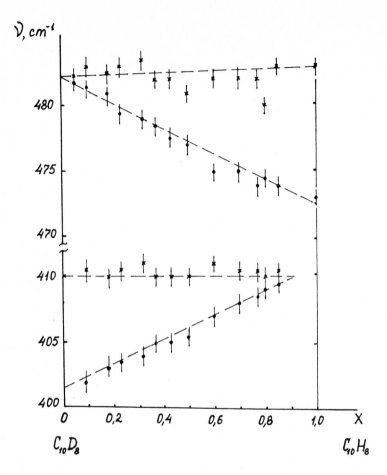

FIG. 8. Concentration dependence of the maximum positions of the Davydov components of the ν_{31} fundamental in naphthalene+naphthalene-d_8 mixed crystals.

deep "trap" since the ratio $(\Delta_u/\Delta_D) \cong 7$ (Δ_u is the isotope shift). The exciton bands of both molecules are formed almost independently. That is why we have a respectively low intensity redistribution between the doublet components (it does not exceed a factor of about 3).

Each component of the doublet is characterized by an asymmetrical contour. Its halfwidth varies with the concentration variation. So, at small concentrations or in pure crystals, the halfwidth of high frequency components is two times larger than those of the second components. At the 30 to 40% concentration their halfwidths become almost equal. Careful measurements of band shapes do not reveal any fine structure, unlike the situation in the electronic spectra [59]. In an effort to find fine structure in the infrared absorption spectra of 32% naphthalene, samples have been studied down to

90K, but nothing was observed except a frequency increase (about 1.5 cm^{-1}) in the peak positions.

Note that, according to Rich and Dows [75], the magnitude of splitting for the 480 cm^{-1} band in naphthalene may be explained within the framework of the atomic potential model by taking into account intermolecular interactions between the atoms H$\cdots$H and C$\cdots$H only.

Mixed Thiophthene+Deuterothiophthene Crystals

Thiophthene crystals have an orthorhombic lattice with the space group D_{2h}^{15} [76]. Its unit cell contains four molecules occupying the positions with C_i site symmetry. The symmetry of the free molecule is C_{2h}. From the symmetry of the D_{2h} factor group one may expect splitting of A_u and B_u fundamentals of the isolated molecule into four components, three of which (B_{1u}, B_{2u}, B_{3u}) are active in infrared absorption. But since the three corresponding transition moments are mutually perpendicular, when infrared radiation falls on the plane of two of them, the third one will not appear in the spectrum.

Vibrational spectra of thiophthene have been interpreted in Refs. [68] and [77]. Some infrared absorption molecular bands reveal considerable Davydov splitting in the infrared spectrum of the crystal. A molecular vibrational band at 700 cm^{-1} (ν_{25}) shows the most splitting magnitude (about 12 cm^{-1}) in the crystal, which is somewhat larger than in the case of naphthalene (Fig. 9).

To form a mixed crystal isotope analog of thiophthene (T_o), dideuterothiophthene (2,5-d_2-thieno-[3,2-b]-thiophthene ($T_{2,5}$-d_2)) was used which is a substance with the same molecular symmetry. The ν_{25} mode in d_2-thiophthene is placed in the 570 cm^{-1} frequency range; it has about a 9 cm^{-1} splitting. In the infrared spectra of mixed T_o + $T_{2,5}$-d_2 crystals these bands do not overlap with other absorption bands.

The spectra presented in Fig. 9 were obtained for two mutually perpendicular polarizations [66]. The choice of one of them was made on the basis of maximum contrast in the polarization of the single 968 cm^{-1} band in the T_o spectrum and the 975 cm^{-1} band in the $T_{2,5}$-d_2 spectrum. Since the spectra for the two perpendicular polarizations were recorded at different times, particular attention was paid to the reproducibility of the wavenumber calibration which was verified by means of the indene absorption spectrum. The reproducibility value did not exceed 0.2 cm^{-1} per day. In the 600-700 cm^{-1} spectral range the resolution of the spectrometer was about 2 cm^{-1}, i.e., it was much smaller than the halfwidth of the absorption bands under study.

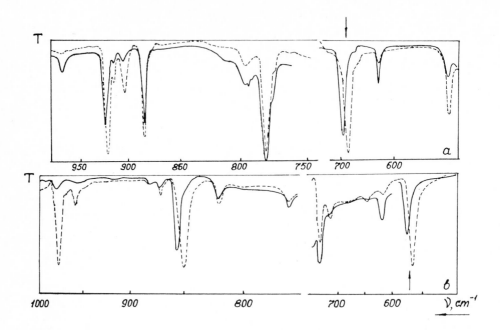

FIG. 9. Portion of the polarized infrared spectra of monocrystalline: (a) thiophthene (thieno-[3,2,-b]-thiophene) and (b) 2,5-d_2-thieno-[3,2-b]-thiophene. Arrows indicate the ν_{25} absorption bands with a larger Davydov splitting magnitude.

The concentration dependences of the Davydov component peak positions in the mixed crystals are presented in Fig. 10. As in the case of naphthalene, each point on the plot is averaged over several measurements and the accuracies of the frequency determination are also presented. As seen in Fig. 10, the splitting is practically absent at the 10% T_0 concentration and at 5% of $T_{2,5}$-d_2. Almost the same picture was observed in methylchloride+ deuteromethylchloride mixed crystals [20] where the resonant interaction disappeared at the 2-3% concentration for the deuteromolecules and the 10% concentration for the undeuterated molecules.

When the concentration varies one may observe a frequency decrease for the high frequency components of the Davydov doublets and an increase for the low frequency ones (Fig. 10). At extreme concentrations Davydov doublets converge into single bands whose frequencies are approximately midway between the initial Davydov components, which is unlike the situation in the electronic exciton spectra and infrared absorption spectra of naphthalene.

Davydov splitting in the infrared spectra of thiophthene may also be observed on other molecular bands, but the out-of-plane ν_{25} mode (the corresponding electric transition dipole moment is perpendicular to the

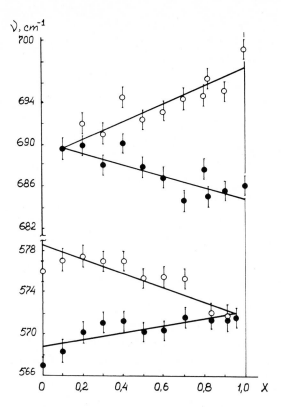

FIG. 10. Concentration dependence of Davydov component maximum positions of the ν_{25} fundamental in thiophthene and deuterothiophthene mixed crystals.

molecular plane) has the largest splitting. These observations confirm Califano's and his coworkers' results [39]: in the vibrational spectra of crystalline anthracene and phenanthrene they have seen the largest splitting for the out-of-plane molecular vibrations.

The results obtained in the vibrational spectra of mixed crystals based on naphthalene and thiophthene allow one to observe the influence of translational disordering, due to an impurity defect, on the magnitude of Davydov splitting. Both studied crystals correspond to the deep "trap" case and, therefore, one could reveal no anomalies in polarization or intensity behavior of Davydov components like those observed in the electronic spectra (Rashba effect).

VI.   USE OF DAVYDOV SPLITTING TO REFINE
INTERMOLECULAR INTERACTION POTENTIALS

In this section, taking ethylene as an example, we shall consider the possibility to refine intermolecular interaction potentials.

From X-ray studies performed by Bunn [78] only carbon atom coordinates and Bravais lattice constants have been found.   Depending on the C-H bond orientation one may suppose several space groups for ethylene crystals. In Ref. [78], the $P_{nnm}(D_{2h}^{12})$ space group has been chosen as the preferable one.   Brecher and Halford have measured the infrared absorption spectra of ethylene single crystals [79].   They pointed out that each molecular vibrational transition (molecular symmetry is D_{2h}) gives rise to a doublet of differently polarized components which are active either only in the Raman spectra or only in the infrared spectra.   Thus, the site symmetry of the ethylene molecules was concluded to be C_i.   This result contradicts that for the D_{2h}^{12} space group.   Brecher and Halford suggested changing the molecular arrangement in the crystal unit cell rather than the carbon atom positions. For this purpose one must rotate both molecules about the C-C bond axis by an angle of 27°.   In this case one should obtain a C_{2h}^5 space group (Table 1).

These conclusions have been completely confirmed in Dows' work [80]. Besides new deuteroethylene crystal infrared absorption data, he also made a significant addition to resolve the ethylene crystal structure problem.   The point is that there are two different space configurations of the molecules depending on the relative rotation direction of molecules about the C-C bond axis.   If rotations have the same direction, one obtains a $P2_1/n11$ structure; in the opposite way, one obtains a $P12_1/n1$ structure (Fig. 11).   In Ref. [80] these structures have been called "a axis version" and "b axis version" according to which of the twofold screw molecular axes holds well in the crystal.[*]

In order to make a choice between these two structures, it has been suggested to use Davydov splitting magnitudes in the infrared spectra of ethylene and ethylene-d_4 crystals [80].   It was supposed to calculate Davydov splitting magnitudes for both structures and to make a choice between them on the basis of the best coincidence between experimental and calculated values.

---

[*] Ethylene molecules occupy centrosymmetrical sites with $(0,0,0)$ and $(1/2, 1/2, 1/2)$ coordinates; the C-C bond molecular axis makes a 36° angle with the orthorhombic crystal axis.

TABLE 1

Correlation Diagram for Crystalline Ethylene

Molecular point group D_{2h}		Site group C_i	Factor group C_{2h}	
ν_1, ν_2, ν_3	$3A_g$		$6A_g$	Raman
ν_5, ν_6	$2B_{1g}$	$6A_g$	$6B_g$	Raman
ν_8	$1B_{2g}$			
ν_4	$1A_u$		$6A_u$	infrared
ν_7	$1B_{1u}$	$6A_u$		
ν_9, ν_{10}	$2B_{2u}$		$6B_u$	infrared
ν_{11}, ν_{12}	$2B_{3u}$			

Lattice modes

R_z	$1B_{1g}$		$3A_g$	Raman
R_y	$1B_{2g}$	$3A_g$	$3B_g$	Raman
R_x	$1B_{3g}$			
T_z	$1B_{1u}$		$3A_u$	infrared + one acoustical mode
T_y	$1B_{2u}$	$3A_u$		
T_x	$1B_{3u}$		$3B_u$	infrared + two acoustical modes

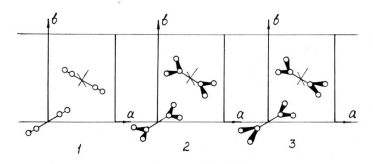

FIG. 11. Projection of possible structures of crystalline ethylene on the a,b plane, (1) structure suggested by Bunn - P_{nnm}, (2) $P2_1/n11$, and (3) $P12_1/n1$.

It has been found from the calculation performed in the dipole-dipole approximation (the experimental band intensity results have also been taken into account) that Davydov splitting magnitudes do not exceed 0.3 cm^{-1} while the experimental values were $2\text{-}8 \text{ cm}^{-1}$ (Table 2). Significantly better agreement with experimental results may be achieved by taking into consideration hydrogen atom repulsion. Although from this calculation the "a version" has been chosen as the preferable one, the dichroism calculation performed by the same author has nevertheless favored the "b version". We shall return to the structural problems in the following sections. For us, in this way, it is important to note that the dipole-dipole approximation calculation does not explain the experimental results until we do not take into account the hydrogen atom repulsion. Dows expressed the same considerations when he studied the fine structure in the vibrational spectra of methylchloride and its deuterium analog. It has been shown that four of the nine Davydov splitting magnitudes may be explained by the hydrogen atom interactions, two others can be explained by the dipole-dipole interaction, while the other three cannot be accounted for within the framework of any model [81,82].

Elliott and Leroi [83] have used ten atom-atom potential versions (between H-H, H-C, and C-C atoms) in the form

$$U_{ij} = -A(r_{ij})^{-6} + B(r_{ij})^{-D} \exp(-Cr_{ij}), \tag{21}$$

where r_{ij} is the distance between the jth and the ith atoms; A, B, C, D are the constants; in most cases we have D equal to zero. To determine atom positions in the crystal unit cell, they minimized the potential interaction energy in the crystal (most stable structure). The low frequency Raman spectrum for such equilibrium atom positions has been calculated [83]. Most parts of the used potentials were rejected during this procedure because imaginary phonon frequencies appeared. Others gave frequencies considerably different from the experimental values with the exception of the "b" structure where the agreement was satisfactory. When using the Eq. (21) potential, one obtained the following parameter values for the interactions: H-H interaction: A = 36.0, B = 4000.0, C = 3.74; H-C interaction: A = 136.0, B = 9410.0, C = 3.67; C-C interaction: A = 535.0, B = 74460.0, C = 3.60. Finally, Elliott and Leroi [83] preferred the "b version" on the basis of the intensity calculations of Raman active librational modes which are less dependent on the potential parameters.

The next step to calculate the ethylene vibrational spectrum using more accurately determined potentials has been performed by Krasyukov and

TABLE 2

Vibrational Spectra of Crystalline Ethylene

Frequencies of Vibrational Bands (cm^{-1})		Davydov splitting magnitude (cm^{-1})
Molecular spectrum	Crystal spectrum	
	infrared absorption [80]	
3105.5	$B_{2u}(\nu_9)$	3089.5	
2989.5	$B_{3u}(\nu_{11})$	2973.5	
1443.5	$B_{3u}(\nu_{12})$	1440.5 1437.0	3.5
1027	$A_u(\nu_4)$	1042 1036	6
949	$B_{1u}(\nu_7)$	949 941	8
810	$B_{2u}(\nu_{10})$	825.5 819.5	6
		110 [83] 73 [83]	
	Raman spectra [83]	
3103	$B_{1g}(\nu_5)$	3086.5 3066	2.5
3026	$A_g(\nu_1)$	3006.0 2995.5	10.5
1623	$A_g(\nu_2)$	1621.5 1614.5	7
1342	$A_g(\nu_3)$	1348.0 1328.5	19.5
1222	$B_{1g}(\nu_6)$	1226.5 1222.5	4
943	$B_{2g}(\nu_8)$	951.5 941.5	10
		177 167 114 97 90 73	

Mukhtarov [84]. Unlike Dows [80] who determined Davydov splitting in the infrared spectrum, and Elliott and Leroi who calculated a phonon Raman spectrum [83], these authors performed detailed simultaneous calculations of the Raman and infrared phonon spectra and Davydov splitting of the intra-molecular vibrations for both the C_2H_4 and C_2D_4 crystals. They have taken into account the experimental spectral data on the whole and minimized the divergences between them and the calculated data.

It is necessary to note that, in spite of the centrosymmetrical crystal structure, many Raman active Davydov doublets are characterized by the splitting magnitudes which exceed those for infrared active doublets in the same frequency region. This fact contradicts the dipole-dipole interaction Davydov splitting model (see above).

VII.   DAVYDOV SPLITTING IN THE VIBRATIONAL SPECTRA
OF CRYSTALS WITH COMPLEX IONS,
CRYSTALS WITH LAYER AND CHAIN STRUCTURE

In recent years many studies have been devoted to the investigation of the vibrational properties of ionic crystals and crystals with complex ions [75, 85-91]. In these investigations it has been found that the Davydov splitting magnitude (often about 50-100 cm^{-1}) cannot be accounted for by the dipole-dipole interaction between polar optical modes. In particular, in Refs. [89-91] the Sheelite type structure crystals MXO_4 (M = Ca, Sr, Ba; X = Mo, W) have been examined. There are two formula units in their tetragonal primitive unit cell (space group C_{4h}^6). Davydov splitting has been observed on the Bethe components (B and E) of the $(XO_4)^{2-}$ ion internal vibrations:

$$
\nu_3^0(F_2)
\begin{cases}
\nu_3(B) \begin{cases} B_g \\ A_u(z) \end{cases} \\
\nu_3(E) \begin{cases} E_g \\ E_u(x,y) \end{cases}
\end{cases}
$$

It has been shown that besides the multipole-multipole interaction there is also a contribution from non-Coulomb interactions (dispersion attraction and interchange energy repulsion forces) which exist between molecules before their vibrational excitation.

By taking into account these additional interactions, one may achieve much better agreement with the experimental Davydov splitting magnitudes ($10\text{-}50 \text{ cm}^{-1}$). The corresponding potential may be represented as a sum of central-force interactions between atom pairs which are not bound through binding forces.

The long-wavelength vibrational properties of a vast family of crystals with a layer or chain structure are analogous to those of molecular crystals [92]. In this case macroscopic layers play the role of immense molecules. Thus, in layered (chain) crystals with two or more layers (chains) per primitive unit cell the following zone-center optical phonons occur: Davydov multiplets corresponding to intralayer modes of an isolated layer (chain) and interlayer rigid-layer modes which are near translational motions of layers constituting the primitive unit cell. Such natural separation of modes may be accounted for by the anisotropy of the intralayer $\varkappa_0$ (covalent or ionic bonding) and interlayer $\varkappa_1$ (usually van der Waals type bonding) forces which are characterized by a factor $\beta = (\varkappa_1/\varkappa_0)$.

It has been found for a number of layered crystals (graphite, MoS_2, GaSe, As_2S_3, As_2Se_3) that this parameter varies from 0.01 to 0.07 [93]. For molecular crystals, where $\nu_1 \sim 100 \text{ cm}^{-1}$ and $\nu_0 \sim 1000 \text{ cm}^{-1}$ (see Eq. (1)), $\beta = (\varkappa_1/\varkappa_0)$ is of the same order of magnitude. Thus, in layered crystals the Davydov splitting magnitudes are about 0.5% to 4% of an intralayer mode frequency.

Rigid-layer modes which provide the most direct information on the strengths of the interlayer forces in layered crystals were of special interest in many experimental studies. For this purpose, investigation of the Raman spectra [93-96] including Raman spectroscopy under pressure [9,97-98] is the most convenient method. In addition, neutron-scattering data have yielded rigid-layer mode frequencies in the important case of graphite which is the most obvious layered crystal [99-100]. It has been determined that the rigid-layer mode frequency is from 15 cm^{-1} up to 100 cm^{-1} which demonstrates the hierarchy of interlayer force constants. In this way Lisitsa [92] has noted a gradual change for the worse for the "molecular" approximation in the series of crystals BiI_3, SbI_3, AsI_3. In AsI_3 crystals, such an approximation is probably true because of the interlayer interaction increase in this crystal compared to that of BiI_3 and SbI_3. For the case of AsI_3 one may also distinguish pairs of modes with suitable symmetry properties and consider them as Davydov pairs. But their interpretation becomes more difficult because of large splitting magnitudes and admixture effects between modes of the same symmetry.

As to Davydov splitting observations in the vibrational spectra of layered crystals, one must note that the number of reliable experimental

studies is very limited. This is due to essential difficulties which hinder the
observation of Davydov splitting components simultaneously and, thus, prevent reliable interpretation (see Sec. V). High anisotropy of force constants
in layered compounds provides a variety of polytypes, for each single layer
atom arrangement, which are characterized by slightly different frequencies of
corresponding intralayer modes. A real crystal is always a mixture of these
polytypes and thus additional difficulties, specific for layered crystals, exist
which prevent correct interpretation. But several relatively simple crystals
are well suited for Davydov multiplet observations and reliable interpretation.

Among them is indium sulphide, InS, which belongs to chalcogenide
semiconductors of the A^3B^6-type. Its crystal structure has much in common
both with layered crystals (GaSe, GaS, InSe, GaTe) and TlSe-type chain
crystals. In spite of that, InS is less anisotropic than the crystals mentioned
above and its Raman spectra are well interpreted on the basis of a simple
molecular crystal model [9] (Sec. I).

The InS crystal forms an orthorhombic lattice with the space group D_{2h}^{12}
(P_{mnn}). Its unit cell contains four formula units. The atom layers with an
S-In-In-S atom sequence are perpendicular to the OX axis (c axis, which can
be singled out in the crystal structure) forming "packets" of four atoms. A
unit cell contains two such "packets". In InS the In atoms have a tetrahedral
coordination (three S atoms and one In atom); however, the two S atoms and
one In atom are in the same "packet" whereas the third S' atom is in the
neighboring "packet". In this relation InS may be considered as an analog of
a molecular crystal with two differently oriented In_2S_4 "molecules" in the unit
cell. It is known that the distances In-S and In-S' are very close to each
other and, thus, from the nearest-neighbor atom interaction model it is
difficult to assume considerable difference between intrapacket and interpacket
interactions.

The group-theoretical analysis results obtained by the general method
[15] are presented in Table 3. The displacement atom vectors for all modes
are summarized in Table 4. The mode pairs, A_g and B_{1g}, B_{2g} and B_{3g}, A_u
and B_{1u}, B_{2u} and B_{3u}, which are conjugated with respect to the symmetry
operations interchanging translationally nonequivalent packets, form Davydov
doublets. The "intermolecular" B_{2u}, B_{3u} and A_u modes, among which B_{2u}
and B_{3u} are infrared active and A_u is inactive in both the infrared and
Raman spectra, are conjugated with the acoustic B_{3u}, B_{2u} and B_{1u} modes.
It is important that both differently polarized components of some Davydov
doublets are either active only in the Raman spectra or only in the infrared
spectra. Thus, it is possible to observe Davydov doublets using the data
obtained by one of these two methods alone. The Davydov splitting in the
Raman spectra has been studied by Faradzhev et al. [101]. It has been

TABLE 3

Vibrational Modes of the InS Lattice[a]

D_{2h}^{12}	n_i	T'+T	T'	T	R'	n_i'	infrared and Raman activity
A_g	4	0	0	0	1	3	$\alpha_{xx}, \alpha_{yy}, \alpha_{zz}$
B_{1g}	4	0	0	0	1	3	α_{xy}
B_{2g}	2	0	0	0	2	0	α_{zx}
B_{3g}	2	0	0	0	2	0	α_{zy}
A_u	2	1	0	1	0	1	
B_{1u}	2	1	1	0	0	1	T_z
B_{2u}	4	2	1	1	0	2	T_y
B_{3u}	4	2	1	1	0	2	T_x

[a] Vibrational modes: n_i - the number of external and internal modes; T - translational (with the exception of acoustical modes); T' - acoustical modes; R' - librational modes; n_i' - internal modes.

shown that the Raman spectra reveal doublet nature. Moreover, the frequencies of the "intermolecular" modes forbidden in the Raman spectra have been evaluated using Eq. (3).

The infrared reflection spectra of InS [102] were recorded in a linearly polarized light in the frequency range from 50 cm^{-1} to 400 cm^{-1} at room temperature with a resolution of about 1 cm^{-1}. The "twinning" of the InS crystal results in indistinguishable $\overline{OY}$ and $\overline{OZ}$ axes. Therefore, it is impossible to record polarized spectra at $\overline{E}||\overline{OY}$ and $\overline{E}||\overline{OZ}$. At both polarizations identical spectra were obtained where four reststrahlen bands were observed, i.e., four modes prove to be active simultaneously: $B_{1u} + 3B_{2u}$. At $\overline{E}||\overline{OX}$ three reststrahlen bands corresponding to the B_{3u} modes were observed.

The results obtained in processing the infrared reflection spectra of the InS crystal are presented in Table 5. As is seen, the pairs of modes $^1B_{2u}$-$^1B_{3u}$ and $^3B_{2u}$-$^3B_{3u}$ which are close to each other in frequency may be considered as Davydov doublets. (The upper index -- mode number of crystal lattice.) Then the low-frequency modes $^2B_{2u}$ and $^4B_{3u}$ correspond to the rigid-layer modes.

The validity of such separation into Davydov pairs follows from the group theoretical analysis. If one imagines the unit cell of InS as two packets or, in other words, as two differently oriented In_2S_4 molecules, then

TABLE 4

Symmetry Displacements for Vibrational Modes of InS

		In 1	In 2	In 3	In 4	S 1	S 2	S 3	S 4
A_g	1	x	-x	x	-x				
	2					x	-x	x	-x
	3	y	-y	-y	y				
	4					y	-y	-y	y
B_{1g}	1	x	-x	-x	x				
	2					x	-x	-x	x
	3	y	-y	y	-y				
	4					y	-y	y	-y
B_{2g}	1	z	-z	z	-z				
	2					z	-z	z	-z
B_{3g}	1	z	-z	-z	z				
	2					z	-z	-z	z
A_u	1	z	z	-z	-z				
	2					z	z	-z	-z
B_{1u}	1	z	z	z	z				
	2					z	z	z	z
B_{2u}	1	x	x	-x	-x				
	2					x	x	-x	-x
	3	y	y	y	y				
	4					y	y	y	y
B_{3u}	1	x	x	x	x				
	2					x	x	x	x
	3	y	y	-y	-y				
	4					y	y	-y	-y

with the aid of symmetrized displacements (Table 4) it is possible to consider the high-frequency B_{2u} and B_{3u} modes as internal stretching and bending vibrations of the In_2S_4 molecule. It is just the stretching and bending vibrations ($^1B_{2u}$-$^1B_{3u}$ and $^3B_{2u}$-$^3B_{3u}$) that form the Davydov doublets. In this case the low-frequency $^2B_{2u}$ and $^4B_{3u}$ modes (Table 5), however, should be assigned to the motions of one In_2S_4 molecule with respect to the other one molecule, i.e., intermolecular or rigid-layer modes. This is also confirmed by the relatively low value of the LO-TO splitting of these modes.

Such complete information on the phonon spectrum of InS has allowed one to predict a new phase transition at high pressures (about 7.0 GPa) and a structure of this high-pressure high-symmetry phase on the basis of the Raman spectrum changes which have been observed when the pressure is in-

TABLE 5

Assignment and Frequencies (cm^{-1}) of the Transverse and Longitudinal Optical Modes of InS

	$^1B_{2u}$	$^2B_{2u}$	$^3B_{2u}$	$^4B_{2u}$	^1A_u	^2A_u
ν_{TO}	288	100	210	acoustic	inactive	inter-molecular
ν_{LO}	314	106	235			inactive

	$^1B_{3u}$	$^2B_{3u}$	$^3B_{3u}$	$^4B_{3u}$	$^1B_{1u}$	$^2B_{1u}$
ν_{TO}	278	acoustic	225	78	247	acoustic
ν_{LO}	316		244	81	268	

creased (up to 1.2 GPa) [103]. When the phase transition occurs, the In and S atoms must shift to new equilibrium positions until linear chains of atoms extended along the $\overline{OX}$ axis are formed. At the phase transition point the lattice constants a and b become equal. The phase transition is accompanied by a decrease, by a factor of two, in the number of atoms in the primitive unit cell because all the S'-In-In-S' "molecules" become congruent. Thus, in the high-pressure phase, Davydov splitting must be absent. The existence of a phase transition in InS at 7.0 GPa has been recently confirmed, but for the present there are no spectral data on it.

There have been several attempts to calculate Davydov splitting values in layered crystals. Most of them are based on the nearest-neighbor approximation. This approach leads to expressions like Eqs. (1-3) which prove to be useful in many cases (see, for example, Ref. [94]). However, for several chalcogenide layered crystals, frequency inversion of the Davydov components [104,105] has been observed in contrast to the Davydov components' relative frequency position predicted from the simple nearest-neighbor approximation. According to Kuroda and Nishina [105] the dipole-dipole deformation charge interaction does not explain the inversion of such modes. It is necessary to take into account the chalcogenide atoms' static dipole moments which play a significant role in the lattice dynamics of layered compounds.

VIII.   THE USE OF DAVYDOV SPLITTING FOR THE STRUCTURAL ANALYSIS OF MOLECULAR CRYSTALS

X-ray and neutron analyses are the main methods for crystal structure determinations.   But in many cases their direct application is either very difficult or quite impossible.   This remark concerns crystals of powder-like compounds, substances with a large number of hydrogen atoms, and those cases when the same X-ray pattern corresponds to different crystal structures.   Among these substances which are difficult for standard structural determination methods are the low-temperature phases of some crystals, including molecular organic crystals.   In this case the spectral methods may be either the only chance to obtain the desired structure information or a significant addition to the standard methods [106].   The vibrational spectra of molecules under the static crystalline field may be considered as the basis of this information.   The details of the interpretation of molecular crystal vibrational spectra were described in Sec. II.   In this section we shall consider several examples of the crystal structure determination using vibrational spectral analysis.

The study of fine structure of vibrational bands and polarization properties of Davydov components, as well as the analysis of transformations of the vibrational selection rules due to the transition from a liquid or gas into a crystal, allows one to ascertain the site symmetry, the factor group and the number of molecules per primitive unit cell.   The reliability of such determination may be rather high, since usually there are many vibrational transitions of a certain symmetry for relatively complex molecules.   If, for some reason, the splitting of some of these species is not large enough to be revealed, then one may observe it on other species where the splitting is larger.

In Sec. VI we have already discussed an example of ethylene; the information on its structure was essentially improved by the use of vibrational spectra examination.   The second successful example is the work performed by Marzochi and Manzelli [107], where CH_2Cl_2 and CD_2Cl_2 crystals have been studied.   The symmetry of the molecules is C_{2v}.   In the infrared absorption spectrum of the crystals there is a band corresponding to the ν_5 (A_2) "forbidden" fundamental.   This result means that the molecular site group should be C_2, C_s or C_1.

By the method of solid solutions (Sec. III) it has been found that there are no Davydov multiplets for most of the fundamentals except two B_2 species and one B_1.   Since the vibrations of the A_1 and A_2 species are not split in the crystals, while the B_1 and B_2 species are split, the authors concluded that the site group is C_2. All polar optical modes of these crystals may be distinguished by their polarization properties depending on the symmetry of corresponding fundamentals: A_1 and A_2 species (a-polarization); one component of the B_1 and B_2 species (b-polarization); another component of the B_1 and B_2 species (c-polarization). Thus, one must conclude that the crystal under study has an orthorhombic lattice and so C_{2v}, D_2 or D_{2h} point groups. Additional data about the crystal structure were obtained from mode intensity measurements in polarized light using the oriented gas model. In this way it is possible to determine the angle between translationally nonequivalent molecules which is formed between the planes which are perpendicular to the C_2 molecular axis. The authors evaluated this angle to be about 29°. On the other hand, from the space group D_{2h}^{14} and the closed packing principle, the authors obtained the same angle to be 27°. On the whole it is possible to believe that the space group of CH_2Cl_2 and CD_2Cl_2 crystals has been obtained from infrared absorption data only.

Spectroscopic Determinations of the Symmetry of Thiophthene, Thiophane and Thiirane

The molecular symmetry of thiophthene (thieno-[3,2-b]-thiophene) is C_{2h}. When the liquid-solid phase transition occurs, several thiophthene fundamentals split into doublets or even triplets (Fig. 9). These multiplets, as well as the corresponding fundamentals, obey the mutual exclusion principle for infrared and Raman active vibrations [108]. In Sec. V we have already shown that these multiplets have the Davydov nature. Thus, one may conclude that thiophthene molecules occupy C_i symmetry positions and, depending on the factor group C_{2h} or D_{2h}, there may be two or four molecules per unit cell. These results are in good agreement with the X-ray results according to which the thiophthene space group is D_{2h}^{15} [76].

For the case of thiirane (ethylene thiooxide) [109], in the infrared spectrum of the second (low temperature) crystalline modification which appears to be due to the phase transition at 166K [110], quartets were observed instead of single bands in the spectrum of the liquid. Measurements of the polarization properties of the 824 cm^{-1} quartet corresponding to the B_2 fundamental [110] showed that the polarization of two components differs from that of the other two (Fig. 12a). Such quartets have also been observed for the B_1 fundamentals with the following frequencies: 668, 1050 and 1435 cm^{-1}. Simultaneously the 1033 (A_1), 1109 (A_1), 1173 (A_2) and 1456 cm^{-1} (A_1) fundamentals have been found to be split into doublets with the same polarization of the components (Table 6). Multiplets with such polarization properties

TABLE 6

Davydov Multiplets in the Infrared Spectrum of Crystalline Thiirane

Frequency of fundamental (gas), cm^{-1}	Frequencies of Davydov components, cm^{-1}	Polarization[a]	Magnitude of splitting, cm^{-1}	Oscillator strength, $f \times 10^7$
620	600 m[b] 604 s	a b	4	
627 (A_1)	608 s 614 m	b a	6	2.4
668 (B_1)	634 sh 641 m 644 m 646 m	a a b b	12	2.8
824 (B_2)	818 m 824 m 828 m 836 m	a b a b	18	1.4
890 (A_2)	899 w 901 w 904 w 910 w	b b a a	11	1.6
944 (B_2)	940.5 sh 946 m 952 m	b a b	11.5	3.1
1033 (A_1)	1028 vw 1030 vw	a a	2	2.5
1050 (B_1)	1044.5 m 1052 m 1058 m 1063 m	a b b a	19.5	11
1109 (A_1)	1114.5 m 1118 w	a	3.5	0.6
1173 (A_2)	1171 m 1174.5 m	a a	3.5	0.95
1435 (B_1)	1423.7 s 1426.8 m 1428.8 s 1436 m		7.7	
1456 (A_1)	1438.5 1445.1		6.6	

[a] a - corresponds to the dotted line in Fig. 12, b - corresponds to the solid line.

[b] mode intensities: s - strong, m - medium, w - weak, sh - shoulder, vw - very weak.

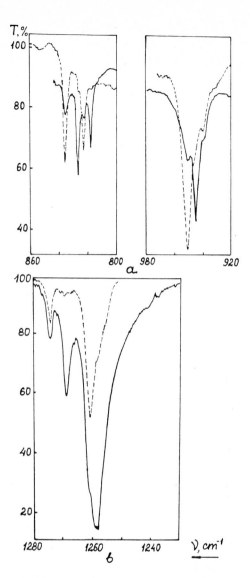

FIG. 12. Portions of the polarized infrared absorption spectra of low temperature crystalline (a) thiirane and (b) thiophane for two different polarizations of infrared radiation.

may be possible only for the C_2 site group, the C_{2v} factor group, with four molecules per unit cell (Table 7). Thus, the low temperature thiirane crystal modification should be assigned to the orthorhombic lattice with one of the space groups C_{2v}^{11}, C_{2v}^{16}, C_{2v}^{17}, C_{2v}^{22}, the choice of which could be made with the help of the potential function symmetry method [111].

The thiophane molecule at crystallization (168K) forms an anisotropic

TABLE 7

Correlation Diagram for Crystalline Thiirane[a]

Molecular point group, C_{2v}	Site group C_2	Factor group C_{2v}, z = 4
A_1	A	$NA_1 = 2$, μ_z doublets
		$NA_2 = 2$, inactive
		$NB_1 = 0$
		$NB_2 = 0$
A_2	A	identical with the previous case
B_1	B	$NA_1 = 0$
		$NA_2 = 0$
		$NB_1 = 2$, μ_x
		$NB_2 = 2$, μ_y quartets
B_2	B	identical with the previous case

[a] NA_1 - number of A_1 modes, etc.; μ_z, μ_y, μ_x - vibrational transition moments of the unit cell (z axis coincides with the C_2 symmetry axis).

crystal, as evidenced by the infrared band dichroism and by the absence of the bandwidth temperature dependence in the crystalline phase [112]. X-ray data for this compound, as well as for thiirane, are most probably absent.

In the infrared spectra of crystalline thiophane many bands show doublet structure, while its solid solution spectra in CCl_4 reveal the resonant nature of these doublets [113] (see also Table 8). The Davydov components are polarized along the mutually perpendicular directions. For example, in Fig. 12b a fragment of the infrared spectrum of crystalline thiophane at 90K is presented.

As shown in Ref. [113], in the infrared spectrum of crystalline thiophane several new bands appear which correspond to forbidden A_2 fundamentals (molecular symmetry is C_{2v}). Thus, the site symmetry may be C_2, C_s or C_1. From the fact that the A_1 and B_1 fundamentals split into Davydov doublets (see above) one may infer that the factor group is C_{2v}, the site group is C_s and there are two molecules per unit cell.

In Tables 6 and 8, together with the data on the Davydov splitting magnitudes, the oscillator strengths of vibrational transitions are also presented. Comparison of the magnitudes of splitting and the oscillator strengths shows that the direct proportionality is absent, although it should be expected assuming the dipole-dipole interaction model.

TABLE 8

Davydov Multiplets in the Infrared Spectrum of Crystalline Thiophane

Frequency of fundamental (gas), cm^{-1}	Frequencies of Davydov components, cm^{-1}	Polarization[a]	Magnitude of splitting, cm^{-1}	Oscillator strength $f \times 10^7$
820 (A$_1$)	822 827	a b	5	10.9
882 (A$_1$)	884 888	a b	4	19.8
1035 (A$_2$)	1035 1039.5	a b	4.5	4.2
1194 (B$_1$)	1189 1196	a b	7	5.5
1256 (B$_1$)	1259 1261	a b	2	16.9
1272 (A$_1$)	1269 1274.5	a b	5.5	3.1
1322 (A$_1$)	1319.5 1324	a b	4.5	2.6

[a] a - corresponds to the dotted line in Fig. 12; b - corresponds to the solid line.

Symmetry Analysis of Crystalline Thiophene Modifications

From heat capacity measurements [114] it is known that crystalline thiophene undergoes three phase transitions at 171.6K (I-II), 138K (II-III) and 112K (III-IV). However, X-ray data [115] have been obtained only for the 212K structure, i.e., for the crystal I. The crystal IV (103K) has been studied in less detail [116,117]. A tetragonal lattice cell has been found to be the most suitable for this thiophene crystal modification.

X-ray experimental studies of crystalline thiophene [115] do not allow one to make an unambiguous choice between D_{2h}^{18} and C_{2v}^{17} (four molecules per unit cell). A possible site symmetry for each molecule should be C_2 (for the C_{2v}^{17} symmetry) or C_{2v} (for the D_{2h}^{18} symmetry). However, bands corresponding to the A$_2$ fundamentals are apparent in the infrared spectra of thiophene crystals [118]. Thus, one may conclude that the lower site symmetry group (i.e., C_2, C_s or C_1) should be realized rather than the C_{2v} one. Depending on the symmetry properties of the multiplets corresponding to the different symmetry species of the fundamentals (their correct assignment was made in

Refs. [118-120]), one may derive the desired site group. For example, if the site symmetry of the thiophene molecule is C_2, then the A_1 and A_2 modes have polarization which differ from those of the B_1 and B_2 modes. Thus, the problem is reduced to the polarized spectra analysis from which a certain site symmetry may be obtained.

These conclusions may be made more precise by examination of the vibrational spectra of 2,5-dideuterothiophene. It is assumed that these two crystals have analogous lattice symmetry. It is also known that the 2,5-dideuterothiophene molecule has C_2 symmetry and therefore there are only two possibilities for the site symmetry: C_s, when the molecular symmetry plane coincides with the crystal one, and C_1 for general molecular sites.

The polarized infrared absorption spectra of polycrystalline thiophene, 2,5-d_2-thiophene and 2-d_1-thiophene films have been measured in the same temperature range (173 - 233K) as in an X-ray study [117]. An oriented film of 30 μm thickness was grown in a metallic cryostat between KBr windows at slow cooling.

The infrared absorption spectra of crystalline thiophene for two mutually perpendicular polarizations are presented in Fig. 13a. Vibrational bands of the A_2 species, ν_9 (920 cm^{-1}) and ν_{11} (565 cm^{-1}), reveal clear polarization behavior. The 870 cm^{-1} band, assigned in Ref. [121] as nearly degenerate in frequency with the B_2 (ν_{20}) and B_1 (ν_{17}) fundamentals, is also polarized, but in a less extent than the previously cited bands. The remaining thiophene bands do not reveal any visible polarization behavior. The infrared absorption spectrum of an oriented crystalline 2,5-d_2-thiophene film is polarized to a greater extent (Fig. 13b).

Therefore, one could conclude that the ν_9 (895 cm^{-1}) and ν_{11} (533 cm^{-1}) bands of the A_2 species and the ν_{19} (586 cm^{-1}) and ν_{20} (819 cm^{-1}) bands assigned to the B_2 species all have the same polarization, while bands assigned to the in-plane vibrations of A_1 and B_1 species have perpendicular ones. The infrared polarized spectrum of oriented 2-d_1-thiophene film also reveals a clear dependence of band polarization from the symmetry species of the fundamentals (Fig. 13c). Assigned to A" species, the 555, 568 and 708 cm^{-1} bands have perpendicular polarization compared to those of A' species, the interpretation of which has been obtained from IR spectra (gas phase) and Raman spectrum band polarizations [123,124].

Thus, using an oriented gas model, one could conclude that the polarization properties of the vibrational bands, assigned to the A_2 and B_2 fundamentals of thiophene and 2,5-d_2-thiophene, differ from those of the A_1 and B_1 fundamentals. The same correlation is true for the A' and A" fundamentals of 2-d_1-thiophene. From this, one may conclude that the site symmetry of the thiophene molecule in the crystal lattice is C_s. However, this

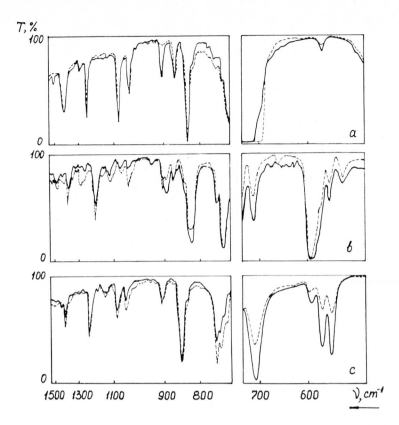

FIG. 13. Infrared absorption spectra of crystalline (a) thiophene, (b) 2,5-d_2-thiophene, and (c) 2-d_1-thiophene films for two different polarizations of incident radiation at 200K.

result is not in agreement with the space group C_{2v}^{17} or D_{2h}^{18} as suggested in Ref. [117] for the thiophene crystal I.

In order to obtain more precise results it is necessary to examine the multiplet fine structure of the vibrational bands and the polarization properties of the multiplet's components [122].

Figure 14 represents a part of the infrared vibrational spectra of thiophene crystals (I-IV). In the spectrum of crystal I, almost all bands of A_1 and B_1 species are split into the Davydov components with mutually perpendicular polarization, while all bands of A_2 and B_2 species are singlets. After the phase transition into crystal II, splitting of some singlet bands of the B_2 (454 cm^{-1} and 720 cm^{-1}) and A_2 (569 cm^{-1}) species into two components occurs.

The crystal III spectrum does not differ from that of crystal II; only a small Davydov component's frequency change has been found.

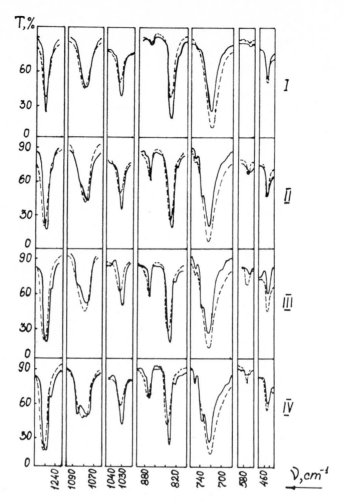

FIG. 14.   Transformation of the infrared spectrum of thiophene by the phase transitions.   Portions of the polarized spectra of four crystalline phases are presented.

In the vibrational spectrum of the thiophene crystal IV, an additional splitting of the B_1 and B_2 molecular species (1250, 1506 and 722 cm^{-1}) appears, while the 833 and 1033 cm^{-1} bands of A_1 species are broadened by about 1-2 cm^{-1}.   It is possible to believe that this broadening is caused by an additional small splitting which remains unresolved in the vibrational spectra.   Thus, most of the bands of the B_1 and B_2 molecular species are triplets or quartets with clear polarization for some part of the components.   Bands of A_1 and A_2 species are distinct doublets, with polarizations perpendicular to each other, although some components are anomalously broadened.

Measurements of a number of bands and the polarization state of the components of the Davydov multiplets confirm the previous conclusion [120] about the C_s site symmetry of the thiophene molecule in crystal I. As mentioned above, this site symmetry for Z = 4 is not consistent with the space groups D_{2h}^{18} and C_{2v}^{17} suggested in Ref. [117]. However, this site symmetry is consistent with the D_{2h} factor group for Z = 4 or the C_{2v} factor group for Z = 2.

From the fact that the A_2 and B_2 fundamentals split into doublets with perpendicular polarizations of the components, one may conclude that the site symmetry is lowered to C_1. If the number of molecules per primitive unit cell is not changed, then the factor group of crystal II would be C_{2v} or still lower. Additionally, the intermolecular infrared bands of crystal IV split. This fact is evidence for a much greater modification in the III-IV phase transition rather than the II-III transition.

It is possible to draw more readily some conclusions about the crystal IV symmetry since vibrational band dichroism clearly exists in this crystal phase. It is very likely that vibrational bands corresponding to the A_1 and A_2 fundamentals are multiplets which definitely have complex unresolved structure, rather than doublets. Thus, the number of molecules per primitive unit cell of crystal IV increases compared to that of crystal III. In this way the most probable factor groups for this phase are C_s or C_2, i.e., the crystal IV has a monoclinic crystal lattice cell. These results have been confirmed by the work of Paliani et al. [123].

The phase transitions I-II and II-III in crystalline thiophene are order-disorder phase transitions. The X-ray study [117] of crystal I showed that the molecular orientations are disordered over space. But this disordering is partial and related probably only to molecular turns in their own plane so that, in infrared absorption, vibrational band dichroism is still observed. Moreover, in the long wavelength infrared region a clear vibrational lattice band was observed [122], the result of which is not consistent with the framework of a completely disordered crystal. A larger amount of details may be observed in the low frequency infrared spectrum for crystal IV, where the molecular orientation ordering is the highest (Fig. 15).

Symmetry Analysis of the Low Temperature Crystalline Cyclopentane Phase

Crystalline cyclopentane is known to exist in three modifications: crystal I - 138 to 181K; crystal II - 122 to 138K; crystal III - lower than 122K. The X-ray data are known for crystal I only. Post et al. [124] have found that the crystal I structure is a close-packed hexagonal with two

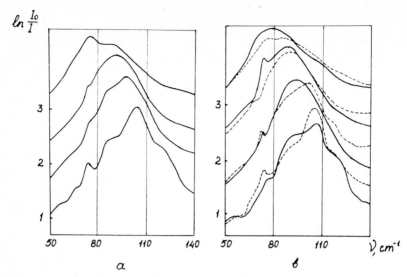

FIG. 15. Low-frequency infrared absorption spectra of four crystalline thiophene phases in (a) unpolarized and (b) polarized radiation. Optical density scales are indicated for the bottom curves at 85K.

TABLE 9

Correlation Diagram for Crystalline Thiophane[a]

Molecular point group, C_{2v}	Site group C_s	Factor group C_{2v}, Z = 2	
A_1	A'	$NA_1 = 1$	
		$NA_2 = 0$	doublets
		$NB_1 = 1$	
		$NB_2 = 0$	
B_1	A'	identical with the previous case	
		$NA_1 = 0$	
A_2	A"	$NA_2 = 1$	inactive
		$NB_1 = 0$	
		$NB_2 = 1$	singlets
B_2	A"	identical with the previous case	

[a] NA_1 - number of A_1 modes, etc.

TABLE 10

Expected Multiplicity of the Infrared Bands of Cyclopentane in the
1400-400 cm^{-1} Region

Site symmetry	Crystalline system	Factor group	Multiplicity
C_s, C_2	Orthorhombic	D_{2h}, C_{2v}, $D_2{}^a$	11 singlets + 11 doublets
C_1			22 triplets
C_s, C_2	Monoclinic	C_{2h}, $C_2{}^a$, $C_s{}^b$	22 singlets
C_1			22 doublets
C_1	Triclinic	C_i, C_1	22 singlets

[a] C_s site symmetry is not compatible with this group.
[b] C_2 site symmetry is not compatible with this group.

molecules per unit cell. The space group should be chosen among the D_{6h}^4, C_{6v}^4 and C_{3h}^4 groups. The spectra of the two high-temperature forms of cyclopentane remain similar to the spectrum of the liquid [125]. Thus, it is possible to assert that both these crystalline phases belong to high-symmetry lattice systems and may be considered as plastic modifications due to the high rotary activity of the molecules.

The infrared spectrum of crystal III is completely different from that of the other phases [125-128], i.e., all the bands present in the other phases split into several very sharp components. The investigators of this subject [125-129] have related this fact with pseudorotation [130] stoppage and close-packed structure formation, where the molecules are in a packed C_2 or C_s conformation.

Observations performed with a polarizing microscope [124] showed that at 122K, when the phase transition II → III occurs, the films transform into a mosaic of dichroic crystallites. Therefore, the investigators of Ref. [126] have excluded the possibility that the crystals remain uniaxial in phase III.

The most important point obtained from the vibrational spectra of crystals is that Davydov splitting into two and only two components in the infrared [126] and Raman [129] spectra occurs for the majority of the observed bands. This splitting has been proved to be true Davydov splitting by the mixed-crystal technique (Fig. 16) [126].

The possible combinations of the site and factor group symmetries are shown in Table 10. The table shows that only a monoclinic factor group with molecules in general positions is consistent with the number of observed

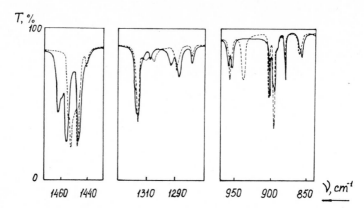

FIG. 16. Infrared absorption spectra of pure crystalline cyclopentane (solid line) and a mixed crystal of cyclopentane-d_0 in cyclopentane-d_{10} (dotted line) in a molecular ratio 1:10, form III.

bands and their multiplicity. The only two close-packed monoclinic space groups for molecules in general positions are the C_{2h}^5 and the C_2^2 groups. The definite choice between them may be made by a comparison of the infrared and Raman band frequencies. They are found to be slightly different from each other. Thus, each nondegenerate fundamental yields four unit cell modes, two of which are infrared and two are Raman active. Therefore, the C_{2h}^5 space group is favored over the C_2^2 space group [126] on the basis of the spectroscopic data.

Symmetry Analysis of Low Temperature Crystalline Cyclohexane Phase

Both the stable and one metastable modification of cyclohexane are known. Crystal I is stable between 279.8 and 186.1K, and all thermodynamic properties are characteristic of a plastic crystal [131]; this conclusion is supported by other experimental results. Stable anisotropic phase II exists below 186.1K. The metastable crystal III will not be considered here. In the gaseous state, the cyclohexane molecule has D_{3d} symmetry. Because of a considerable freedom of molecular reorientation, crystal I may be interpreted on the basis of a cubic cell [131]. Symmetry lowering in phase II leads to the splitting of infrared and Raman bands (Fig. 17) [132,133] and to the appearance of low-frequency lattice modes [134]. Mixed-crystal technique measurements [135] and polarized infrared spectral investigations [132,136] showed that the degenerate E_u fundamentals correspond to the quartet of infrared active modes in phase II, two of which have perpendicular polarization to the other pair. The splitting arising in phase II is a combination of Bethe and Davydov splittings.

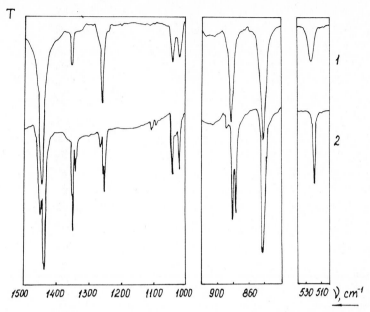

FIG. 17. Infrared absorption spectra of crystalline cyclohexane: 1 - plastic crystal; 2 - crystal II.

These spectroscopic experimental results are in agreement with X-ray measurements [137,138] where the cyclohexane crystal II structure has been attributed to a monoclinic system with the C_{2h}^6 (C2/C) space group. In this case there are two molecules in the primitive unit cell on C_i symmetry sites.

A corresponding diagram of symmetry species is shown in Table 11. Note that, in the case of cyclohexane as well as cyclopentane, lowering of the molecular symmetry appears in the lower symmetry phase [137]. In the latter case, the corresponding molecular distortion is significantly smaller, which is essentially indicated by the smaller magnitude of the Bethe splitting (1-3 cm^{-1} [132] compared to 20-35 cm^{-1} [126]).

In many cases, conclusions about the symmetry parameters of the unit cell, provided by infrared or Raman spectral measurements, can make an essential addition to the X-ray data analysis particularly when a crystal or one of its modifications exists only at low temperature. Until recently, investigations performed simultaneously by crystallographers and spectroscopists were very rare [138,139], but one may hope for an increase of such studies.

The spectral data accuracy about a crystal symmetry depends considerably on the possibility of resolving the multiplet structure of the infrared vibrational bands. It is clear that the multiplet is better resolved when the component bandwidth is smaller (low temperatures) and when the instrumental distortions are lower (high resolution). Thus sensitive, high

TABLE 11

Correlation Diagram for Crystalline Cyclohexane

Molecular point group, D_{3d}	Site group C_i	Factor group C_{2h}, $Z = 2$	
$6A_{1g}$ $2A_{2g}$ $8E_g$	$24A_g$	$24A_g$ $24B_g$	Raman Raman
$3A_{1u}$ $5A_{2u}$ $8E_u$	$24A_u$	$24A_u$ $24B_u$	infrared infrared

Lattice modes

R_z	A_{2g}	$3A_g$	$3A_g$	Raman
$R_{x,y}$	E_g		$3B_g$	Raman
T_z	A_{2u}	$3A_u$	$3A_u$	infrared + one acoustic mode
$T_{x,y}$	E_u		$3B_u$	infrared + two acoustic modes

resolution spectral equipment is needed for this purpose. Note that correct polarization measurements need sample illumination by weakly converged beams (~1:10). It is very likely that such conditions may be satisfied for interference spectrometers with axial symmetry or with tunable lasers [140]. For the investigation of fine structure in the infrared spectra of molecular crystals (with a usual spectral region from 200 to 3000 cm^{-1}), instruments with as large as possible spectral region are of great interest. Fourier transform spectrometers are among them. As modern experience has shown up to now, a spectrometer resolution of about 0.1 cm^{-1} would be sufficient for this purpose in the spectral region 200-5000 cm^{-1}. In the case of classical slit spectrometers, equivalent results demand the creation of instruments with a resolution limit of about 0.05 cm^{-1}. Raman scattering spectra obtained with the aid of laser excitation essentially supplement infrared data on the crystal structure.

IX. OBSERVATION OF THE PHASE TRANSITION IN PHENANTHRENE BY THE MEASUREMENT OF RELATIVE FREQUENCY SHIFTS OF DAVYDOV COMPONENTS

During the heat capacity measurements of crystalline phenanthrene [141] a small anomaly (about 400 cal) in the heat capacity was found which was revealed over the temperature range a few degrees in the vicinity of 72°C. This anomaly corresponds to the heat capacity increment of 0.8×10^{-16} erg/degree, which is equivalent to the excitation of two new degrees of freedom.

A temperature examination of phenanthrene single crystal under a polarizing microscope did not show any anomaly over the temperature range 25-85°C. X-ray studies [142-143] seem to indicate that phenanthrene crystals have the C_2^2 space group (two molecules per unit cell). Temperature X-ray measurements showed only slight anomalies over the temperature range 60-80°C; linear and angular variations of the unit cell parameters have also been indicated. Chiang et al. [143] have assumed that in this temperature interval a phase transition from one monoclinic lattice into another one occurs, which leads to a small crystal structure modification only.

The phase transition has been confirmed by measurements of the temperature dependence of the dielectric constant and the thermal expansion constant, and by neutron scattering measurements [144]. The thermal expansion constant remains invariable along the "a" axis, while it diminishes from 0.74 to 0.64 along the "b" axis and from 2.2 to 1.8 along the "c" axis (in $10^{-4}/°C$ units) (Fig. 18).

From neutron scattering data one can see that near the 72°C temperature some decrease of the low frequency phonon state density occurs which differs from that expected from the usual temperature dependence of phonon spectra. This decrease continues up to 85°C. Spielberg et al. [144] relate the phase transitions to the activation of the out-of-plane vibrational motion, shown in Fig. 19, by assuming that, in the low temperature modification, only harmonic vibrations with a small amplitude compared to the equilibrium plane molecular conformation exist.

Since one concerns the deformation of the molecule, then a vibrational spectral analysis is naturally suitable. In the first temperature study of the infrared absorption spectra of polycrystalline phenanthrene, a hardly visible shift of the 498 cm^{-1} band was observed, while in the other spectral regions no variation was observed.

The single crystal infrared absorption studies [145,146] provide considerably clearer pictures characterizing the phase transition. Figure 20

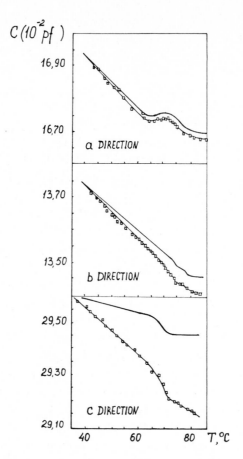

FIG. 18. Temperature dependence of capacitor capacitance, where phenanthrene has been used as a dielectric for three mutually perpendicular polarizations. □ are experimental points, full lines are drawn with the correction on the temperature expansion of a dielectric. Used by permission [143].

shows the temperature dependence of the Davydov splitting magnitude Δv_D for the 498 cm^{-1} fundamental in the temperature range between liquid nitrogen and 368K. We can easily see that Δv_D varies significantly only near the phase transition point. The state of polarization and the relative intensity of the components are practically unaffected by the transition.

Analogous but less expressed changes have been observed for the 850 cm^{-1} band. Both of these bands (498 and 850 cm^{-1}) have been attributed [147] to the out-of-plane B_2 fundamentals which are usually characterized by a considerable Davydov splitting value. The Δv_D value was measured during heating as well as during cooling. No hysteresis effect was observed (Fig. 20) which would have been expected for the first order phase transition.

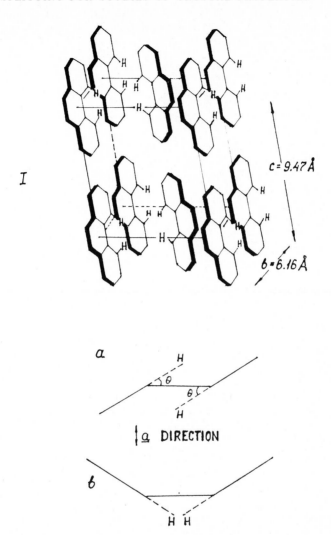

FIG. 19.   Disposition of phenanthrene molecules in the unit cell (I).   (a) Conformation of the phenanthrene molecule in the low-temperature modification.   (b) Expected second conformation in the high-temperature modification. Arrow indicates the "a" axis direction.   Used by permission [143].

The retention of the infrared dichroism in both crystalline modifications and during the transition itself, as well as the delicate character of changes we observed in the infrared spectra, allow one to suggest that the changes in the molecular and crystal structure are very small.   However, from these data it is very difficult to say anything about the nature of the phase transition under study.   One may assume both relative ordered molecular reorientation ($\Delta\nu_D$ changes due to the angular dependence of the dipole-dipole molecular interaction) and partial disordering in the high temperature phase ($\Delta\nu_D$

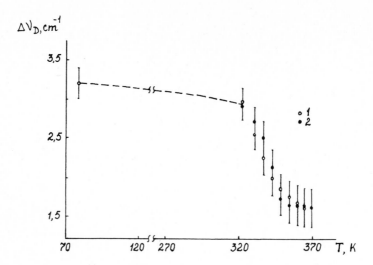

FIG. 20. Temperature dependence of Davydov splitting magnitude in the vicinity of the phase transition (1 - direct run, 2 - reverse run).

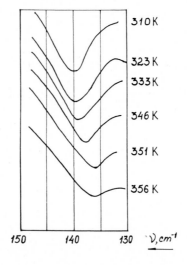

FIG. 21. Temperature dependence of the 140 cm^{-1} infrared absorption band shape.

changes due to the concentration dependence of the Davydov splitting magnitude). In the latter case molecules disordered in space may be considered as defects, which are analogous to isotopic impurities.

This experiment shows that Davydov splitting in the vibrational spectra would be related to the order parameter of the phase transition which was introduced for the first time by Landau and Lifshitz [148,149].

The phase transition in phenanthrene was also indicated by the frequency shift of the 140 cm^{-1} lattice mode. In the high temperature phase this mode softens by 4 cm^{-1} with respect to its frequency in the low temperature phase (Fig. 21).

X. CONCLUSION

The examples which we examined in this review show that the study of Davydov splitting of vibrational transitions in crystals provides valuable structural information, allows one to indicate delicate structure phase transitions, and may serve as an indicator of the mixed crystal composition. Certainly, these examples do not exhaust the whole variety of Davydov splitting's manifestations, but they indicate the great possibilities of their use.

The Davydov splitting phenomenon remains uncertain in some aspects which open interesting directions for further investigations. Therefore, Davydov splitting magnitude as a rule is not explained by dipole-dipole interaction, and this value probably should be related to the whole interference of contributions from which one usually explains the vibrational state dispersion in crystals. Such an approach would be fruitful for band shape studies in disordered systems (mixed crystals), due to its relation to the vibrational state density function.

REFERENCES

1. A. S. Davydov, _Theory of Molecular Excitons_, Plenum Press, New York, 1971.

2. V. M. Agranovich, _Exciton Theory_, Nauka, Moscow, 1968.

3. H. Bethe, _Ann. der Phys._, _3_, 133 (1929).

4. D. F. Hornig, _J. Chem. Phys._, _16_, 1063 (1948).

5. H. Winston and R. S. Halford, _J. Chem. Phys._, _17_, 607 (1949).

6. J. C. Decius and R. M. Hexter, _Molecular Vibrations in Crystals_, McGraw-Hill, New York, 1977.

7. H. Poulet and J. P. Mathieu, _Spectres de Vibration et Symétrie des Cristaux_, Gordon and Breach, Paris, 1970.

8. D. Dows, _Physics and Chemistry of the Organic Solid State_, Inter-science, New York, 1963; Mir, Moscow, 1967.

9. R. Zallen, _Phys. Rev. B_, 9, 4485 (1974).

10. M. A. El'yashevich, _Atomic and Molecular Spectroscopy_ (in Russian), Fizmatgiz, Moscow, 1962.

11. Yu. K. Khokhlov, _Tr. Fiz. Inst. im. P. N. Lebedeva Akad. Nauk SSSR_, 59, 221 (1972).

12. G. C. Pimentel, _J. Chem. Phys._, 19, 1536 (1951).

13. R. S. Halford, _J. Chem. Phys._, 14, 8 (1946).

14. E. B. Wilson, J. C. Decius, and P. C. Cross, _Molecular Vibrations_, 1955, McGraw-Hill, London.

15. S. Bhagavantam and T. Venkatarayudu, _Theory of Groups and Its Application to Physical Problems_, 3rd edition, 1962, Academic Press, New York.

16. G. C. Pimentel, A. L. McClellan, W. B. Person, and O. Schnepp, _J. Chem. Phys._, 23, 234 (1955).

17. D. F. Hornig and G. L. Hiebert, _J. Chem. Phys._, 27, 752 (1957).

18. J. Hollenberg and D. A. Dows, _J. Chem. Phys._, 39, 495 (1963).

19. L. I. Maklakov, V. N. Nikitin, and A. V. Purkina, _Opt. Spektrosk._, 15, 332 (1963); _Opt. Spectrosc._, 15, 178 (1963).

20. M. E. Jacox and R. M. Hexter, _J. Chem. Phys._, 35, 183 (1961).

21. M. M. Denariez, _J. Chem. Phys._, 62, 323 (1965).

22. D. A. Dows, _J. Chem. Phys._, 63, 168 (1966).

23. J. L. Duncan and D. C. McKean, _J. Mol. Spectrosc._, 27, 117 (1968).

24. A. Théorêt and C. Sandorfy, _Spectrochim. Acta_, 23A, 519 (1967).

25. A. B. Pasquier, C. Sourisseau and M. L. Josien, _C. R. Acad. Sc. Paris_, 268B, 1366 (1969).

26. S. Califano, _Mol. Phys._, 5, 601 (1962).

27. I. D. Mikhailov, V. A. Savelev, N. D. Sokolov, and N. D. Bokii, _Phys. Status Solidi (B)_, 57, 719 (1973).

28. R. J. Jakobsen, J. W. Brasch, and Y. Mikawa, _J. Mol. Struct._, 1, 309 (1967).

29. R. M. Hexter, _J. Chem. Phys._, 33, 1833 (1960).

30. M. P. Lisitsa and N. P. Kharchenko, _Zh. Prikl. Spectrosk._, 8, 667 (1968).

31. O. V. Fialkovskaya, _Proceedings of Tenth Conference on Spectroscopy_ (in Russian), 1957, p. 151.

32. O. V. Fialkovskaya, Opt. Spektrosc., 17, 397 (1964), (Opt. Spectrosc. 17, 211 (1964)).

33. W. B. Person, G. C. Pimentel and O. Schnepp, J. Chem. Phys., 23, 230 (1955).

34. S. Zwerdling and R. S. Halford, J. Chem. Phys., 23, 2221 (1955).

35. S. Califano, J. Chem. Phys., 36, 903 (1962).

36. L. W. Daasch, J. Mol. Spectrosc., 8, 86 (1962).

37. L. Colombo, J. Chem. Phys., 39, 1942 (1963).

38. J. Freund and R. S. Halford, J. Chem. Phys., 43, 3795 (1965).

39. V. Schettino, N. Neto and S. Califano, J. Chem. Phys., 44, 2724 (1966).

40. A. Bree and R. Zwarich, J. Chem. Phys., 51, 912 (1969).

41. A. Bree and R. A. Kydd, J. Chem. Phys., 51, 989 (1969).

42. D. M. Hanson and A. R. Gee, J. Chem. Phys., 51, 5052 (1969).

43. E. Gazis, P. Heim, and Ch. Meister, Spectrochim. Acta, 26A, 497 (1970).

44. C. Pecile and B. Lunelli, J. Chem. Phys., 46, 2109 (1967).

45. T. M. K. Nedunghadi, Proc. Ind. Acad. Sci., 15, 376 (1972).

46. A. R. Gee and G. W. Robinson, J. Chem. Phys., 46, 4847 (1967).

47. I. I. Kondilenko, P. A. Korotkov, and G. S. Litvinov, Opt. Spectrosk., 30, 437 (1971).

48. V. Schettino and M. P. Marzocchi, J. Chem. Phys., 57, 4225 (1972).

49. M. Ito, Spectrochim. Acta, 21, 2063 (1965).

50. I. I. Kondilenko, P. A. Korotkov, and G. S. Litvinov, Advanced Problems in Optics and Nuclear Physics (in Russian), Naukova Dumka, Kiev, 1974, p. 109.

51. E. F. Sheka, Izv. an SSSR, Ser. Fiz., 27, 503 (1963).

52. V. L. Broude and S. Kochubei, Fiz. Tverd. Tela, 6, 34 (1964).

53. E. I. Rashba, Opt. Spektrosk., 2, 576 (1963).

54. V. L. Broude, E. I. Rashba, and E. F. Sheka, Dokl. Akad. Nauk SSSR, 139, 1085 (1961).

55. V. L. Broude and E. I. Rashba, Fiz. Tverd. Tela, 3, 1941 (1961), (Sov. Phys. Solid State, 3, 1415 (1961)).

56. O. A. Dubovskii and Yu. V. Konobeev, Fiz. Tverd. Tela, 12, 405 (1970), (Sov. Phys. Solid State, 12, 321 (1970)).

57. O. A. Dubovskii, Candidate's Thesis, FEI, Obninsk, 1970.

58. H. K. Hong and G. W. Robinson, J. Chem. Phys., 52, 825 (1970).

59. J. Hoshen and J. Jortner, Chem. Phys. Lett., 5, 351 (1970).

60. Y. Onodera and Y. Toyozawa, J. Phys. Soc. Jpn., 24, 341 (1968).

61. F. Yonezawa and T. Matsubara, Prog. Theor. Phys., 35, 357, 759
     (1966); 36, 695 (1966).

62. V. L. Broude, A. V. Leiderman, and T. G. Tratas, Fiz. Tverd. Tela,
     13, 3624 (1971).

63. O. A. Dubovskii, Fiz. Tverd. Tela, 15, 205 (1973).

64. V. L. Broude and A. V. Leiderman, Zh. ETF Pis. Red., 13, 426 (1971).

65. V. M. Agranovich, Uspekhi Fizicheskich Nauk, 112, 143 (1974).

66. G. N. Zhizhin, M. A. Moskaleva, and E. B. Perminov, Opt.
     Specktrosk., 30, 1047 (1971); (Opt. Spectrosc., 30, 562 (1971)).

67. G. N. Zhizhin and M. A. Moskaleva, Opt. Spectrosk., 37, 99 (1974).

68. M. A. Moskaleva, Candidate's Thesis, Moscow State Univ., Moscow,
     1974.

69. V. M. Kozin and A. I. Kitaigorodskii, Zh. Fiz. Khim., 29, 897 (1971).

70. M. A. Kovner, E. P. Krainov, and N. I. Davydova, Proceedings of
     Commission on Spectroscopy,   Academy of Sciences of the USSR   (in
     Russian), Nauka, Moscow, 1, 114 (1965).

71. N. Neto, M. Srocco, and S. Califano, Spectrochim. Acta, 22, 1981
     (1966).

72. A. Bree and R. A. Kydd, Spectrochim. Acta., 26A, 1791 (1970).

73. N. J. Harrick, Internal Reflection Spectroscopy, John Wiley and Sons,
     New York, 1967.

74. H. Yamada and K. Susuki, Spectrochim. Acta, 23A, 1735 (1967).

75. N. Rich and D. A. Dows, Mol. Cryst., 5, 111 (1968).

76. E. G. Cox, R. J. J. Gillot, and G. A. Jefrey, Acta Crystallogr., 2,
     356, (1949).

77. Ya. M. Kimelfeld, M. A. Moskaleva, G. N. Zhizhin, V. P. Litvinov, S.
     A. Ozolin, and Ya. L. Goldfarb, Opt. Spektrosk., 28, 1112 (1970), (Opt
     Spectrosc., 28, 599 (1970)).

78. C. W. Bunn, Trans. Faraday Soc., 40, 23 (1944).

79. C. Brecher and R. S. Halford, J. Chem. Phys., 35, 1109 (1961).

80. D. A. Dows, J. Chem. Phys., 36, 2833, 2837 (1962).

81. D. A. Dows, J. Chem. Phys., 32, 1342 (1960).

82. D. A. Dows, J. Chem. Phys., 35, 282 (1961).

83. G. R. Elliott and G. E. Leroi, J. Chem. Phys., 59, 1217 (1973).

84. Yu. N. Krasyukov and E. I. Mukhtarov, Opt. Spektrosk., 55 (5), 870 (1983).

85. A. N. Lazarev, N. A. Mazhenov and A. P. Mirgorodskii, Zh. Neorg. Materialy, 14, 2107 (1978).

86. N. A. Mazhenov and A. P. Mirgorodskii, and A. N. Lazarev, Zh. Neorg. Materialy, 15, 495 (1979).

87. N. A. Mazhenov and A. N. Lazarev, Zh. Neorg. Materialy, 15, 504 (1979).

88. A. N. Lazarev, Kolebatelnye Spectry i Stroenie Silikatov (Vibrational Spectra and Structure of Silicates), Nauka, Leningrad, 1968.

89. Yu. P. Tsyashchenko and G. E. Krasnyanskii, Opt. Spektrosk., 47, 911 (1979).

90. I. I. Kondilenko, G. E. Krasnyanskii, and Yu. P. Tsyashchenko, Opt. Spektrosk., 47, 519 (1979).

91. G. E. Krasnyanskii and Yu. P. Tsyashchenko, Phys. Status Solidi, 101, K151 (1980).

92. M. P. Lisitsa, Zh. Prikl. Spektrosk., 27, 589 (1977).

93. R. Zallen and M. L. Slade, Phys. Rev. B, 9, 1627 (1974).

94. R. Zallen, M. L. Slade, and A. T. Ward, Phys. Rev. B, 3, 4257 (1971).

95. A. Polian, K. Kunc, and A. Kuhn, Solid State Commun., 19, 1079 (1976).

96. T. J. Wieting, Solid State Commun., 12, 931 (1973).

97. E. A. Vinogradov, G. N. Zhizhin, N. N. Melnik, S. I. Subbotin, V. V. Panfilov, K. R. Allakhverdiev, S. S. Babaev, and V. F. Zhitar, Phys. Status Solidi (B), 99, 215 (1980).

98. A. Polian, J. C. Chervin, and J. M. Besson, Phys. Rev. B, 22, 3049 (1980).

99. G. Dolling and B. N. Brockhouse, Phys. Rev., 128, 1120 (1962).

100. R. Nicklow, N. Wakabayashi, and H. G. Smith, Phys. Rev. B, 5, 4951 (1972).

101. F. E. Faradzhev, N. M. Gasanly, B. N. Mavrin, and N. N. Melnik, Phys. Status Solidi (B), 85, 381 (1978).

102. N. M. Gasanly, F. E. Faradzhev, A. S. Ragimov, V. M. Burlakov, A. F. Goncharov, and E. A. Vinogradov, Solid State Commun., 42, 843 (1982).

103. F. E. Faradzhev, A. S. Ragimov, A. F. Goncharov, and S. I. Subbotin, Solid State Commun., 39, 587 (1981).

104. T. J. Wieting and J. L. Verble, Phys. Rev. B, 3, 4257 (1971).

105. N. Kuroda and Y. Nishina, Phys. Rev. B, 19, 1312 (1979).

106. D. K. Arkhipenko and G. B. Bokii, Kristallografiya, 22, 1176 (1977).

107. M. P. Marzochi and P. Manzelli, J. Chem. Phys., 52, 2630 (1970).

108. G. N. Zhizhin, Doctor's Thesis, Inst. Chem. Phys., Moscow, 1974.

109. G. N. Zhizhin, G. M. Kuziants, M. A. Moskaleva, and A. Usmanov, Opt. Spektrosk., 33, 903 (1972).

110. V. T. Aleksanyan and G. M. Kuziants, Zh. Strukt. Khim., 12, 266 (1971).

111. P. M. Zorkii, Doctor's Thesis, Moscow State Univ., Moscow, 1973.

112. A. Usmanov, Candidate's Thesis, Moscow State Univ., Moscow, 1971.

113. Ya. M. Kimelfeld, A. Usmanov, G. N. Zhizhin, and V. P. Litvinov, Zh. Strukt. Khim., 12, 864 (1970).

114. C. Wadington and J. W. Knowlton, J. Amer. Chem. Soc., 71, 797 (1949).

115. S. C. Abrahams and W. N. Lipscomb, Acta Crystallogr., 5, 93 (1952).

116. G. Bruni and G. Nata, C. R. Acad. Lincoi, 11, 929 (1930).

117. G. Bruni and G. Nata, Recl. Trav. Chim. Pays-Bas, 48, 860 (1929).

118. V. T. Aleksanyan, Ya. M. Kimelfeld, N. N. Magdesieva, and Yu. K. Yuriev, Opt. Spektrosk., 22, 216 (1967); (Opt. Spectrosc., 22, 116 (1967)).

119. V. T. Aleksanyan, Ya. M. Kimelfeld, N. N. Magdesieva, and Yu. K. Yuriev, Optics and Spectroscopy, Suppl. 3, Molecular Spectroscopy II, USSR Academy of Science, 1967, p. 168; (Optical Society of America, 1968, p. 65).

120. Ya. M. Kimelfeld, M. A. Moskaleva, G. N. Zhizhin, and V. T. Aleksanyan, Zh. Strukt. Khim., 11, 656 (1970).

121. M. Rice, J. M. Orza, and J. Morcillo, Spectrochim. Acta, 21, 689 (1965).

122. G. N. Zhizhin, M. A. Moskaleva, and V. N. Rogovoi, Zh. Strukt. Khim., 14, 656 (1973).

123. G. Paliani, A. Poletti and R. Cataliotti, Chem. Phys. Lett., 18, 525 (1973).

124. B. Post, R. S. Schwartz, and J. Fankuchen, J. Amer. Chem. Soc., 73, 5113 (1951).

125. G. N. Zhizhin, Yu. E. Lozovik, M. A. Moskaleva, and A. Usmanov, Dokl. Acad. Nauk SSSR, 190, 301 (1970), (Soviet Physics-Doklady 15, 36 (1970)).

126. V. Schettino, M. P. Marzocchi, and S. Califano, J. Chem. Phys., 51, 5264 (1969).

127. A. LeRoy, C. R. Acad. Sci., Paris, 264B, 1087 (1967).

128. S. Lifson and A. Warshel, J. Chem. Phys., 49, 5116 (1968).

129. V. Schettino and M. P. Marzocchi, J. Chem. Phys., 57, 4225 (1972).

130. J. E. Kilpatrik, K. S. Pitzer and P. Spitzer, J. Am. Chem. Soc., 69, 2483 (1947).

131. J. G. Aston, G. L. Szasz and H. L. Fink, J. Am. Chem. Soc., 65, 1135 (1943).

132. G. N. Zhizhin and Kh. E. Sterin, Opt. Spektrosk., 19, 55 (1965), (Opt. Spectrosc., 19, 28 (1965)).

133. M. Ito, Spectrochim. Acta, 21, 2063 (1965).

134. Kh. E. Sterin, A. V. Bobrov, and G. N. Zhizhin, Opt. Spektrosk., 18, 905 (1965); (Opt. Spectrosc., 18, 509 (1965)).

135. M. A. Moskaleva and G. N. Zhizhin, Opt. Spektrosk., 36, 916 (1974).

136. G. N. Zhizhin, M. A. Moskaleva and Kh. E. Sterin, Optics and Spectroscopy, Suppl. 4, Molecular Spectroscopy, USSR Academy of Science, 1969, p. 146.

137. R. Kahn, R. Fourme, D. André, and M. Renaud, Acta Cryst., B29, 131 (1973).

138. R. Kahn, R. Fourme, D. André, and M. Renaud, C. R. Acad. Sc., Paris, 271B, 1078 (1970).

139. Ch. Menard and A. Meller, C. R. Acad. Sc., Paris, 271B, 1181 (1970).

140. N. I. Bagdanskis, V. S. Bukreev, G. N. Zhizhin, and M. N. Popova, Advanced Tendencies in Spectroscopy Technique (in Russian), Siberia Department of USSR Academy of Science, Novosibirsk, 1982, p. 153.

141. R. A. Arndt and A. C. Damask, J. Chem. Phys., 45, 755 (1966).

142. S. Matsumoto and T. Fakuda, Bull. Chem. Soc. Jpn., 40, 743 (1967).

143. K. Chiang, P. Forsyth, L. Morrison, J. B. Cohen, and J. W. Kauffman, Phys. Lett., 30A, 531 (1969).

144. D. H. Spielberg, R. A. Arndt, A. C. Damask, and J. Lefkowitz, J. Chem. Phys., 54, 2597 (1971).

145. G. N. Zhizhin, E. L. Terpugov, M. A. Moskaleva, N. I. Bagdanskis, E. I. Balabanov, and A. I. Vasil'ev, Fiz. Tverd. Tela, 14, 3612 (1972), (Soviet Physics-Solid State, 14, 3028 (1973)).

146. G. N. Zhizhin and N. I. Bagdanskis, Opt. Spektrosk., 34, 1150 (1973).

147. V. Schettino, N. Neto, and S. Califano, J. Chem. Phys., 44, 2724 (1966).

148. L. D. Landau and E. M. Lifshitz, Statistical Physics, Pergamon Press, Oxford, 1969.

149. N. G. Parsonage and L. A. K. Staveley, Disorder In Crystals, Clarendon Press, Oxford, 1978.

Chapter 6

RAMAN SPECTROSCOPY ON MATRIX ISOLATED SPECIES

H. J. Jodl

Fachbereich Physik, Universitat Kaiserslautern
6750 Kaiserslautern, W. Germany

I. INTRODUCTION

Almost thirty years have passed since the first paper on matrix isolation spectroscopy (MIS) was published. Nowadays an enormous number of publications appear each year concerning MI in physics and chemistry. Every well established technique like Mössbauer analysis, optical spectroscopy, NMR and ESR, etc., has been successfully applied to matrix isolated (MI)-species. This publication rate (hundred per year) demonstrates the importance of this subject. In no sense does this chapter encompass the whole field or answer all the questions one might ask about optical properties of matrix isolated (MI)-systems.

Instead we review the experimental results on light scattering by prototypic systems (mainly diatomic molecules) isolated in rare gas matrices (RGM). The motivation for this apparently severe restriction was the desire to simplify the interpretation of experimental data and the theoretical analysis. Although a careful attempt has been made to omit no work of outstanding importance, it must be recognized that in a field as large as this it is impossible to include everything. This is a review of experimental work and is written primarily for experimentalists.

A. Ideas, Advantages and Limitations of MI-RS

The usual approach in MIS is to gather spectroscopic data like transition energies, intensities, depolarization ratios, zero phonon lines and sidebands, etc. of the MI-species as a function of different parameters like temperature, pressure, excitation frequencies, matrices, etc. and to compare these data with those in the free or gaseous state.

This knowledge, in turn, allows a deeper insight into the static and dynamic interactions of the defect with the host matrix. The method can also be turned around to deduce information on species (atoms, ions, free radicals, etc.) not readily accessible in the free state. As a consequence MIS renders information about the defect, the matrix itself and the defect matrix system depending on the physical consideration, and to the methodical procedure.

There are many advantages to using MIS (matrix isolation spectroscopy) in general and MI-RS (matrix isolation Raman spectroscopy), in particular MI-IRS (matrix isolation infrared absorption spectroscopy):

(i)   Although similar in principle to IR-absorption spectroscopy, Raman scattering (RS) is inherently more sensitive. Basically, in RS one looks for small signals in the absence of any background while in IR absorption one looks for a small change in an otherwise very large continuum. As a consequence one needs rather thick samples for MI-IRS (~1 mm and quick sample condensation, 10 mmol/h) and thin samples for MI-RS (~10 μm and very slow spray on, ~100 μmol/h) with low light scattering qualities.

(ii) With a symmetrical molecule like N_2 or $XeCl_2$ the Raman effect offers the opportunity, normally denied in IR-absorption, of observing totally symmetric vibrational modes which are commonly observed as the strongest and sharpest bands in the Raman spectrum.

(iii) Polarization measurements are easily made in Raman spectroscopy and, in principle, depolarization ratios may be obtained for species randomly oriented in the host matrix.

(iv) The complete vibrational spectroscopic range (10 to a few 10^3 cm^{-1}) can be covered in one scan without changing spectrometer components and without the complication of atmospheric bands common in IRS. Furthermore, there is easy access to atomic lines for frequency calibration in the case of RS, whereas the situation in IRS is much more complicated if one needs wavenumber accuracy of <1 cm^{-1}.

(v) MI-RS can also be used to clarify the ambiguities which often arise in MI-IRS and the determination of vibrational spectra and molecular structure of new species.

(vi) The application of lasers has enabled the observation of resonance Raman spectra. This provides harmonic and anharmonic vibrational constants from overtone bands, permits photolysis to create new chemical species, and enables aggregation of defects as a consequence of local heating of the matrix material.

(vii) With a laser one can monitor different areas of the matrix by scanning the laser beam over the surface of the matrix in order to check the uniformity of the matrix.

(viii) The determination of Raman data for matrix-isolated high temperature species which are normally not accessible because of temperature and pressure, and of special species whose fluorescence dominates and obliterates vapor phase Raman spectra.

(ix) The reason to use RGS as matrix material is physically well understood. The extension to non-RGS as host crystals allows the investigation of further physical aspects, e.g., N_2-matrix for reasons of symmetry and for the use of a Raman scattering standard, CO_2 matrix for additional intra- and intermolecular modes, CO matrix for permanent dipole moments, CCl_4, CH_4, etc.

There are several factors which mitigate against the success and the applicability of MI-RS. These include:

(i)   the light scattering properties of the matrix,

(ii)   inhomogeneous broadening of Raman lines which limits spectral resolution,

(iii)   fluorescence from the MI species themselves and from contaminants,

(iv)   the problem of distinguishing between resonance Raman and resonance fluorescence,

(v)   local heating by the incident radiation,

(vi)   the need for relative high matrix-solute ratios (M/S) because of the weakness of Raman effect, and

(vii)   the inferior resolution imposed by the large spectral slit-widths which are normally required to achieve an acceptable signal-to-noise ratio.

Most experiments have been carried out by chemists or physicists. This fact influences, via a feedback, the way samples are chosen and the experimental design. Chemists are, for example, interested in the search for new compounds and their spectroscopic constants (XeF_6), in the isolation of reactive species (NO, NO_2), and the investigation of chemical reactions (diffusion of defects and aggregation, photolysis and thermolysis, charge transfer reactions). Further aims include the examination of high temperature molecules (C_4, for example) and radicals (CH_3) and the investigation of chemistry in excited states. Physicists are more interested in comparing experimental data with theoretical models. They therefore investigate the influence of parameters like temperature and the type of matrix and explore relaxation processes like energy transfer from one defect to another one or to the lattice. Also of interest is the discussion of different anharmonic contributions from the defect (e.g. anharmonic potential parameter), the matrix (e.g. lattice anharmonicities) and the defect-matrix system (e.g. coupling of defect transitions with lattice modes, Huang-Rhys-factor). The influence of different symmetrical surroundings to the defect (like N_2 in a N_2-lattice or in a RGM), the importance of impurity induced phenomena (e.g. phase transition in doped RGS), the contribution of dimers or polymers to monomeric states (Ag, Ag_2, Ag_x in RGS), the transition of MI-systems to alloys (Cu-atoms in RGM; Cu-sublattice), the combination of MI-impurity with a lattice defect, i.e. a vacancy (C_6H_6 in Ne) and the search for applications of MIS (e.g. eximer laser like KrF) are other important topics.

B. Shortcomings of the Theory

In the theory of MI-systems it has been usual to make a considerable number of approximations. It is good to be aware of them when discussing the capability of theoretical models and when planning new experiments to test the theory.

The Born Oppenheimer approximation is used to separate nuclear and electronic transitions. The large difference between high frequency internal molecular (solute S) and low frequency intramolecular vibrations of the matrix (matrix M) often implies the rigid molecule approximation. It is common to describe the behavior of the nuclei as harmonic. One therefore often ignores anharmonic contributions (non-linear terms in the mechanical and electrical parts of the defect-matrix interaction). In some cases the center of mass and the center of interaction does not coincide for the defect in the cage; considering this fact the coupling of different modes (rotation/translation, vibration/translation, etc.) can be described. Temperature dependent effects are incorporated by a mere deformation or volume part and by an interaction or a multiphonon part. Usually no term in the potential energy refers to more than two particles (M-M, M-S, i.e. pair approximation). Different mathematical functions are chosen to fit the interaction potential. Combination rules are used to generate the potential parameters for the defect-RG-atom interaction from the potential M-M and S-S. Symmetry arguments and dimensionality of the model (linear chain model) are incorporated to simplify the mathematical procedure. Most theory refers to one defect space approximation where interactions between trapped defects S are negligible (low concentration limit M:S $\geq$ 100:1). In this case one assumes a perfect RG-lattice which contains only a single defect on a substitutional lattice site and which is confined to a cage formed by its neighbors. The defect is treated by mass changes and force constant changes in the case of atomic impurities, or considered as rigid rotators or free vibrations in the case of molecular impurities. The change in local polarization enters theory via effective charges and inductive effects. The assumption that the local situation around a defect can be described by the overall picture, for example, relative changes in bondlength of the defect is assumed the same as the relative change in lattice constants, or force constants of the interaction potential are related to the bulk compressibility. In addition there are the rather crude assumptions concerning the change of polarizability, effective charge, quadrupole moment by the substitutional defect.

A possible way to test assumptions concerning the situation around a defect is by EXAFS-measurements (extended X-ray absorption fine structure).

Nowadays it is possible to determine local changes in lattice constants on the % or %0 level, when a defect S is trapped in a RGM.

C. Historical Development

During the very beginning of MIS the opinion was common that the theory of RGS was easy because of the simple crystal structure and closed shell character, whereas classic solid state experiments like single crystal growth, neutron scattering, heat conductivity, optical spectroscopy, etc., turned out to be enormously difficult. In the course of time this situation has changed. As the experimental difficulties were overcome the more theoretical problems arose. A certain success was achieved for topics like matrix shifts when comparing theoretical predictions with experimental data, but not for others like the phase stability at T = 0K or at low impurity concentration.

Historical beginning of MI-RS: The use of matrices was introduced by Pimentel and coworkers in 1954. The first experiments in laser Raman matrix isolation were executed on thin films by Shirk and Claassen [1], Ozin [2], and Nibler [3], on doped single crystals by Jodl et al. [4] and on free radicals and ionic species by Andrews and coworkers [5]. Technical inventions also influenced the development of MIS. For MI-RS the advent of different laser types, closed cycle cryostats, triple-monochromators and photo-electronic detection systems have been important.

It is worthwhile to point out new trends in MIS in general as a feedback to MI-RS or as a demonstration of the capability of this method (only a restricted collection is mentioned).

(a) physical questions: such as compressive metallization of the inert gas solids; like optical spectroscopy of MI-clusters; the transition from atom to bulk and from clusters to microcrystals; such as the mixing of different binding mechanisms, for example van der Waals interaction for MI-Na in RGS and covalent interaction for Na_x in RGS; like relaxation processes and time dependent spectroscopy; chemisorption and reactive processes of MI-species.

(b) sample preparation: slow spray on or pulsed spray on condensation; like supersonic molecular beams and RG-atoms being solidified; such as collecting sputtered atoms in a cryogenic noble gas matrix; doping of a RG-crystal via nuclear reactions ($Ar^{40}(n,\beta) K^{41}$); like new sample holders or special wave guides for light propagation in a frozen gas matrix.

(c) measuring techniques: the application of every standard spectroscopic method from synchrotron radiation to Far-IR, including tunable lasers and time-dependent spectroscopy; laser induced photochemistry; like simultaneous spectroscopic and non spectroscopic investigations in the self same matrix; like Far-IR Fourier transform spectroscopy; combination of high pressure and low temperatures with MIS; like magnetic circular dichroism spectroscopy, application of attenuated total reflection (ATR) in conjunction with IR-spectroscopy and photoconductivity; like ESR, NMR and Mössbauer analysis.

(d) new species in MIS: in addition to the classical MI-substances like RG-atoms in RG-solids, or alkali atoms in RGS, or gaseous di- and tri-atomic molecules new species were chosen: Metal atoms and aggregates up to four particles (Ag, Au, Cu, Ta, W, Mo), transition elements isolated and halogenated like Ce, Pr, Sm, Eu, Tb, Th, Sc, Y, La, Nd, Gd, carbon-hydrogen combinations and new compounds like C_4 and UF_6.

(e) application of MIS to other fields: laser induced isotope enrichment in a RGM; such as diagnostic investigations of low pressure plasma by MIS; the measurement of chemical composition of the atmosphere using a modified MI-technique; the determination of sputtering yields via MIS.

D.   Reviews Concerning MIS

Perhaps the most significant recent development in the field of MIS has proven to be the laser-Raman matrix technique. This chapter will cover portions of the literature appearing in the past decade as far as MI-RS is concerned but will include all the main reviews dealing with MIS as a whole.

Since the appearance of the three books, Low Temperature Spectroscopy (Meyer [6]), Vibrational Spectroscopy of Trapped Species (Hallam [7]), and Cryochemistry (Moskovits and Ozin [8]), it is no longer appropriate to deal at length with experimental aspects of the subject such as generation of species, nature and preparation of the matrix, spectroscopic properties of matrix systems and spectroscopic method generally. Almost each year at least one review has appeared dealing with MIS, MI-IRS or MI-RS. We list them in chronological order below.

Impurities and Alloys - Jodl, Bruno and Lüscher [9]. Production and measurements/ Point defects and single impurities/ impurity clusters and compounds. A review in this early stage could cover only very general the beginning problems of the whole field of MI.

Matrix Isolation and Molecular Spectroscopy - Downs and Peake [10]. This is a very thorough, critical and complete review. Different classes of MI-species and their generation are discussed. The nature and preparation of the matrix and their physical and technical parameters, spectroscopic properties of matrix-trapped systems, and spectroscopic methods are included. The conclusion calls for strict scientific procedures in MIS. An appendix of fifty pages contains a data collection (1960-1972).

Raman Matrix Isolation Spectroscopy - Nibler [11]. Some efforts at solving the experimental difficulties and applying polarized RS to MI are presented and the value of this technique for a few chemical problems is illustrated. Structural aspects and vibrational assignment, superposition of fluorescence and RS are considered.

Matrix-Isolation Laser Raman Spectroscopy - Ozin [12]. The advantages and limitations of MI-RS are discussed in the light of the Raman technique and in the special case of MI-systems. MI-RS is applied to selected examples like microwave discharge reaction products, high temperature species, highly reactive and thermally unstable compounds.

Infrared and Raman Spectra of Unique MI-Molecules - Andrews [13]. Reviews between 1971-75 are discussed. The advantages of MI-RS in comparison to MI-IRS are counted and the physical relevant parameters are quoted. In combination with IR spectroscopy RS results for the class of MI-species, metal dihalogen molecules, are presented.

Characterization of the Products of Metal Atom-Molecule Cocondensation Reactions by MI-IRS and MI-RS - Moskovits and Ozin [14]. Normally unstable chemical species are stabilized by immobilizing them in an RGM. The application of MI-IRS and MI-RS yield information about the identity of the products, coordination number, mode of ligand attachment, molecular structure and thermodynamic and bonding properties.

Infrared and Raman Spectroscopic Studies of Alkali-Metal-Atom Matrix Reaction Products - Andrews [15]. The experimental goal is extensively described to bring alkali metal atoms like Li, Na, K, Rb, Cs and reactive molecules (NO^-, O_2^-, O_3^-, Cl_2^-, ClO_2^-, N_2^-, N_2O^-, NO_2^-) together long enough for primary reactions to take place and then quickly trap them in a matrix for spectroscopic studies by IR and RS.

Laser Excitation Matrix-Isolation Spectroscopy - Andrews [16]. The laser excitation technique offers several advantages over the IR experiment: (1) the possibility of obtaining new chemical species by laser photolysis, (2) the observation of resonance Raman spectra which provide harmonic and anharmonic vibrational constants from overtone bands, and (3) the possible observation of fluorescence from new chemical species. Specific examples are discussed in combination with experimental tips for the chosen species considering sample preparation and Raman investigations.

The Study of the Shapes of Inorganic Molecules Using Vibrational IR and R-Spectroscopy - Barraclough, Beattie and Everett [17]. Application of

MI-IRS and MI-RS to high temperature gases like M^+F_5 (M^+ = Nb, Sb, Ta) in combination with non spectroscopic measurements. The difficulties and advantages of MI studies are also mentioned.

Vibrational Energy Levels in Matrix Isolated Species - Burdett, Poliakoff, Turner, and Dubost [18]. Only selected specific problems involving vibrational energy levels of trapped species are discussed along with the determination of molecular constants of triatomic molecules, the study of transition metal carbonyl in matrices and IR energy processes in matrices.

Spectroscopic Identification and Characterization of Matrix Isolated Atoms - Gruen [19]. The correlation between gaseous atomic spectra and the absorption spectra of atoms isolated in noble gas matrices is presented and applied to about forty elements belonging to each chemical group I-VII.

Matrix Isolation in Vibrational Spectroscopy - Barnes and Hallam [20]. Experimental technique of MI-IRS and MI-RS and matrix effects like energy shifts, rotation/vibration of species, multiple trapping sites, aggregation are described and applied either to simple cases (e.g. HCl) or very complicated ones (e.g. C_2H_5OH).

International Conference on MIS (Berlin 1977-1978) [20a]. Presentations covered the whole area of MIS, the generation of reactive species and their isolation in matrices, spectra of metal atoms and cluster formation, stable molecules in matrices, RS and IR spectroscopy, high temperature molecules, reactive matrices, relaxation phenomena in matrices.

Spectroscopy of Transient Species and Molecular Ions in Matrices - Andrews [21]. MIS is useful for the study of transient species and molecular ions and is applied here to selective topics like isotopic effects of $Li^+O_2^-$, resonance Raman scattering of $Cs^+Cl_2^-$, laser photolysis of OF etc.

Matrix Isolation in "Molecular Spectroscopy" - Chadwick [22]. This report completes the discussion of research involving molecular matrices published between 1974 and 1977 with a helpful table of data in the appendix. In addition, the main trends in MIS during this period are pointed out and well documented by selectively chosen publications.

Matrix Isolation Raman Spectroscopy - King and Stephenson [23]. MI-RS is used for testing field standards and measurement techniqus used to detect trace amounts of small quantities of complex organic molecules and to analyze them spectroscopically.

Absorption and Laser - Excited Fluorescence Spectra of Matrix-Isolated Metal van der Waals Dimers - Miller and Andrews [24]. The electronic structure, bonding mechanism and molecular constants of alkaline earth dimers (Be_2, Mg_2, Ca_2, Sr_2, Ba_2) and group IIB dimers (Zn_2, Cd_2, Hg_2) are extensively discussed. The well known matrix effects, such as solvent effects, site effects, vibrational relaxation, are applied to these species. The summary contains further open questions and suggests interesting experiments.

Third International Meeting on Matrix Isolation - (Nottingham 1981). The presentations were arranged around the invited lectures and mirror in a way the scientific program: Astrophysics and Matrix Isolation, Ions in Matrices, Organic Photochemistry at Low Temperatures, IR Laser Photochemistry, Matrix Effects, Energy Transfer in Matrices, Application of Matrix Isolation to Study of Surfaces, Liquid Matrices, Radiation Chemistry, NMR in Matrices.

Raman Studies of Molecules in Matrices - Downs and Hawkins [25]. This article contains primarily chemistry oriented questions and appropriate species: introduction to and purpose of these experiments/advantages and disadvantages of the application of RS to matrix experiments/practical aspects like equipment, preparation, problems associated with excitation, technique for measuring/specific applications like monomeric halides, dimeric species, mixed carbonyl dinitrogen complexes, metal dimers and clusters/table containing about 180 MI molecules.

A concise textbook for newcomers to the subject is Matrix Isolation: A Technique for the Study of Reactive Inorganic Species - Cradock and Hinchcliffe [26].

E. Scope of the Review

Because MIS has been so extensively reviewed, the material in this chapter deals exclusively with MI-RS where more than one hundred original papers have appeared up to now. It is the aim of this chapter to describe the recent progress in the area of MI-RS and in doing so we focus on four main topics. First, we consider the necessary theoretical background and the physical parameters used in this technique like laser frequency, M/S ratio, polarization, isotopes, temperature, etc. Second, different experimental techniques are reviewed such as Raman geometries, number and types of problems involved in MI-RS and new approaches. Third, we present a complete summary of recent matrix Raman results on di-, tri- and polyatomic molecules embedded in RGS. Finally, we present some more detailed results on a few selected MI-species (our own field of interest).

II. BACKGROUND

A. Theory

The theory of light scattering from free molecules has been discussed by Sushchinskii [27], for solids by Loudon [28], Cowley [29] and Sushchinskii [30], for pure RGS by Werthamer [31] and [32] and the case of Raman, resonance Raman scattering and resonance fluorescence by Rousseau and Williams [33]. Therefore, the primary purpose of this section is to present the theoretical equations which will be used in the analysis of the experimental data rather than to reiterate the theory in detail. For the purpose of this article we shall designate as Raman effects all those inelastic light scattering phenomena in which the scattering mechanism produces a change in the polarizability tensor of the system as a whole (defect plus host lattice).

The total scattering intensity I_S in photons per Raman active particle per second for a transition from the initial ground state, $|G>$, to the final state $|F>$, may be given by the standard expression, for example (Rousseau and Williams [33])

$$I_S = \frac{8\pi \, w_S^4 \, I_L}{9 \, c^4} \sum_{\rho\sigma} |(\alpha_{\rho\sigma})_{GF}|^2 \tag{1}$$

Here c is the velocity of light, I_L is the incident intensity at frequency w_L, and w_S is the scattered frequency. $(\alpha_{\rho\sigma})_{GF}$ is the polarizability tensor for the transition from $|G>$ to $|F>$ with the incident and scattered polarizations indicated by σ and ρ, respectively. The second order perturbation expression for $(\alpha_{\rho\sigma})_{GF}$ is

$$(\alpha_{\rho\sigma})_{GF} = \frac{1}{\hbar} \sum_I \frac{<F|p_\rho|I> <I|p_\sigma|G>}{w_{GI} - w_L} + \frac{<I|p_\rho|G> <F|p_\sigma|I>}{w_{IF} + w_L} \tag{2}$$

In this expression w_{GI} and w_{IF} are energy spacings, w_L is the laser frequency and p is the electron momentum operator. This equation is completely general and is expected to correctly treat both Raman and resonance Raman (resonance fluorescence) - processes within the limits of perturbation theory. If the incident laser energy is very much less than the energy needed for a real transition, i.e. $w_L \ll w_{GI}$, this represents off-resonance light scattering (normal RS). In the limit of the laser frequency

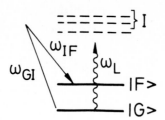

FIG. 1.

coinciding with a real transition ($\omega_L \cong \omega_{GI}$) for some $|I>$, independent of whether or not the levels are discrete or continuous, a situation arises which is completely different from non resonance case (discrete and continuous resonance RS). This general approach combines very elegantly the empirically found marked differences between RS, resonance RS and resonance fluorescence.

The time dependence of the polarizability α necessary to scatter light inelastically can generally be attributed to fluctuation in the local electric field $\vec{E}_{loc}$ at the scattering particle

$$\vec{d}_{ind.} = \alpha \, \vec{E}_{loc.} = \tilde{\alpha}_{effect.} \, \vec{E}_{ext.} \tag{3}$$

The effective polarizability tensor $\tilde{\alpha}_{eff}$ is then a function of the instantaneous coordinates of all particles in a medium and can be expanded in a power series of the relative displacement r of the inherent Raman active mode

$$(\alpha_{\rho\sigma}) = \alpha_{\rho\sigma}^{(o)} + \sum_{\mu} \left(\frac{\partial\alpha_{\rho\sigma}}{\partial r_\mu}\right)_{r=o} r_\mu + \sum_{\mu\nu} \left(\frac{\partial\alpha_{\rho\sigma}^2}{\partial r_\mu \partial r_\nu}\right)_{r=o} r_\mu r_\nu + O(r^3) \tag{4}$$

The first nonvanishing term, that changes the polarizability or the symmetry of the states $|G>$ and $|F>$, mainly determines the feature of the Raman spectrum.

In the case of matrix-isolated species different modes are Raman active in principle according to the kind of doping material (Fig. 2.). Polyatomic molecules in RGS do not reveal any new aspects but the number of Raman active internal modes increase rapidly.

1.   High Frequency Raman Spectrum

The electronic ($\Delta\nu \sim 10^4$ cm^{-1}), vibrational ($\Delta\nu \sim 10^3$ cm^{-1}) and rotational ($\Delta\nu \sim 10$ cm^{-1}) Raman effect of the matrix isolated defect may be described

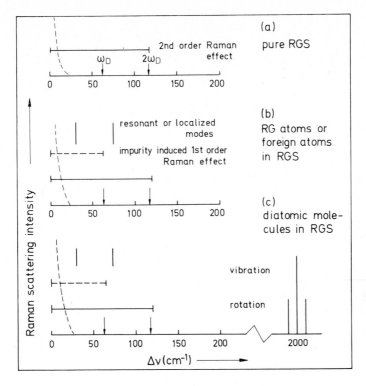

FIG. 2. Raman active modes of the impurity/RGS-system.

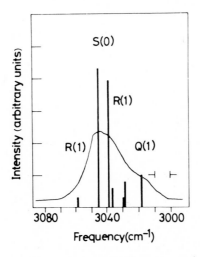

FIG. 3. Calculated spacings and Raman scattering intensities for the vibration-rotation states of CH_4 molecule in a Kr matrix. The curve is the experimental result. Used by permission [35].

similarly by substituting the relevant states |G> and |F> in Eq. (2) by electronic, vibrational or rotational wavefunctions and by using the appropriate relative displacements according to the different cases. Group theory predicts the number of transitions which are Raman or IR active for all molecules provided that the shape of the molecule or its point group is known (Fateley et al. [34]). As a representative example of the literature we quote here the case of matrix isolated CH_4-molecules discussed by Kobashi et al. [35]. In order to derive the selection rules for the Raman scattering of the rotationally split ν_3 and ν_4 vibration they utilized the full symmetry of the Hamiltonian and the symmetry properties of the rotational and vibrational-rotation states. The relative intensities and spacings of the Raman spectra were calculated and successfully compared with experiment.

It is interesting to mention here the "vibrational overtone" Raman effect. The selection rule for a vibrational Raman effect is $\Delta \nu = \pm 1$; an overtone Raman effect is a process in which $\Delta \nu = \pm 2$. This transition can arise from two different causes, either the mechanical anharmonicity of the harmonic oscillator or a non linear term in the polarizability, i.e. $(\partial^2 \alpha / \partial r^2) \neq 0$, the electric anharmonicity. If the rotational motion of the matrix isolated molecule in the RGS is free, nearly free or hindered, depending on the ratio of potential barrier to rotational constant, the observable frequencies in RS are markedly different. In the case of polyatomic molecules, combinations of vibrational modes are also detectable. The following sections deal mainly with intramolecular modes where some examples will be described in detail.

2.   Raman Effect in RGS

The theory of light scattering from RGS by assuming a simple model for their dielectric properties is developed by Werthamer and coworkers [31,32,36]. In the case of hcp and fcc RGS the conservation of momentum and energy and symmetry arguments reduce the complicated structure of the Raman scattering tensor. This fourth rank tensor then has only three independent elements. If this tensor is expanded as a power series in the displacements, the first non vanishing term in an ideal fcc RGS corresponds to the two phonon Raman scattering or second order Raman effect. If the translational symmetry of the crystal is disturbed we can also expect one phonon Raman scattering or impurity induced first order Raman effect. Only those phonons which transform like the Raman tensor contribute to the Raman scattering. The continuous spectrum of the second order Raman effect is proportional to the weighted phonon density of states; the weighted coefficients being regulated by the electron-phonon interaction.

Klein and Koehler [37] compared the theoretical approach of Werthamer with the recently reported experiments on light scattering from solid Xe, Kr and Ar (Fleury et al., [38]; see also Crawford et al., [39]). The RGS can crystallize either in the fcc or hcp structures. The components of the polarizability tensor for O_h and D_{6h} symmetry are listed in Table 1 (Sushchinsky [30]).

TABLE 1

fcc O_h

$$
\overset{A_{1g}}{\begin{pmatrix} a & . & . \\ . & a & . \\ . & . & a \end{pmatrix}}
\overset{E_g}{\begin{pmatrix} b & . & . \\ . & b & . \\ . & . & b \end{pmatrix}}
\overset{E_g}{\begin{pmatrix} b & . & . \\ . & b & . \\ . & . & b \end{pmatrix}}
\overset{F_{2g}}{\begin{pmatrix} . & . & . \\ . & . & d \\ . & d & . \end{pmatrix}}
\overset{F_{2g}}{\begin{pmatrix} . & . & d \\ . & . & . \\ d & . & . \end{pmatrix}}
\overset{F_{2g}}{\begin{pmatrix} . & d & . \\ d & . & . \\ . & . & . \end{pmatrix}}
$$

hcp D_{6h}

$$
\overset{A_{1g}}{\begin{pmatrix} a & . & . \\ . & a & . \\ . & . & a \end{pmatrix}}
\overset{E_{1g}}{\begin{pmatrix} . & . & . \\ . & . & c \\ . & c & . \end{pmatrix}}
\overset{E_{1g}}{\begin{pmatrix} . & . & -c \\ . & . & . \\ -c & . & . \end{pmatrix}}
\overset{E_{2g}}{\begin{pmatrix} . & d & . \\ d & . & . \\ . & . & . \end{pmatrix}}
\overset{E_{2g}}{\begin{pmatrix} d & . & . \\ . & -d & . \\ . & . & . \end{pmatrix}}
$$

The energy and full description of the two phonon contributions to the Raman active modes at critical points of the Brillouin zone are gathered in Table 2.

TABLE 2

Point[b]	Phonon contributions	Reduced Raman active combinations	energy cm^{-1} for ^{36}Ar[a]
X	$\pm$ 2 T A	$A_{1g} + E_g + F_{2g}$	96
	$\pm$ L A $\pm$ T A	F_{2g}	115
	$\pm$ 2 L A	$A_{1g} + E_g$	134
L	$\pm$ 2 T A	$A_{1g} + E_g + F_{2g}$	64
	$\pm$ L A $\pm$ T A	$E_g + F_{2g}$	101
	$\pm$ 2 L A	$A_{1g} + F_{2g}$	138

[a] Fujii et al. [40], see also Klein and Koehler [37] (Fig. 23).
[b] Klein and Venables [41], p. 596.

3.   Low Energy Raman Spectrum

If the kind of defect is such that it changes neither force constants nor crystal structure we have to take into account only breaking of the translational invariance of the lattice.   As a consequence, energy and momentum conservation is no more valid and phonons of the whole Brillouin zone will contribute to the scattering process.   For this impurity induced first order Raman effect one would expect a broad spectrum which resembles the form of the one phonon density of states distribution (see for example Klein and Venables [41], Vol. II p. 969).   The interpretation of the spectra is complicated somewhat by the numerous modes throughout the Brillouin zone for which activity is induced, even when the dispersion relations for the perfect crystal are available.

TABLE 3

Raman Allowed Symmetry of the Phonon Branch in Doped f_{cc}-Structure, Partially the L-point

phonon branch	phonon mode	space group	Reduced Raman active combinations	energy (cm^{-1}) for ^{36}Ar
L - point	TA	L_3	$E_g + F_{2g}$	32
L - point	LA	L_1	$A_{1g} + F_{2g}$	69

The Raman efficiencies for impurity induced one phonon and two phonon scattering are of the same order and depend mainly on powers of u, the amplitude of motion about the equilibrium atomic position at spacing a, on a itself, and on α_o, the atomic polarizability.   Werthamer et al. [36] calculated these scattering efficiencies for Ne and Ar and Werthamer et al. [42] that of He.   Fleury et al. [38] estimated experimental values of the Raman-intensities of the heavier RGS and Slusher and Surko [43] that of solid He.

Although solid Rare Gases are found to crystallize in fcc-structure, small amounts of N_2 or O_2 can modify the fcc structure and can lead to hcp admixtures (Barrett and Meyer [44], Barrett et al. [45], Meyer [46]).   As a consequence we have to consider the Raman activity of hcp-phase with D_{6h}-symmetry also.   At the Γ-point of the hcp-Brillouin zone the irreducible representation of the phonons are A_{2u}, B_{1g}, E_{2g} and E_{1u}.   The uneven modes belong to the acoustic branches, the even ones can be considered as the quasioptic TO and LO phonons.   Only one of these modes, the doubly degenerated E_{2g} quasioptic TO-phonon, is Raman active (Table 1).   The corresponding mode in fcc structure is the TA phonon at the L-point of the

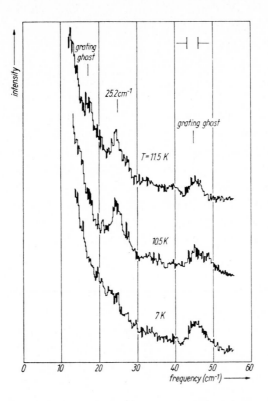

FIG. 4. Unpolarized Raman spectra from a neon crystal at three temperatures during heating. Resolution 3 cm-1. Used by permission [47].

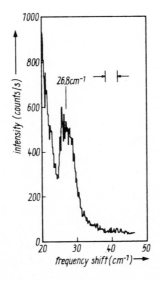

FIG. 5. Raman spectrum of an Ar crystal at T $\sim$ 39 K. Resolution 3.5 cm-1. Used by permission [48].

Brillouin zone. If only neighboring planes are considered the two corresponding modes will have exactly the same energy since a B-plane vibration
within the stacking sequence ABA for hcp is equivalent to that within the
sequence ABC for fcc. The difference between fcc and hcp-structures is in
the third nearest neighbors. Even if these are included the relative change
in the force constants is very small. From this point of view we would
expect an additional Raman peak on top of the impurity induced first order
Raman spectrum; this Raman peak is generated by the structure and is
independent of the kind of the matrix isolated-substance. The following
experimental results are known: a sharp peak at 9.4 cm^{-1} for hcp He
(Slusher and Surko [43]), one at 25.2 cm^{-1} for hcp-Ne (Schuberth et al.
[47]) and one at 26.8 cm^{-1} for hcp-Ar (Schuberth [48]).

When an otherwise perfect lattice is locally disturbed by an impurity, we
have to recalculate the lattice dynamical problem to get the phonon modes.
Even in the simplest case, a linear chain with atoms of mass m and force
constants k with one defect of mass m' and force constant k', all resonant or
local modes may appear. Barker and Sievers [49] studied such modes in
doped alkali halides and semiconductors and Cohen and Klein [50] applied
existing theoretical results to study rare gas mixtures like Ar: Ne, Ar: Kr,
Kr: Ar and Kr: Xe. These new modes were first seen by IR absorption and
so most of the theory considers only this activity. Until recently only one
Raman experiment was known (Brunel et al. [51]), although many groups try
to measure impurity induced MI-RS. The laser light was focused at a glacing
angle onto the thin film, collected in a double grating monochromator in
conjunction with an iodine filter to suppress scattered radiation, and the
spectra (Fig. 6) recorded between 10 and 100 cm^{-1}. The authors tentatively
assign the peaks at 20, 22 and 27 cm^{-1} in Xe, Kr and Ar, respectively, as
phonon band modes induced by the guest molecules (here HCl). The mode at
70 cm^{-1} in Xe is ascribed to a localized mode and the peak at 54 cm^{-1} in Xe
may correspond to the edge of the phonon band. Mannheim [52] pointed out
that localized modes may be produced not only when the mass of the impurity
is smaller than the mass of the host but also when the force constants
increase. The distinction between in band and localized modes must be
carefully examined.

Very recently Nosé and Klein [53] presented the results of a molecular
dynamics calculation on HCl molecules matrix-isolated in solid fcc argon.
From their considerations it is immediately apparent that the zone boundary
TA phonon at the L-point of the Brillouin zone is exactly the frequency seen
in the Raman spectrum (for comparison Table 3 mode energies at 10 K, Fig. 6
27 cm^{-1} in Ar at 22K).

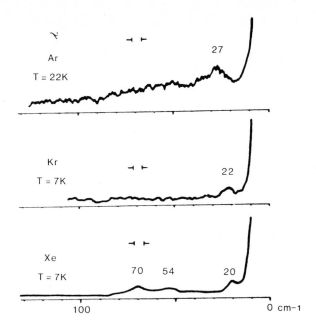

FIG. 6. Low frequency Raman spectra of HCl in argon, krypton and xenon matrices HCl/Ar = 2/100, HCl/Kr = 2/100, HCl/Xe = 2/100. Used by permission [51].

Besides the above described experiment on HCl in Ar, Kr, Xe, a further experiment is documented concerning the low frequency Raman spectra of single crystals of argon doped with O_2, N_2 or CO. These spectra show two distinct features superimposed on a sloping background: one peak at 10 cm^{-1} for Ar-O_2, 15 cm^{-1} for Ar-N_2 and 22 cm^{-1} for Ar-CO; the other one appears at the same frequency in all the spectra at about 55 ± 3 cm^{-1} for the crystals at 80K (Rich et al., [54]).

To close this section we would like to emphasize the information we can deduce from Raman scattering, like

- energy spacings, energy shifts and fine structure
- full width half maximum and lifetime data
- intensities, changes in polarizability, depolarization ratios, number of active centers, occupation number
- anharmonicities and overtone frequencies
- symmetries of different modes,
- detection of new species,
- additional information by combination with IR-spectroscopy data.

But when scanning through MIRS-literature, some unnecessary diffi- culties arise. For example the Raman data are not consistently defined, used and published: when an observed Raman line ν_i is given or quoted it is unclear whether the fundamental frequency ω_o or its anharmonic corrected value ω_e (see Table 5 in Sec. IV) is meant. The formal interrelation is well known in standard literature. Similar problems may arise, when it is not mentioned how the experimental spectroscopic data were calibrated or whether the Raman values are vacuum corrected. The severest fault to complain about is that when unwanted impurities are detected but not specific physical statements are made about the MI-system: e.g. molecule A in RGM, N_2-Raman line in the spectra. This will be discussed later in Sec. III. The most sophisticated experimental design, the highly developed apparatus and the most interesting physical problem is of no use if unscientific methods are apparent in their publication.

B. Physical Parameters

The relevant parameters used in MI-RS can be separated into those of technical character (e.g. influence on sample preparation or on Raman geometries - see Sec. III) and physical ones:

- concentration ratio (M/S),
- kind of matrix,
- sample growth and preparation,
- temperature, pressure and other external influences,
- depolarization ratio,
- isotopic effects and
- laser frequency.

To fulfill the single defect approximation the M/S ratio must exceed 100:1 which ideally implies a defect-defect separation of about 20 Å. Although the Raman scattering cross-sections of the MI-species are very small, they are detectable by modern Raman technology (Table 4). Figures 7-9 demonstrate the increase in Raman intensity with defect concentration. Sometimes the monomeric lines broaden and, in addition, Raman signals due to polymers appear.

Generally, MIS uses either inert spherically symmetric matrices (Ne, Ar, Kr, Xe) or inert oriented matrices (e.g. N_2) or reactive matrices (e.g. CO) depending on the physical question to be considered or on the technical equipment available (e.g. lowest accessible temperature or the type of cryostat). Neon as a matrix produces samples of very good spectroscopic quality but delivers additional difficulties like multiple trapping sites and the beginning of quantum-mechanical behavior. One uses argon, krypton and

TABLE 4

Common Atmospheric Gases[a]

Molecule	Raman shift (cm^{-1})	Raman scattering-cross section
SO_2	1151	5.5
H_2S	2611	6.6
N_2	2331	1.0
O_2	1556	1.2
O_3	1103	4.0
H_2	4160	2.2
CO	2145	0.9
CO_2	1388	1.5
NO	1877	0.6
NH_3	3334	3.1
H_2O	3652	2.5
CH_4	2914	8.0

[a]Used by permission [55].

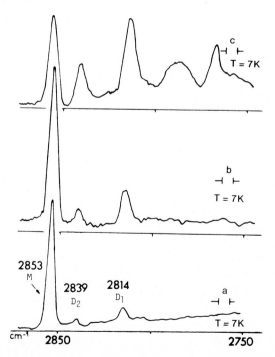

FIG. 7. Effect on the Raman spectra of HCl concentration in Kr matrices. (a) HCl/Kr = 1/100; (b) HCl/Kr = 2/100; (c) HCl/Kr = 10/100. All samples deposited at 7K with a low deposition rate. Used by permission [51].

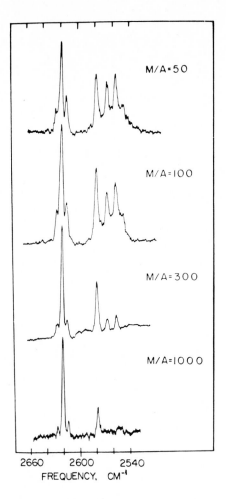

FIG. 8. Raman spectrum of H_2S isolated in Ar at 20 K; M/A = 50, 100, 300, 1000. All spectra were measured with a spectral slit-width < 3 cm^{-1}. Used by permission [56].

xenon as a series if one is interested in the matrix shift of a special impurity. Nitrogen is sometimes favored as a matrix because the N_2-fundamental vibration may be used as a Raman line standard for the impurity lines and because of possible crystal field splitting of the Raman line of the MI-species due to the T_h factor group of the nitrogen crystal structure (Fig. 10, inset d).

Other kinds of matrices have been used such as CO_2, CO, CCl_4, etc. which complicate MIS because of their additional properties (like electric dipolmoment, chemical reactivity or intra-molecular matrix modes) but this will not concern us here (Smith et al. [59]; Barnes and Hallam [20]).

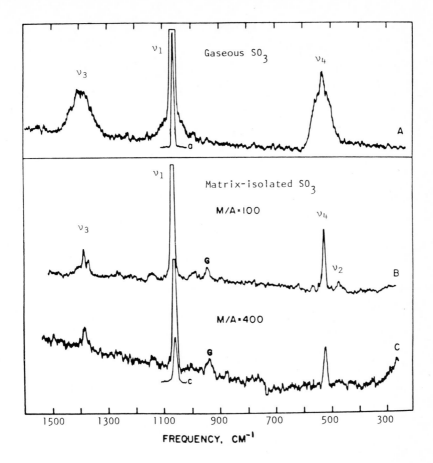

FIG. 9. Raman spectra of gaseous and matrix-isolated SO_3, slit-width (sensitivity in cps): A = 8 cm^{-1} (300), a = 5 cm^{-1} (3K); B = 8 cm^{-1} (1K); C = 11 cm^{-1} (3K); G grating ghost. Used by permission [57].

It is perhaps superfluous to stress that sample purity, generation, and preparation determine decisively the quality of MIS. Slow or pulsed spray on, thin film condensation, crystal growth from the liquid of the impurity/host matrix mixture or other kinds of solid state doping and annealing effects have been employed. This aspect will be discussed in the next section. In general MI-RS is carried out using condensed thin films and almost no systematic investigations are dedicated to complex matters like the comparison of surface and bulk effects, and inhomogeneous line broadening due to solid state conditions etc.

Temperature is used as a parameter in MI-RS for two reasons. First, it affects the deposition and annealing procedure (see Sec. III) and second there is the effect on the Raman spectra of MI-species. The temperature-

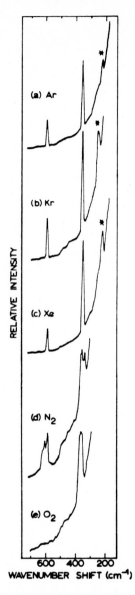

FIG. 10.   Rotational Raman spectrum of hydrogen in various matrices at
12K.   Sample concentrations are M/R = 50/1; spectra scanned at 50 cm-^{1}/min
with slit-widths = 500μ/500μ/500μ.   Spectrum (a) argon matrix, scanned on 10
× 10-^{9}A range, 4880Å excitation (700 mW); (b) krypton matrix, 3.0 × 10-^{9}A
range, 5145 Å excitation (800 mW); (c) xenon matrix, 3.0 × 10-^{9} A range,
4880 Å excitation (700 mW); (d) nitrogen matrix, 10 × 10-^{9}A range, 5145 Å
excitation (800 mW); (e) oxygen matrix, 10 × 10-^{9}A range, 5145 Å excitation
(800 mW).   Used by permission [58].

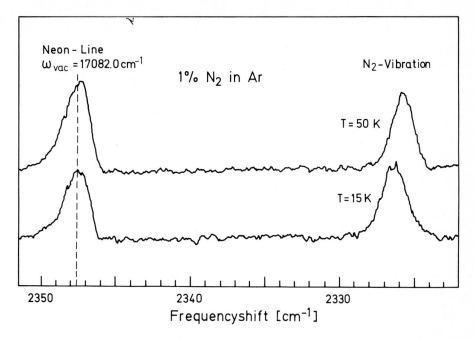

FIG. 11. The Raman spectrum of 1 mol % N_2 in Ar at various temperatures. Used by permission [60].

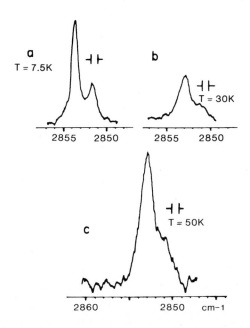

FIG. 12. Temperature effect on the Q branch of the Raman spectra of HCl in krypton. T_d = 7K, HCl/Kr = 1%. (a) T = 7.5K; (b) T = 30K; (c) T = 50K. Used by permission [51].

shift of Raman lines is of the order of -0.05 cm^{-1}/K (Jodl and Bolduan [60]) and is only detectable against frequency standards (Fig. 11 and Sec. V). Very few papers deal with this interesting anharmonic phenomenon effects. In general MI-RS measurements are executed at fixed temperatures lying between 5 and 20 K.

Varying the temperature changes the halfwidth, the monomer/polymer ratio, and the rotational occupation number possibly giving rise to additional Raman lines to be observed. Unfortunately, this parameter is only variable within limits of 5K and 50K because of evaporation of matrix material and therefore the measured absolute temperature shift is of the order of 1 to 2 cm^{-1}.

Pressure as a parameter has a certain importance in the category of external influences on MI-species like electric or magnetic fields. The separate variation of temperature and pressure deliver parameters which characterize the anharmonic behavior of the impurity-lattice system (see Sec. V). Nevertheless, the combination of MI-RS with high pressure and low temperature is extremely difficult from the experimental point of view (Jodl and Holzapfel, [61]). Gas pressure cells allow a variation from 1 bar to $\leq$ 10 kbars at 5K (Jodl and Holzapfel, [62]). This pressure variation generates a typical frequency shift of $+1$ cm^{-1}/kbar in RGM (Jodl and Bolduan, [60]) and was again only detectable against a frequency standard with sufficient accuracy (Fig. 13).

According to theory, the depolarization ratio $\rho = (I_{\perp}/I_{\parallel})$ for symmetric vibrations of free molecules and should be less than 3/4. The determination of this quantity is a means to distinguish between symmetric and asymmetric vibrations in the case of polyatomic molecules. Applying this definition of ρ in the gaseous or liquid case to the MI-case implies that the guests are assumed to be a perfectly random collection of non interacting species in a weakly interacting host. This parameter is easy to measure in MI-RS and delivers additional information about the impurity/matrix system (symmetry assignment, crystal quality, orientation of the defect, comparison of ρ values for the free molecule, MI-molecule or pure solid molecular crystal). Figure 14 shows an example where $\rho = 0.5$ (Ar/$^{16}O_2 = 100$), whereas $\rho = 0.1$ for Ar/N_2 $= 100$ (Jodl and Bolduan, [60]). For comparison $\rho = 0.1$ for pure solid N_2 is shown in Fig. 15.

The use of isotopes is a classical method in the structural analysis of molecules. This procedure is well known in literature. Hence, it is no surprise that this technique was applied to MIS from the very beginning. Isotope splittings range from a few cm^{-1} (halogens), to about 40 cm^{-1} (N_2) and ~1000 cm^{-1}(H_2). However, judging from literature this technique has, in practice, been limited to easily accessible species.

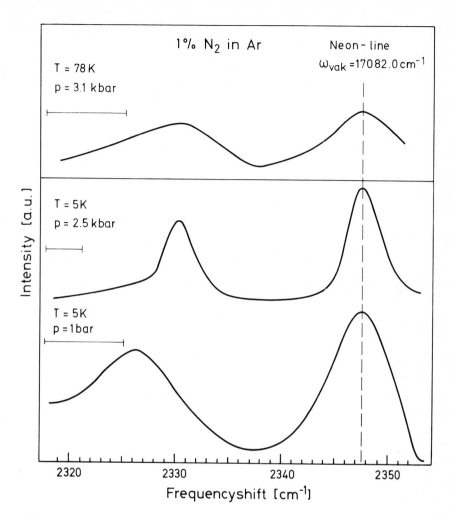

FIG. 13. The Raman spectrum of 1 mol % N_2 in Ar at various pressures and different temperatures. Apparatus conditions were: excitation with 5145 Å, 300 mW and spike filter, slit-width of the triple grating monochromator is indicated by horizontal bars, full scale scattering intensity of about 300 c/s and signal to noise ratio between 1.3 and 2. Used by permission [60].

Friedman et al. [65], for example, monitored simultaneously the isotopic composition of MI-Br_2 using both Raman scattering and relaxed fluorescence as the excitation is tuned over a sharp resonance. Moskovits and Dilella [66] compared the relative abundance of $^{58}Ni_2$ and ^{60}Ni ^{58}Ni with the isotopic split Raman intensities of MI-nickel. As a consequence the Raman lines were ascribed to Ni_3.

Isotopic effects of the halogen atoms, the zinc atom or both (Givan and Loewenschuss [67]), of halogenated mercury (Givan and Loewenschuss [68]),

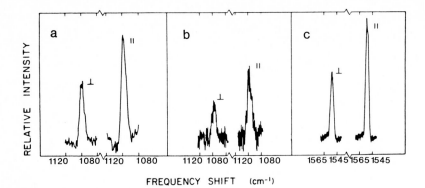

FIG. 14. Raman polarization data for the $^{16}O_2$ - $^{16}O_2$ fundamentals using a 8 mM sample of $^{16}O_2$ gas diluted in argon (M/R = 100) deposited with sodium atoms at 16°K. Excitation = 4880 Å, resolution = 5 cm^{-1}, scanning speed = 10 cm^{-1}, rise time = 3 sec. Spectrum (a) 10 scans, time averaged, scanning speed (per single scan) = 20 cm^{-1}/min, range = 0.3 × 10^{-9}A, ρ = 0.52; (b) single scan, range = 0.3 × 10^{-9}A; (c) single scan $^{16}O_2$ fundamental, range 1.0 × 10^{-9}A, ρ = 0.53. Used by permission [63].

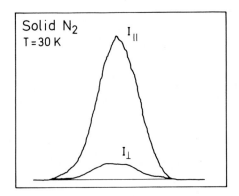

FIG. 15. Raman polarization data for solid N_2 at 30K. Used by permission [60].

of $FeCl_3$ and Fe_2Cl_6 (Givan and Loewenschuss [69]) and of $InCl_3$ (Givan and Loewenschuss [70]) were resolved in the stretching mode bands. These detailed isotopic analyses in conjunction with computer calculation delivers results about assignment of bands, molecular shape, interaction force constants and thermodynamic considerations of the MI-species. Lesiecki and Nibler [64] concluded from the results of isotopic split MI-IRS and MI-RS data (Fig. 16) that the molecules Tl_2F_2 and Tl_2Cl_2 have a planar rhombic structure rather than a linear configuration as previously proposed on the basis of non-spectroscopic investigations. Beattie confirmed by the characteristic Raman line distributions (9:6:1) of the species $^{35}ClXe^{35}Cl$, $^{37}ClXe^{35}Cl$ and

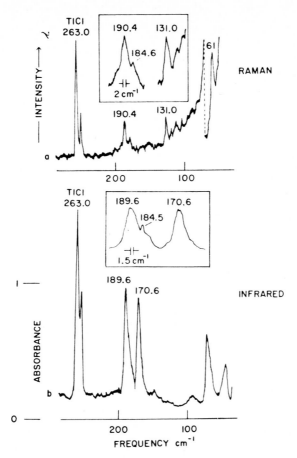

FIG. 16. Vibrational spectra of TlCl/Tl$_2$Cl$_2$ isolated in an argon matrix at 14K. Raman a: 5145 Å, 300 cps, SBW = 2 cm-^{1}; inset: 5145 Å, 100 cps, SBW = 2 cm-^{1}. Infrared b: SBW = 2 cm-^{1}, inset SBW = 1.5 cm-^{1}. Used by permission [64].

^{37}ClXe^{37}Cl the natural abundance of ^{35}Cl/^{37}Cl (3:1) and the presence of just two chlorine atoms in this molecule (Fig. 17; Beattie, work is mentioned in Barraclough et al. [17], Fig. 10).

Familiar to each Raman experimentalist is the necessity to observe the Raman lines of a given species at identical energy intervals off several exciting laser frequencies. A nice example for MI-RS in conjunction with IR measurements is shown in Fig. 18. But in addition sometimes astonishing effects appear, if different laser lines are used; these resonant processes increase the Raman scattering intensity to a much greater extent than expected from the ω^4-law of nonresonance Raman scattering (Fig. 19).

Now a large variety of different lasers whose frequencies range from the beginning of the UV to the near infrared have become available for MI-RS.

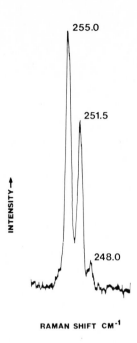

FIG. 17. The Raman spectrum of matrix-isolated xenon di-chloride. Used by permission [17].

Optimal intensities must be chosen individually for each MI-system and range from 100 mW to over 1 W. Problems inherent in laser light scattering from thin films will be discussed in Sec. III and cover local heating, straying light behavior, absorption and scattering processes by the impurity etc.

C. Enhanced Raman Scattering

The availability of new laser types with different or tunable frequencies open the field of MI-species to investigations of the transition from off-resonance Raman to pre-resonance Raman and resonance Raman scattering, of the separation between resonance Raman and resonance fluorescence or of the distinction between discrete resonance and continuous resonance Raman scattering. From the experimental point of view other mechanisms, which drastically enhance Raman intensities, are known like anharmonic effects or Fermi resonance. These are successfully applied to MI-RS by taking advantage of intrinsic physical parameters of the MI-impurity like excitation profiles, mechanical and electric anharmonic interactions or accidental degeneracy of vibrational states, to augment the rather weak Raman scattering cross sections. As shown in Fig. 20 surprisingly low laser power is necessary to obtain a distinct Raman spectrum.

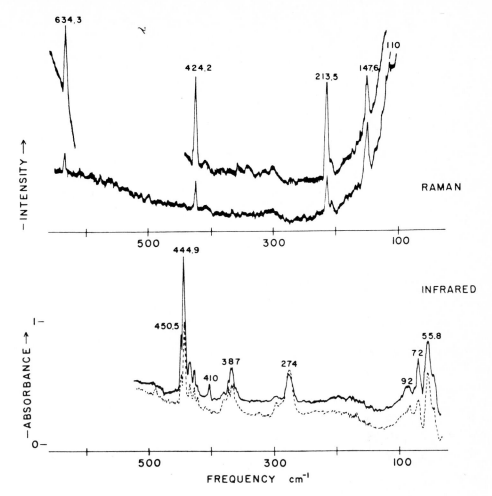

FIG. 18. Vibrational spectra of MgI_2 in Ar at 14K. Raman: 3h at 700°C, 300 cps, SBW = 4 cm^{-1}; upper, 5145 Å; lower, 4880 Å. Infrared: 1.5 h at 700°C, SBW = 2.0 cm^{-1}; 14K - - -, after annealing at 35 - 40K. Used by permission [71].

The overtone intensity pattern and excitation wavelength dependence characterize the Cl_2^- observations as resonance Raman spectra and were the first observation of this phenomena in inert gas matrices. Many investigations on other species followed (see Table Sec. IV).

This resonant enhancement of the Raman cross sections is a well documented effect and the distinction between fluorescence and resonance Raman scattering in the gas phase is the subject of considerable experimental and theoretical attention (Kiefer [74] and Behringer [75]). The present state of this subject is reviewed by Rousseau and Williams [33].

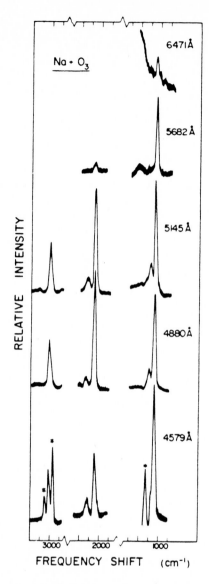

FIG. 19. Emission spectra observed from products of matrix reactions of sodium atoms (Ar/Na ~ 200) codeposited at 16K with ozone (Ar/O$_3$ = 100). All spectra were scanned at 20 cm^{-1}/min using a 3-sec rise time. 250 μ slits. 0.1 × 10^{-9}A range and 5 Å pass dielectric filters except where otherwise noted: 6471 Å, approximately 45 mW of power at the sample; 5682 Å, 30 mW, 1-sec rise time: 5145 A, 120 mW, 0.3 × 10^{-9} range; 4880 Å, 20 mW, 1 × 10^{-9} range, 1-sec rise time; 4579 Å, 30 mW, no dielectric filter, Ar$^+$ fluorescence bands denoted by *. Used by permission [72].

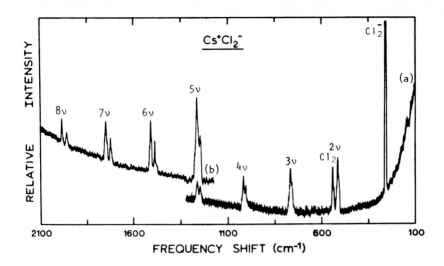

FIG. 20. Resonance Raman spectrum of matrix-isolated $Cs^{+}Cl_2$ using 75 mW of 457.9 nm excitation with dielectric "spike" filter. Scan (a) on 0.3×10^{-9}A range; scan (b) on 0.1×10^{-9}A range. Used by permission [73].

High intensity resonance Raman overtone progressions out to $v = 25$ have been observed with different excitations for monomeric iodine isolated in an argon matrix (Fig. 21). The position of the laser lines relative to the absorption spectrum is also graphed for comparison.

I_2 is a model substance for discrete resonance Raman scattering (Rousseau and Williams [33]). Typical features displayed in the MI-RS spectra are investigated, such as energy of overtones, intensities of progressions, change in polarization ratio, broad band fluorescence, laser line dependence, etc. So MIS in combination with resonance Raman studies delivers additional information about the several possible relaxation processes and damping constants.

If the exciting laser is in resonance with discrete vibronic transitions of Ca_2 in solid Kr the Raman scattering is found to be also strongly enhanced. The intensity pattern changes drastically if the laser is tuned over the sharp resonance (Fig. 22).

Resonant absorption is the first step in the observed processes; while a fraction of the excited molecules emits light coherently giving rise to the observed Raman signal, most of them do interact with the solid environment, leading to a change in the character of the scattered radiation and redistribution of the energy. Friedman et al. [65] studied how the cross section of the resonance scattering in Br_2 is modified by dissipative interactions with an argon matrix by tuning over the sharp resonance. An example for continuous

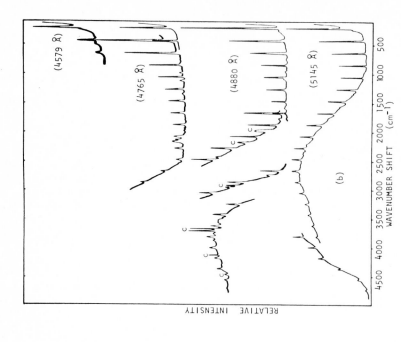

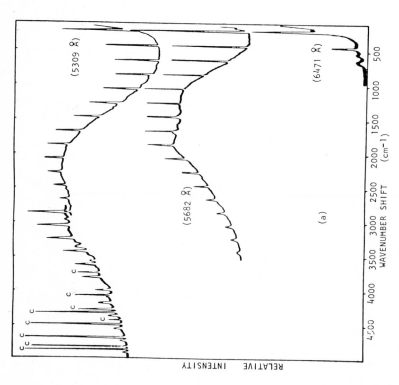

FIG. 21. (a) Raman spectra of monomeric iodine obtained as, in situ, photodecomposition product of TiI₄ isolated in an argon matrix at 16K for the major lines of a Kr ion laser. General instrumental parameters: Spectral slit width = 10 cm⁻¹ at exciting line, scan speed = 10 to 50 cm⁻¹ min⁻¹, rise time = 0.3 to 3s. Components of narrow band fluorescence series and laser plasma emission lines are labelled C. (b) Raman spectra of same samples for the major exciting lines of an Ar ion laser.

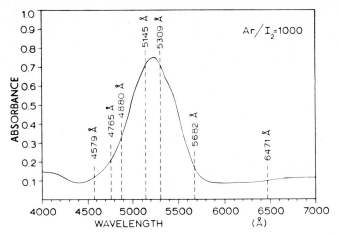

FIG. 21 (c). Absorption spectrum of a sample of pure I_2 isolated in an Ar matrix at 4 K. Used by permission [76].

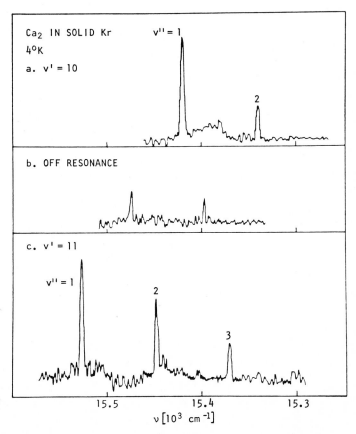

FIG. 22. Ca_2 Raman spectra in solid Kr at 4K. (a) v' = 10 excitation at 15599 cm-1. (b) Excitation of resonance at 15656 cm-1 in the v' = 10 phonon sideband. (c) v' = 11 excitation at 15714 cm-1. Used by permission [77].

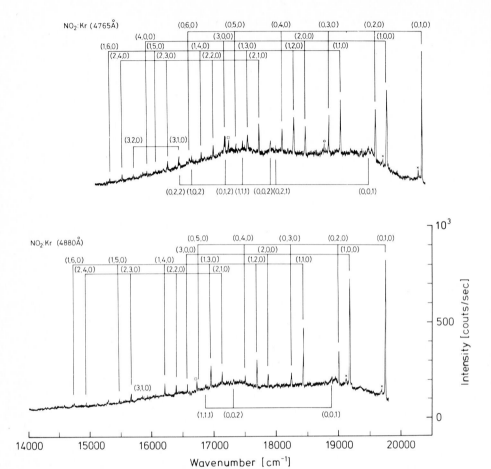

FIG. 23. Continuous resonance Raman spectrum of Kr/NO_2 at 15K excited by two different laser lines. Used by permission [78].

resonance Raman scattering will be discussed in detail in Sec. V. Figure 23 shows well resolved overtone progressions of all three fundamental vibrations of $MI-NO_2$; no drastic changes in spectral features can be seen when comparing the excitation with either 4765 Å or 4880 Å laser lines.

The observation of overtone progressions in Raman spectra deliver the possibility to calculate anharmonic coefficients of MI-species, e.g. Sn_2 (Teichman et al., [79]) and Ni_3 (Moskovits and Dilella, [66]). Recently, Moskovits and coworkers observed also a progression on the anti-Stokes side of the exciting line (Fig. 24). The intensities of these relative to the Stokes resonance Raman features were investigated as a function of laser frequencies and power, and metal/argon concentration: Fe_2 and NiFe (Moskovits and Dilella, [80]) and Ti_2 and V_2 (Cossé et al., [81]). According to this work

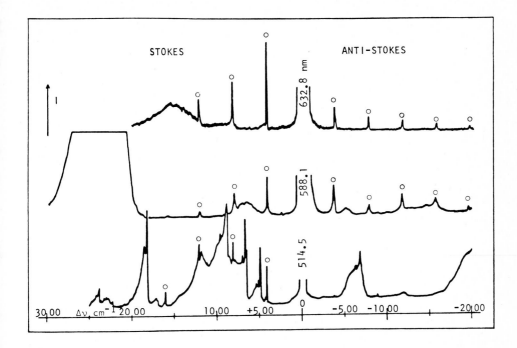

FIG. 24.   Resonance Raman scattering (marked with 0) of Ti_2 in argon matrix at 15K obtained with the three following excitations: 632.8 nm (30 mW) He-Ne laser (6 cm-1 slits); 588.1 nm (10 mW) 6G dye laser excited with Ar+ (6 cm-1 slits); 514.5 nm (10 mW) Ar+ ion laser (5 cm-1 slits).   The Ti to Ar ratio was approximately 1/1000.   Used by permission [81].

laser irradiation populated excited vibrational states of the ground state producing an anti-Stokes resonance Raman progression.   These consecutive two photon scattering processes are efficient because of very long vibrational relaxation lifetimes in the case of these Mi-species.   Fermi resonance is a further effect, which enhances Raman intensities and is treated in literature (Sushchinski [27]).   The classical example CO_2, in which Fermi found his effect in 1930, has been investigated by MI-RS too.   The Raman spectrum and corresponding IR spectra are presented in Fig. 25.   Two of three fundamental vibrations, v_2 at about 660 cm^{-1} and v_3 at about 2340 cm^{-1}, are IR active according to symmetry considerations. v_1 is Raman active and form together with $2v_2$ the well known Fermi doublet.   The energy splitting of this doublet is a measure of molecular anharmonicities and can be systematically investigated as a function of different matrices, temperature, pressure, concentration (Jodl and Kobel [83]).   Figure 26 (inset a) shows the RS of MI-OF_2 and besides many other Raman lines the Fermi doublet v_1, $2v_2$ at about 900 cm^{-1} is well resolved; inset b illustrates nicely photolysis of OF_2 to OF and F_2.

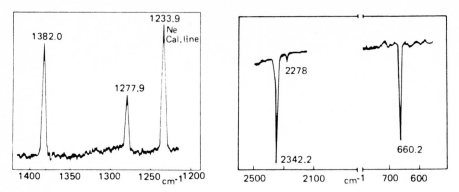

FIG. 25. Raman spectra of CO_2 and infrared spectrum in solid krypton. Used by permission [82].

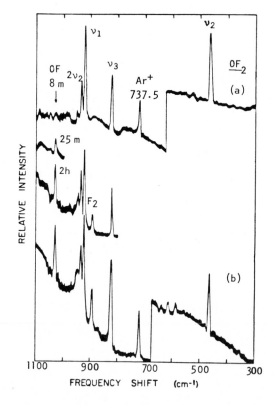

FIG. 26. Raman spectra from 300 to 1100 cm^{-1} for oxygen difluoride in solid argon at 16K using 4880 Å excitation (Ar/OF$_2$ = 100). Spectrum (a) initial spectrum recorded, arrows denote region of spectrum scanned after indicated time of exposure to the 4880 Å laser line. Spectrum (b) sample deposited for 4 hr with simultaneous 4880 Å photolysis for the last 3 hr, initial spectrum recorded. Used by permission [5].

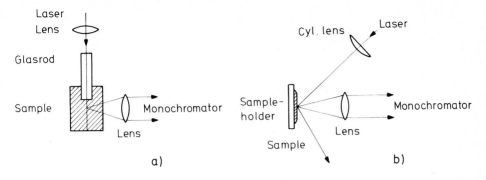

FIG. 27.   Raman scattering by crystals (a) or by thin films (b).

The inevitable progress of laser technology stimulated not only RS investigations but also laser excited emission studies on MI-systems. About one hundred papers covering this new field appear yearly. Some of these are listed (Sec. IV) when they supplement RS data. But on a whole this class of interesting experiments relates to other physical questions in comparison to MI-RS and is omitted here. Some reasons why they got so much attention recently are that old experiments can now be done with new equipment, such as time dependent spectroscopy leading to new discussions of vibrational relaxation, reactive processes stimulated by strong light sources, new laser types for hitherto uninvestigated species.

III. EXPERIMENTAL TECHNIQUES

The aim of this section is to point out only those problems which follow necessarily from the application of RS to MI, e.g. special matrix preparation for Raman studies, various Raman geometries and their advantages and difficulties. The samples may be prepared in several ways. The matrix material can be grown in the form of a crystal, some cm in diameter (Egger et al. [84]; Daniels et al. [85]) or the matrix condensed as a thin film, some 100 µm thick (Andrews [13]; Downs and Peake [10]). The methods for RS are different in principle accordingly (Fig. 27a,b).

Most of the examinations use thin film condensation. The advantages are quick sample preparation (few hours), small amount of matrix/solute material, elaborate polycrystal generation techniques that can be applied to a wide range of species, easy mixing of matrix/solute, suitable cryotips. The disadvantages are in general bad crystal quality, uncertain real M/S ratio, sample heating by laser irradiation, surface effects. Very few investigations

use macroscopic MI-crystals (Jodl et al. [4] and Kiefte et al. [86]) a technique which considering advantages and disadvantages is on the whole
complementary to thin film condensation. At the end of this section some new
techniques of the recent years concerning sample preparation, MI-RS and RS
in combination with other techniques will be reviewed.

A. Parameters of Sample Preparation

Most of the review articles (Sec. I) describe the problems of thin film
sample condensation for MI more on an academic level than for practical purposes. Few manuscripts cover the field in detail. But one should have in
mind that each apparatus for MI-RS has its own technical parameters and the
differences between the equipment of various scientific groups are enormous.
Rochkind [87] invented the pulsed deposition technique. Shirk and Claassen
[1] reported the first laser Raman scattering of MI-species at high dilution
and gave clear experimental and technical hints. Nibler and Coe [3] and
Nibler [11] described experimental observations on illumination geometries,
laser heating of samples, fluorescence problems, M/S ratio and depolarization
measurements. Perutz and Turner [88] compared by different criteria slow
spray on and pulsed matrix isolation as a technique with the aid of IRS-data.
Barnes et al. [89] investigated the effects on the quality of MI-RS by varying parameters such as deposition conditions and substrate metal, and discussed the optimum conditions. Dubost [90] obtained a controlled ratio of
monomers to polymers in MI-IRS by a very slow spray on technique.

In what follows, the technical parameters for sample preparation are
discussed. The matrix/solute ratio is different in the first stage (gas
container, or atoms from a furnace, etc.), from that in between, and in the
last stage (thin film in the cryostat). No systematic work is known. Many
people quote the M/S ratio of the first stage, which is only an upper limit,
since clustering and demixing may happen, impurities may enter etc. Mann
and Behrens [91] made some thermodynamic considerations concerning the
demixing effects due to different masses of the constituents during matrix
preparation. According to our studies (Rödler [92]) the geometrical arrangement for condensation, like the distance from gas capillary to target, diameter
of capillary, angle of capillary and target, has no influence but rather characterizes the use of the apparatus. Figure 28 shows the known influence of
deposition temperature, which depends on the kind of matrix used and on the
desired monomer/polymer ratio. The more stable the temperature is maintained ($\Delta T < 0.1$ K) during the few hours of condensation the better is the
optical and physical quality of the sample. It is one of the most sensitive

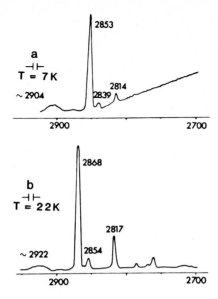

FIG. 28. Effect of deposition temperature (T_d) on the Raman spectra of HCl in matrices. (a) HCl/Kr = 1/100, T_d = 7K, deposition rate: 0.2 mM/h. (b) HCl/Ar = 1/100 T_d = 20 K, deposition rate: 0.2 mM/h. Used by permission [51].

parameters to our mind.

The deposition rate varies in the literature from m mol/h to μ mol/h. We ourselves use 10 to 50 μ mol/h and very thin films (20-40 μm) and obtain reproducible high quality RS with different MI-species. This velocity of condensation is a factor of at least 10 smaller than the one used by Dubost [90] (known as Dubost catastrophe). For example, we investigated systematically the physical question of the ortho/para ratio in MI-H_2 and D_2 as a function of deposition rate (see Sec. V). The comparison between quick spray on, slow spray on, very slow spray on, pulsed spray on, etc., according to IR- or RS-signals is delicate because the intensity can vary within large limits (Fig. 29) if the investigations are not executed under identical circumstances, such as same matrix, same apparatus and so on.

Thickness measurement of MI-thin films is standard nowadays, either by the application of the quartz crystal microbalance (Moskovits and Ozin [93]) or by the use of interference effects by thin films (Groner et al. [94]). The optical quality of a perfect layer can be judged naturally by the Raman signal, if this is already known, or by the diminuation of the amplitudes of interference rings, i.e. the contrast ratio (Fig. 30). Sometimes one observes a larger oscillation in this interference pattern which is due to focussing by the curved film which changes during growth.

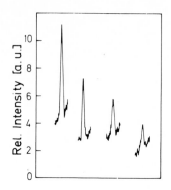

FIG. 29.   RS of ν_2 in NO_2/Xe at 18 K.   The laser is scanned over an area of 5 × 5 mm, under identical circumstances.   Used by permission [95].

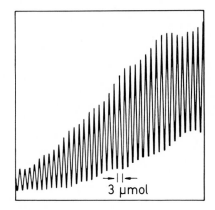

FIG. 30.   Typical interference pattern during sample condensation of 1 mol % N_2 in Ar to control thickness and quality.   Used by permission [96].

The history of the sample before and during measurements is of great importance.   Figure 31 demonstrates a typical annealing procedure common in MIS: sample condensation at 10-15K, warming up to 40-50K followed by recooling to 10-15K.

The values are a little bit different for various matrix material.   On the other hand they mirror the experience of the investigator, the capability of the apparatus or the specific conditions of the MI-system under investigation. Our experience is that this procedure delivers in general "better" Raman signals (Fig. 32).   But for small molecules, H_2 in RGM (see Sec. V), or for reactive species (Fig. 33) annealing effects produce unwanted aggregates by diffusion.

Very few papers describe quantitatively this annealing procedure.   For example, the intensity ratio of monomer to dimer Raman line as a function of

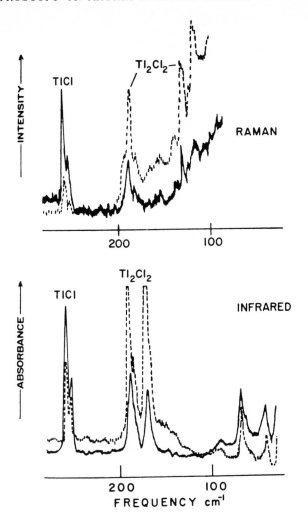

FIG. 31. Effect of diffusion on the vibrational spectra of TlCl/Tl$_2$Cl$_2$ isolated in an argon matrix, ——— 14K ----- after warm up to 40K followed by recooling to 14K. Used by permission [64].

controlled warm up (see NO in Sec. VB). Sudden changes in temperature can be the death of MI-film. We found that temperature variation should not exceed 1K per minute to maintain the optical quality.

To close this section the author would like to emphasize the main technical parameters for thin film condensation. These are deposition temperature and rate and the annealing procedure; for Raman studies it is recommended to use very thin films (20-40 μm) by very slow spray on (10-50 μmol/h) and deposition temperatures of 15K for Ar and Kr and of 20K for Xe.

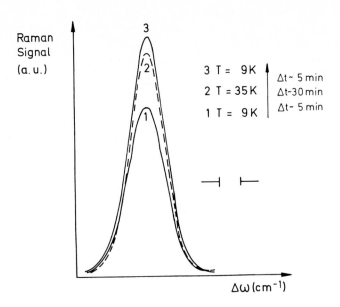

FIG. 32. Annealing effect of the molecule vibration. Apparatus conditions: 1 mol % N_2 in Ar, thickness ~ 150 μm excitation with 5145 Å, 800 mW and spike filter, $3 \cdot 10^3$ c/s. Used by permission [60].

B. Different Raman Geometries

Most laser-Raman apparatus for MIS use a 90° angle or a near grazing angle of incidence - 10° to 20° - as the irradiation geometry and adapt the form of the sample holder to that. The following different ways taken from literature should be judged according to the amount for adjusting, to applicability to different samples, to reproducibility of Raman data, to signal to noise ratio (straylight behavior), to good thermal contact. Figure 34 a to f presents the various standard irradiation geometries used in MI-RS; only different ones are printed, analog modifications are omitted (new approaches are described in III.D).

A variety of multiple reflection cold tips have been tried in an attempt to enhance the Raman signal but it was found that after a single pass through the matrix-film the laser was randomly defocussed and no gain was made on back reflection.

The Raman apparatus usually consists of a laser, a double grating monochromator and a photon counting system; analog using a ratemeter or digitally with a multichannel analyzer. In addition to the rare gas-ion-laser (Ar^+ lines at 4579, 4765, 4880, 5145 Å and Kr^+ lines 3507, 3564, 5309, 5681, 6471 Å; 1W

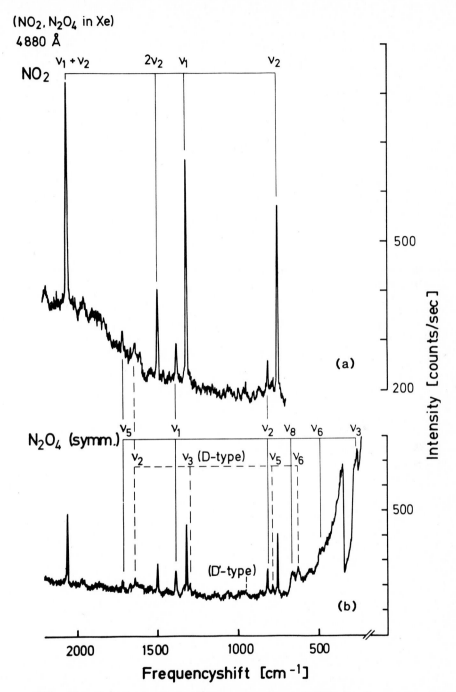

FIG. 33. (a) Part of the Raman spectrum of NO_2 in Xe condensed at 15K and measured at 15K. (b) Same region after warming up to 50 K for 30 min. and recooling to 15 K (details in Sec. V.E). Used by permission [78].

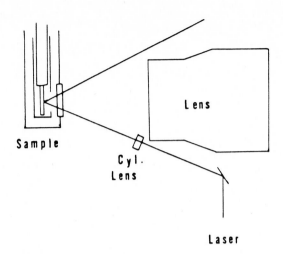

FIG. 34a. Different Raman geometries used in MIS. Arrangement of sample, lens, and laser beam for Raman spectroscopy. Used by permission [1].

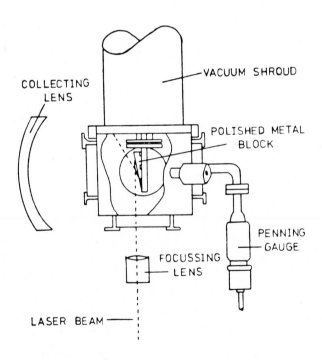

FIG. 34b. Schematic diagram of the experimental arrangement for matrix isolation Raman spectroscopy. Used by permission [89].

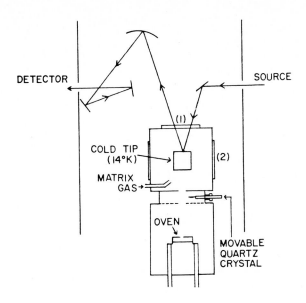

FIG. 34c. Schematic of high temperature furnace and cold cell positioned on reflection assembly compartment of infrared spectrophotometer (viewed from above). Used by permission [64].

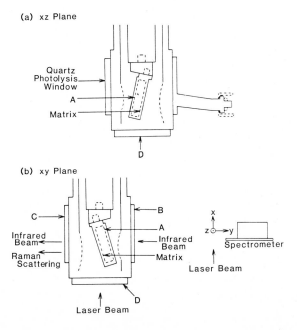

FIG. 34d. Experimental system for the acquisition of infra-red absorption and Raman spectroscopic data from the self-same matrix: (a) Deposition configuration, and (b) Spectral acquisition configuration. Used by permission [97].

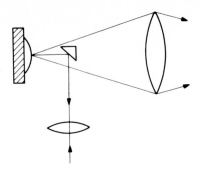

FIG. 34e. Schematic diagram of a back scattering geometry of MI-RS. The laser is focussed and deflected by a small mirror on the MI-film which is condensed on a He-cooled Ag coated copper plate. The Raman light is collected by a camera objective (1:1.2)-viewed from above. Used by permission [61].

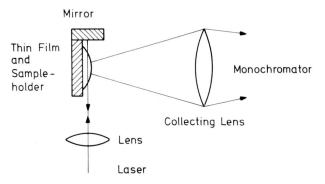

FIG. 34f. Arrangement of a specially chosen Raman geometry for MIS - viewed by the side. Used by permission [96].

of power) dye lasers ($\Delta\lambda$ ~4000 - 6000 A; 1/2W of power) are most helpful to take advantage of resonance Raman effect. Yet the straylight behavior of thin films should not be underestimated: a triple monochromator is very helpful, narrow band filters for each laser line to reject plasmalines, a variable filter in front of the sample is absolutely necessary to avoid the scattering of the dye fluorescence (in detail see next section).

As a result it is absolutely necessary to use a triple monochromator or iodine filter with a double monochromator in order to reduce the straylight as much as possible. We tried most of the different Raman geometries and we would recommend the one from Shirk (Fig. 34a) with a cylindric lens, near grazing angle and a mirrorlike sample holder; this arrangement is to be preferred because it delivers strong RS, it is very flexible, not time consuming and the film surface can easily be scanned.

C. Problems Connected With MI-RS

The problems which appear in applying RS to MIS are partially reviewed by Nibler [11], Ozin [12], Barnes et al. [89] and Andrews [13,15]. About ten papers of the hitherto known one hundred papers concerning with MI-RS give some insight in these kind of problems but no systematic approach to solve them is known. The following paragraph exhibits the inherent experimental difficulties and gives some hints on how to avoid them. As far as the author can judge the occurrence of the following general physical problems are highly connected with the specific Raman apparatus which is used by the different research groups and must be solved individually by them. The consensus of experimental practice is not too high. From this point of view it is absolutely necessary that each article should contain all the relevant parameters of sample production, of sample preparation and of Raman spectroscopy; otherwise there is no way of reproducing or comparing data in MI-RS.

Comparing RS intensities makes sense, to an uncertainty ~10%, if observations are made carefully on the same matrix as a function of temperature (Fig. 11) of depolarization (Fig. 14 or 15), of exciting frequencies (Fig. 23), etc.

It is very difficult to compare RS spectra on different thin film samples by changing M/S ratio (Fig. 7 to 9), matrix material (Fig. 10) or with IR spectra (Fig. 16, 18, 25).

M/S ratios between 100 and 1000 are commonly used and under favorable conditions (for example CO_2 and resonance Raman effect) 100 ppm of solute have been detected. Successful isolation of the impurity can be tested either by the observation of monomer and polymer Raman lines (NO_2 and N_2O_4 in Fig. 33), by varying concentration or by controlled warm up or by crystal field splitting of Raman lines (for example H_2 in N_2).

To avoid sample deterioration the warming up or cooling down procedures should not exceed 1 K/min. The sample temperature should always be smaller than 1/3 of the melting temperature of the matrix material. This value is found empirically (Hallam [7]) and is physically explainable by the onset of diffusion processes in the matrix (activation enthalpies for the migration of interstitials 0.058 eV and of vacancies 0.13 eV in solid Ar (Wehr et al. [98]). If the laser beam hits the sample one can observe a small rise of actual temperature of between 1 and 3K. This depends on the kind of solute, if it absorbs laserlight or not, on the quality of thermal contact between sample and cooling bath (typical values are 10-15K in the case of liquid helium), on the way how and where the temperature of the sample is measured, on laser power and on the way how the laser beam is focussed in the sample (convex or cylindric lens).

It is a real problem to specify the actual absolute temperature of the sample irradiated by the laser even if the thermocouple is attached to the thin film directly. Most of the authors are too optimistic about their temperature values or give no information how and where it is measured. As a consequence an important value for the interpretation of the data is lost and uncontrolled processes do happen like phonon excitation, diffusion processes and polymerization. A way out of this situation could be the determination of the local temperature by comparing the intensities of anti-Stokes to Stokes Raman lines next to the laser frequency: for a rotational transition in MI-molecules, for the E_g librational mode at about 30 cm^{-1} in nitrogen matrices.

The sample holder consists commonly of a copper, silver, aluminum or platinum mirror coated with different materials or of a sapphire disk on which the matrix and the solute beam are condensed. Great care should be taken to carefully clean the highly polished metal surface before each experiment. This procedure determines the optical quality of the sample, because chemical sediments act as nuclei for condensation and hinder perfect sample growing. High thermal conductivity, no laser induced fluorescence (e.g. copper has a broad band at 14,000 cm^{-1}) and mirror like behavior characterize good sample holders for MI-RS.

Almost each matrix has different scattering behavior against laserlight and the straylight becomes evident at various structures in the Raman spectra. This unwanted effect can be diminished by experimenting with different sample preparation techniques, by using various matrix support material and by recording the spectra with different irradiation geometries (see above). This perturbation quality can also be reduced by optical components: Figure 35 demonstrates the influence of grating ghosts, which are avoidable by the use of long pass filters or of holographic gratings. Laser straylight increases enormously the Rayleigh wing which is overcome by a triple monochromator, a double monochromator in conjunction with a third one acting as a filter (Fig. 36), or by an iodine filter-monochromator unity.

If plasma lines from the laser excitation source appear in the spectrum they can be removed by narrow band pass filters (5Å pass, 65-75% transmission; 50Å, 90-95% transmission). On the other side these lines are sometimes used as wavelength standards to ameliorate the frequency accuracy of the monochromator of a few cm^{-1} to ±1 cm^{-1} according to Prochaska and Andrews [58] or to optimize the focussing conditions, if the position of the Raman line is not known until then. (An asterisk (*) denotes a plasma emission line from the laser in Fig. 10.) In the case of a dye laser the matrix scatters an observable amount (about 10^3 to 10^4 counts/s) of the dye fluorescence (continuum up to 2000 cm^{-1} off resonance energy) into the light

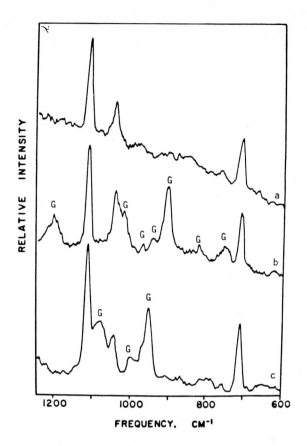

FIG. 35. Raman spectra of matrix isolated O_3 at the ratio of $Ar/O_2/O_3 \cong$ 25/25/1: (a) 5145 Å exciting line using a long pass filter (Oriel No. G-772-5400) to eliminate grating ghosts; (b) 5145 Å exciting line without the long pass filter; and (c) 4880 Å exciting line without a filter. Bands labelled G are due to grating ghosts. All spectra measured at spectral slit widths < 5 cm-1. Used by permission [99].

detection system; a prism monochromator as preselector or a pinhole arrangement is an effective help. Typical Raman values for signal to noise ratio during one scan are <10 c/s dark level, ~200 c/s straylight, ~300 c/s signal (Fig. 37a). Bad optical quality of the matrix can even deteriorate these values. MI-IR spectroscopy has the opportunity to sweep several times over the spectrum to increase the signal by averaging (Fig. 37b).

From the very beginning of MI-RS depolarization measurements have been common (Nibler and Coe [3]). Depending on the optical quality of the film it is possible to gain values with an uncertainty of $\Delta\rho = 0.05 - 0.1$. Care must be taken that the optical equipment itself is not polarizing like window material, gratings, etc.

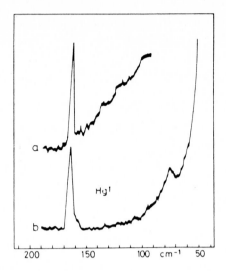

FIG. 36. Comparison of Raman spectrum of HgI_2 in krypton matrix at 20K as obtained without (a) and with the third monochromator (b) (single oven experiments). Used by permission [100].

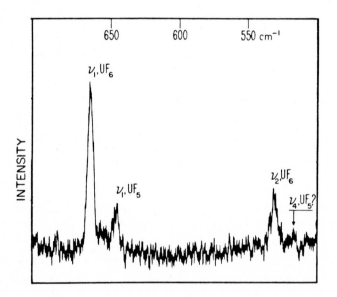

FIG. 37a. Raman spectrum of UF_6 and a small amount of UF_5 in an argon matrix at 10K. This matrix had a ratio of $400/1$ = argon/UF_6 and contained about 200-300 micromoles of UF_6 per square cm. Used by permission [101].

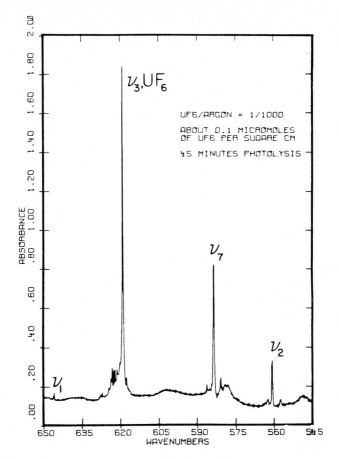

FIG. 37b. Infrared spectrum of UF stretching region for UF_6 and UF_5 in an argon matrix at 10K. The peaks labelled ν_1, ν_2, and ν_7 are the three infrared-active U-F fundamentals of UF_5. This spectrum was recorded at 0.1 cm^{-1} resolution with a Nicolet series 7000 Fourier Transform Infrared Spectrometer.

Raman monochromators have a wavenumber accuracy of ±1 cm^{-1}. However, typical MI-values for matrix shifts of vibrational lines are smaller than 10 cm^{-1} and for temperatures shift are less than 1 to 2 cm^{-1}. A thorough analysis of Raman literature for $MI-N_2$ (Jodl and Bolduan [60]) reveals that the reported matrix shifts range from +1 and -7 cm^{-1}. A brief look at the list in Sec. IV shows the same astonishing results for a series of other species. Therefore it is absolutely necessary to use atomic lines for frequency calibration ($\Delta\omega$ of Raman lines ±0.3 to 0.5 cm^{-1} according to slit-width). The light of a calibrating lamp is easily deflected into the light path by a moveable mirror. Plasma lines for calibration have the disadvantage that they are rare, very strong and sometimes far off the Raman line under investigation. It is economical to use lines less than 10 to 20 cm^{-1} off the Raman line; e.g. Thorium possesses very many lines in the interesting spectral region (see Figs. 11, 13, 25).

According to our experience plasma lines may shift in frequency by more than a few cm^{-1} depending on laser service conditions such as radiation power, radiation duration and wavelength; therefore they are useless for calibrating purposes. From published Raman data it is generally not possible to recognize if they are vacuum corrected or not. It is not necessary to emphasize that only these energies have a physical meaning and are comparable with theory; for example: if the wavenumber values are registered from the monochromator with an inherent uncertainty of ±1 cm^{-1} the vacuum correction in the case of about 2300 cm^{-1} runs up to 0.7 cm^{-1}.

If the optical quality of MI-films is perfect and if the MI-system is chosen favorably it is even possible by MIS to resolve the fine structure (for CNN $\Delta\omega$ < 10 cm^{-1} (Bondybey and English [102]) or the crystal field splitting (for pure N_2 $\Delta\omega \leq 1$ cm^{-1} (Jodl and Bolduan [60]). This demonstrates very well the capability of MI-RS as a technique although some problems remain to be solved.

D. New Approaches

To close this section some remarks concerning new physical and technical methods in sample generation, in Raman scattering, in combining MI-RS with other techniques or in laser induced emission are mentioned very briefly, just to state their existence not to discuss their capability.

Different species like gaseous ones, like high-temperature ones do need their specific technique to produce and isolate them in a matrix. High temperature Knudsen cells of various design have been used by a number of investigators. Recently Scheuermann and Nakamato [103] report the design of a novel miniature oven, which is heated by a laser beam to vaporize small quantities of solid compounds (here for example $ZnCl_2$). Hofmann et al. [104] produce Na and Na_x by a molecular beam oven, control by a mass spectrometer the atom to molecule ratio and deposit them in a RGM for spectroscopic analysis and clustering investigations. Felder and Günthard [105] detect by MI-IRS the freezing of internal rotation temperature in a supersonic jet. Cooling of rotational and vibrational degrees of freedom is well known to occur in supersonic experiments. On the other hand a MI-film produced by an effusive thermal molecular beam reveals spectra which contain different individual conformers of one species according to the thermodynamic condition in the beam (Fig. 38).

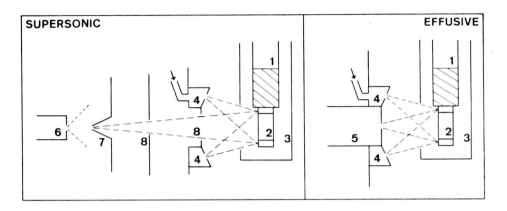

FIG. 38. Molecular beam matrix deposition systems (vertical section). Left side: supersonic beam system; right side: effusive thermal beam. 1: LHe cooled cryostat, 2: target window, 3: radiation shield, 4: matrix gas inlet, 5: heatable Knudsen cell, 6: supersonic nozzle, 7: conical skimmer, 8: collimators. Used by permission [105].

Allamandola et al. [106] describes an apparatus which permits flash photolysis of a pulse-deposited gas mixture in a matrix isolation experiment. The efficacy of this technique for producing, trapping and spectroscopic measuring is demonstrated by different examples (Fig. 39).

New Raman devices apply further laser types by extension to other spectral regions or by the possibility of tuning. For example, Poliakoff et al. [107] uses a tunable cw spin flip Raman laser to obtain spectra with a resolution of <0.01 cm^{-1} and to induce chemical reactions in a matrix. More efficient photo-multipliers, new light detecting systems (Vidicon, OMA), and new electronic procedures (like computer time averaging of commercial available Raman spectrometers or multi scan operation of these) lead to an improvement in signal to noise ratio beyond the best obtainable conventional photon counting systems and in duration of measurements.

Rosetti and Brus [108] ameliorated Raman geometries by investigating laser beam wave guide propagation in frozen gaseous layers. A scattering enhancement of about 30 in a neon matrix is observed in the waveguide mode, as compared with normal front surface illumination; in addition, less local heating of the film by laser scattering is expected (Fig. 40).

By use of a multiple-reflection double-pass-ATR element considerably thinner matrices and shorter deposition times (minutes instead of hours) are required for spectra of comparable intensity (Huber-Wälchli and Günthard [109]).

MI-RS is nowadays successfully combined with further techniques. Boal and Ozin [110] passed a xenon chlorine mixture through a microwave dis-

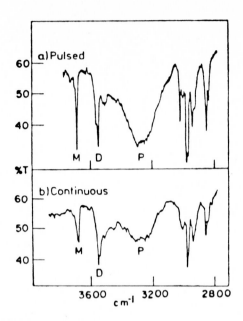

FIG. 39. The O-H stretching region of the infrared spectrum of methanol in Ar at 20K (Ar/CH, OH = 100). (a) Sample prepared by pulsed deposition, 2 pulses/min for 2.5 h. Overall flowrate - 2.8 m mol/h. (b) Sample prepared by a 2-h continuous deposition, flowrate - 2.5 to 3 m mol/h. M-monomer; D-dimer; P-higher polymers. Used by permission [106].

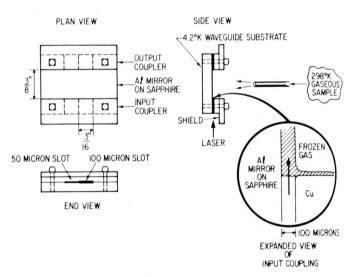

FIG. 40. Schematic diagram of "slot" waveguide device. Used by permission [108].

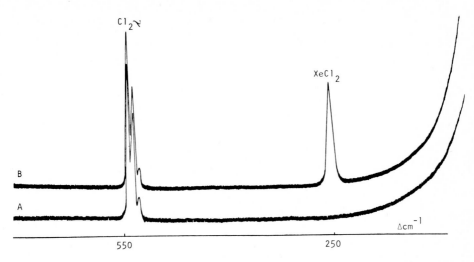

FIG. 41. The Raman spectra of a xenon-chlorine (25:1) matrix at 4.2K. (a) <u>without</u> microwave discharge of the gaseous mixture, showing the presence of only Cl_2; (b) <u>with</u> microwave discharge of the gaseous mixture, showing the presence of both $\overline{XeCl_2}$ and Cl_2. Used by permission [110].

charge and condensed the molecules onto the cold tip of a Raman cell. A strong new band appeared and was attributed to $XeCl_2$ (Fig. 41).

Jodl and Holzapfel [61] developed the technique of MI-RS at high pressures and low temperatures. Effects of pressure and of temperature on vibrational spectra are of significant value in the fundamental understanding of anharmonic effects. This method is applied to different molecular species and postulates precise frequency measurements (smaller 1 cm^{-1}) of temperature, and pressure shifts independently; a specially high pressure cell was developed for this purpose (Fig. 42: see also Sec. V). The capability of MI-RS at high pressure is well documented by Fig. 43 in which the librational Raman spectrum of pure solid CO_2 at different pressures at 25°C is compared to the same one of CO_2 in Ar at various pressures at 20K.

As stated above the comparison of IR-data and Raman-data in MI is delicate unless the measurements are carried out on the self same matrix. Lesiecki and Nibler [64] (see Fig. 34c) and Grzybowski et al. [97] (see Fig. 34d) discovered a special experimental set up for the acquisition of both kinds of spectral data.

The use of several lasers operating simultaneously in MIS opens a wide application area for new experiments like IR double resonance experiments, bi-photon excitation, time resolved double resonance investigations and photolysis; which deliver further information of MI-species about relaxation mechanism, lifetime, high lying electronic states and chemical reactions.

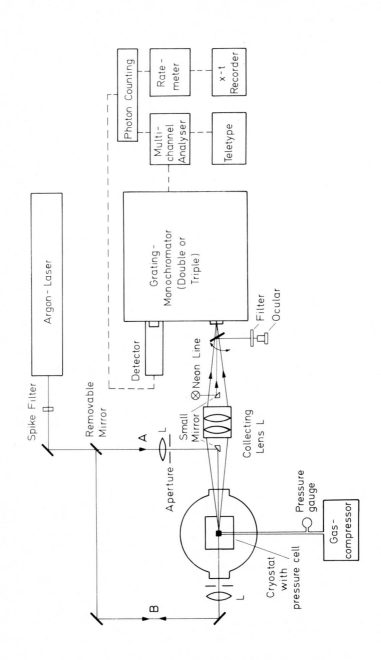

FIG. 42a. Schematic diagram of high pressure Raman apparatus.

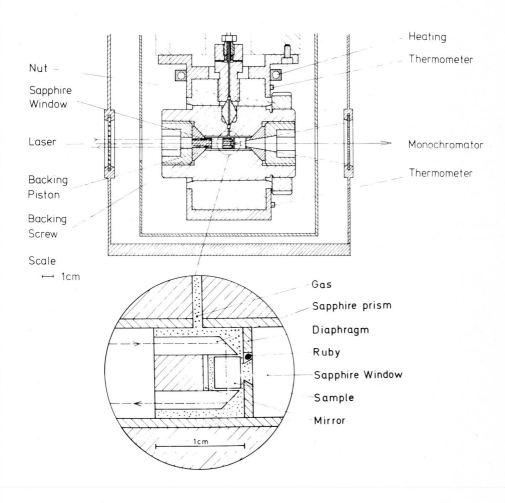

Nut
Sapphire Window
Laser
Backing Piston
Backing Screw
Scale
⊢—⊣ 1cm

Heating
Thermometer
Monochromator
Thermometer

Gas
Sapphire prism
Diaphragm
Ruby
Sapphire Window
Sample
Mirror

1cm

FIG. 42b. High pressure optical cell and sample chamber. Used by permission [62].

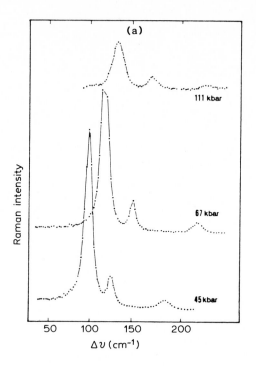

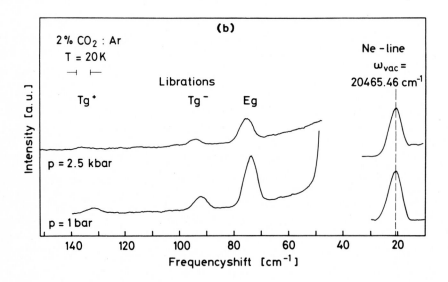

FIG. 43.   (a) Librational mode Raman spectra of solid CO_2 at 25°C and pressures of 45, 67 and 111 kbar.   Used by permission [111].   (b) RS of MI-CO_2.   Used by permission [83].

IV. SUMMARY OF EXPERIMENTAL DATA

A. General Remarks

The aim of this review article is not to quote and discuss each manu-script dealing with MI-RS. Certain relevant ones are described in different sections in detail. However, it is very helpful if a more or less complete data collection is made available. The last published review article (Downs and Hawkins [25]) contains a table of about 180 matrix isolated molecules studied by their Raman spectra during 1976 until 1981: Species, method of production, matrix, temperature, support, exciting laser wavelength and power, supportive techniques, observations, references. The period 1970 to 1982 is considered here.

The collection in this section contains the Raman data from each known diatomic and triatomic species available to the author; some poly-atomic examples are also mentioned, if the species, the technique or the problem considered was of interest. The list contains the impurity (chemical symbol), vibrational frequencies, matrix material and temperature, necessary remarks and references. If available, gas phase values are mentioned in parentheses. In the case of diatomic species sometimes IR-data or laser excited emission data are presented, in addition, if a comparison to Raman results seems meaningful. The quality of Raman spectra can be seen from Fig. 44 and many other examples spread over the whole article. On scanning through the list curious facts come to the eye. For example the discrepancies in vibrational energies reported by different authors or ill defined spectroscopic values like ω_o, ω_e, $\nu(0\text{-}1)$, etc.

The investigations concerning diatomic molecules are stimulated by pre-cise physical questions and honored by well executed measurements. These species may be classified as homonuclear (H_2, N_2, Br_2 ...) or heteronuclear (CO, NO, HCl) gaseous molecules, as old chemical compounds like GeS, ZrO or new ones like XeCl. The number of triatomic species consist of gaseous ones (NO_2, SO_2, CS_2), of metal halogen molecules like $CdBr_2$, CsI_2, of metallic clusters (Ag_3, Ni_3) or new species ($XeCl_2$). Investigations concerning triatomic and polyatomic molecules are mainly stimulated by physical or chemical questions coming from the specific material itself, such as isotopic effect or chemical substitution. Sometimes they have not convincingly considered the method and the technique.

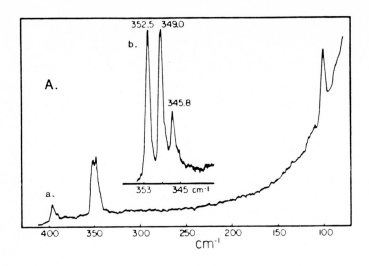

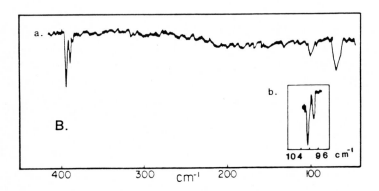

FIG. 44. A. (a) Raman spectrum of $InCl_3$ trapped in solid krypton at 20K. (b) High resolution spectrum of ν_1 band of $InCl_3$, (slit width: 1 cm-1). B. (a) IR spectrum of $InCl_3$ trapped in solid krypton at 20K. (b) High resolution spectrum of the 100 cm-1 band (slit width: 1.5 cm-1). Used by permission [70].

For these examples the aims are the same, namely to detect new chemical species, to analyze them spectroscopically and structurally, or to apply RS as a technique in addition to others. Hardly ever do new, physically important questions enter the scene like atom-cluster-transitions, anharmonic effects, etc.

B. Diatomic Molecules

TABLE 5a

Impurity	Vibrational Frequency cm⁻¹	Matrix	T (K)	Remarks	References
Ag_2	194 (192)	Kr	5	other polymers	Schulze et al. [112]
Ar_2	25.5	gas	103	theory and exp.	Frommhold [113]
Ar_2		gas	40	Raman scattering in a super sonic expansion; Rotation and Rotation-Vibr.	Godfried and Silvera [114]
As_2	429.55	Ne	4	laser induced fluorescence, other polymers	Bondybey et al. [115]
Bi_2	173 (172.71)	Ne	4	laser induced fluorescence, other polymers	Bondybey et al. [116]
Bi_2	122 244 366	Ar		resonance Raman, other polymers	Ahmed and Nixon [117]
Bi_2	173.5 172 169.5 (172.71)	Ar Kr Xe	8 8 8	resonance Raman, other polymers	Manzel et al. [118]
Br_2	317.5 (323.3)	Ar	15	resonance Raman, other isotopes and other RGM	Ault et al. [119]

(continued)

TABLE 5a (continued)

Impurity	Vibrational Frequency cm^{-1}	Matrix	T$_{(K)}$	Remarks	References
Br$_2$	315	Ar	5-10	comparison between Raman scattering and absorption	Friedman et al. [120]
Br$_2$	316	Ar	4-10	resonance Raman	Friedman et al. [65]
BrCl	436.6 (440)	Ar	12	resonance Raman, other isotopes and other RGM	Wight et al. [121]
Ca$_2$	79	Kr	4	resonance Raman	Miller and Andrews [122]; Bondybey and English [77]
Cd$_2$	58	Kr	20		Givan and Loewenschuss [123]
CdO	810	Ar	15	in combination with IR and visible absorption	Prochaska and Andrews [124]
Cl$_2$	543	Xe	4.2		Boal and Ozin [110]
Cl$_2$	554.5 568 (559.7)	Ar Kr	15 15	resonance Raman and other isotopes	Ault et al. [119]
Cl$_2$	554.7 550.0	Ar Kr	4 4	resonance Raman	Bondybey and Fletcher [125]
ClO	850	Ar	16	other halides	Andrews [16]; Loewenschuss et al. [126]

Species		Matrix		Comments	Reference
CO	2140.9 2140.1 2138.2 2134.8 (2143.274)	Ne Ar Kr Xe	10 14 15 20	other polymers	Jodl and Bier [127]
CO	2139.7	N_2	15	CARS	Beattie et al [128]
CO	2138.4	Ar	80		Kiefte et al. [129]
Cu_2	280				Harris and Jones [130]
D_2	2977 (2993.6)	Ar	12	vibration-rotational and pure rotational Raman	Prochaska and Andrews [58]
D_2 (ortho)	2973.1 (2993.548)	Ar	15	and pure rotational Raman	Jodl and Bier [127]
D_2 (para)	2971.1 (2991.446)	Ar	15		
DCl	2079	Ar	7	other isotopes and polymers; comparison between IR and RS spectra	Brunel et al. [51]
F_2	892 (892)	Ar			Ault et al. [119]
Fe_2	299.6	Ar	11	resonance Raman	Moskovits and Dilella [80]
Ga_2	180 (ω_e)	Ar	12	resonance Raman	Schulze and Froben [131]
H_2	4138 (4161.6)	Ar	12	vibration-rotational and pure rotational Raman	Prochaska and Andrews [58]
H_2 (ortho)	4136.9 4118.5 4110.0 (4155.2543)	Ar Kr Xe	15 15 15	vibration-rotational and pure rotational Raman	Jodl and Bier [127]

(continued)

TABLE 5a (continued)

Impurity	Vibrational Frequency cm^{-1}	Matrix	T(K)	Remarks	References
H$_2$ (para)	4143.2 4124.4 4116.0 (4161.1653)	Ar Kr Xe	15 15 15	and other RGM	Prochaska and Andrews [58]
HD	3619 (3632.1)	Ar	12		Brunel et al. [51]
HCl	2868 2853 2838 2888	Ar Kr Xe pure solid	20 7 7		Barnes et al. [132]
HCl	2868	Ar	20	in combination with IR	Givan and Loewenschuss [133]
HgCl HgBr	287 178	Kr	20	resonance Raman overtone progressions and other RGM	Howard and Andrews [134]
I$_2$	213.3 (214.5)	Ar	16	resonance Raman and resonance fluorescence	Grzybowski and Andrews [76]
I$_2$	213.70	Ar	16	resonance Raman, other isotopes and other RGM	Wight et al. [121]
IBr ICl	268.4 (269) 377.3 (384.3)	Ar Ar	12	resonance Raman	Schulze and Froben [131]
In$_2$	118 (ω_e)	Ar	12		Miller and Andrews [135]
Mg$_2$	90.8 (51.12)	Ar	12	laser excited emission	

Species	Frequency	Matrix	K	Notes	Reference
Mg_2	51.1 (ω_e)	Ar	12	collection of about 30 further metallic diatomics in RGM	DiLella et al. [136]
N_2	2323	Ar	81	in combination with IR experiments	De Remigis et al. [137]
N_2	2328	Ar	77		Jodl et al. [4]
N_2	2326.7 / 2327.6 (2329.92)	Ar / pure solid	8 / 8	concentration-temperature-and pressure dependent	Jodl and Bolduan [60]
N_2	2326.7 / 2324.6 / 2322.0	Ar / Kr / Xe	18 / 18 / 18		Jodl and Kobel [83]
N_2	2326.2	Ar	80		Kiefte et al. [129]
NH	3302 (ω_e) / 3284 / 3264 (3289.9)	Ne / Ar / Kr	4	laser excited emission	Bondybey and English [138]
NCl	825.0 (827.0)	Ar	12	laser excited emission	Miller and Andrews [139]
NBr	684.6 (691.75)	Ar	12		
N I	603.4	Ar	12		
NO	1873 / 1872.2 / 1867 (1875.972)	Ar / Kr / Xe	15 / 20 / 20		Jodl and Bier [127]
Ni_2	380.9	Ar	10	laser excited emission	Ahmed and Nixon [140]
O_2	1554	Ar	77		Jodl et al. [4]

(continued)

TABLE 5a (continued)

Impurity	Vibrational Frequency cm^{-1}	Matrix	$T_{(K)}$	Remarks	References
O_2	1552 (1556)	Ar	16		Andrews and Smardzewski [141]
O_2	1551.0	Ar	40		Kiefte et al. [86]
O_2	1551.0 (1556.2)	Ar	80		Kiefte et al. [129]
O_2	1550.5 1554.5	solid N_2	15 15		Jodl and Molter [142]
P_2	780.77	Ne	4	laser induced fluorescence, other polymers	Bondybey et al. [115]
Pb_2	225.0 (256.5)	Ar	10	laser excited emission	Teichman and Nixon [143]
Pb_2	112.2 (109.8)	Ar	6	laser excited emission	Bondybey and English [144]
Pb_2	111 109 (119.1)	Kr Xe	10	dimers and clusters resonance Raman	Manzel et al. [145]
Pb_2	107 (119.1)	Xe	12	dimers and clusters resonance Raman	Stranz and Khanna [146]
PbS	423.2 (429.4)	Ar	11		Teichman and Nixon [147]
S_2	716 (725.68)	Ar	7		Barletta et al. [148]

Species	Wavenumber	Matrix		Remarks	Reference
S_2	725.4 722.4 725.6 725.0 (725.7)	Ne Ar Kr Xe	5 5 5 5	laser excited emission	Bondybey and English [149]
Sb_2	269.9	Ne	4	laser induced fluorescence and other polymers	Bondybey et al. [115]
Se_2	383.7 (385.3)	Ar	4	laser excited emission	Bondybey and English [150]
Sn_2	188	Ar	12	overtone progression and isotopic effect	Teichman et al. [79]
Sr_2	44 43	Ar Kr	12 12	laser excited emission	Miller and Andrews [151]
Te_2	248.3 (ω_e) (247.036)	Ar	10	laser excited emission	Ahmed and Nixon [152]
Ti_2	407.9	Ar	11	resonance Raman	Cossé et al. [81]
Tl_2	80 (ω_e)	Ar	12	resonance Raman	Schulze and Froben [131]
V_2	537.5	Ar	11	resonance Raman	Cossé et al. [81]
YF	632 (ω_e) 625 (636.3)	Ne Ar	15	laser excited emission	Hamilton et al. [153]
Zn_2	80	Kr	20		Givan and Loewenschuss [123]
ZnCl ZnBr	385 198	Kr	20		Givan and Loewenschuss [133]
ZnO	810	Ar	15	in combination with IR and visible absorption	Prochaska and Andrews [124]
ZrO	964 (969)	Ne	4	laser excited emission	Lauchlan et al. [154]

C. Triatomic Molecules

TABLE 5b

Impurity	Vibrational Frequency (cm^{-1})			Matrix	T(K)	Remarks	References
	ν_1	ν_2	ν_3				
Ag$_3$	120.5			Kr	5		Schulze et al. [112]
Br$_3$	197 190	305 ν_{intra} (Br-Br)		Kr Xe	4.2 4.2		Boal and Ozin [155]
Cd Br$_2$	209.1			Kr	20	other isotopes and mixed di-halides	Strull et al. [156]
Cd Cl$_2$	329.8						
Cd I$_2$	155.1						
CdF$_2$	555.0			Kr	20	comparison between IR and RS spectra, other isotopes and polymers	Givan and Loewenschuss [157]
CdFCl	365		610				
CdBrI	305		583.5				
CNN	2824	394	1235	Ar	4	laser excited emission of this free radical	Bondybey and English [102]
CO$_2$	1382.0 (1388.17)	1277.9 = 2ν_2 (1285.41)		Kr	20	comparison between IR and RS spectra other isotopes	Loewenschuss and Givan [82]
CO$_2$	1385.0 1383.3 1380.2 (1388.187)	1280.8 = 2ν_2 1279.2 1276.9 (1285.412)		Ar Kr Xe	15 15 20		Jodl and Kobel [83]

Molecule				Matrix		Remarks	Reference
CS$_2$	655.5 (658.0)	800.8 = 2ν_2 (802.1)		Kr	20		Loewenschuss and Givan [82]
CS$_2$	657.7 658.3 657.8	807 = 2ν_2 804.9 800		Ar Kr Xe	16 16 16		Jodl and Strelau [158]
Cl$_2$O	638	297	678	Ar	16	other isomeric species	Andrews [16]
ClO$_2$	942 940 (955)			Ar,Kr Xe	16 16	comparison between resonance Raman and resonance fluorescence	Chi and Andrews [159]
CuO$_2$	668			Ar	10	laser excited emission	
CuO$_2$	624			Ar	10	resonance Raman	Tevault [160]
GeCl$_2$	390		362	N$_2$	5		Ozin and Voet [161]
H$_2$S	2619 (2611)	1179 (1183)		Ar	20	other polymers	Sheng-Yuh Tang and Brown [56]
HCN	2088.8	721.5	3303.9	Ar	20	comparison between IR and RS spectra other polymers	Barnes and Orville-Thomas [162]
HgCl$_2$	358.4		407.5	Kr	20	other isotopes and other polymers	Givan and Loewenschuss [68]
Hg Br$_2$	225.0		293.8				
Hg I$_2$	163.5	74	237.5				
HgClBr	387.5	91	255.5	Kr	20		Givan and Loewenschuss [68]
HgCl I	378.0	83.5					
HgBr I		66	272.0				

(continued)

TABLE 5b (continued)

Impurity	Vibrational Frequency (cm⁻¹) ν_1	ν_2	ν_3	Matrix	$T_{(K)}$	Remarks	References
HgF_2	567.6			Kr	20	comparison between IR and RS spectra other isotopes and polymers	Givan and Loewenschuss [157]
$HgFCl$	397.5		592.6				
$HgFBr$	265.8		581.5				
$M^+Cl_2^-$	ν_{intra} (Cl-Cl)⁻			Ar	12		Howard and Andrews [73]
Li^+	246						
Na^+	225						
K^+	264						
Rb^+	260						
Cs^+	259						
$M^+I_2^-$	ν_{intra} (I-I)⁻			Ar	12	resonance Raman and other halogenide	Howard and Andrews [163]
Li^+	115						
Na^+	114						
K^+	113						
Rb^+	116						
Cs^+	115						
$M^+O_2^-$		ν_{intra}(O-O)⁻	ν_{intra}(O₂-O₂)⁻	Ar	16	isotopic effects and other dimers	Andrews and Smardzewski [141], Smardzewski and Andrews [63], Smardzewski and Andrews [164]
Li^+		1097	694				
Na^+		1094	391				
K^+		1108	305				

Species				Matrix	K		Reference
Rb^+		298	1110				
Cs^+		287	1114				
MgF_2	841.8	249.0	550.0	Ar	14	comparison between IR and RS spectra	Lesiecki and Nibler [71]
$MgCl_2$	600.8	93.0	326.5				
$Mg\,Br_2$	497.1	81.0	197.9				
$Mg\,I_2$	444.9	55.8	147.6				
NO_2	(1633.4)	752 (750.1)	1325 (1329.6)	Ar	16	resonance Raman, isotopic effects and other polymers	Tevault and Andrews [165]
NO_2	1616.0	751.4	1320.5	Ar	15	resonance Raman, dimers and N_2O_4 in N_e	Jodl and Bolduan [78]
	1615.0	750.3	1321.6	Kr	15		
	1614.0	748.9	1315.8	Xe	15		
	(1618)	(749.6)	(1320.2)				
Ni_3			232.3	Ar	5	resonance Raman overtone progressions and isotopic splitting	Moskovits and Dilella [66]
O_3	1037.9	705.4 (702.1)	1106.4 (1103.3)	Ar	20		Hopkins and Brown [99]
OF_2	825	464	920	Ar	16	Fermi resonance photolysis	Andrews [5]
		$\nu_{intra}(O-F)$	1028.9				Andrews [15]
		$\nu_{intra}(O-O)$	1552				
		$\nu_{intra}(F-F)$	892				

(continued)

TABLE 5b (continued)

Impurity	Vibrational Frequency (cm^{-1}) ν_1	ν_2	ν_3	Matrix	$T_{(K)}$	Remarks	References
Pb$_3$	117			Xe	12	resonance Raman	Stranz and Khanna [146]
PbCl$_2$	322 (314)	103 (99)	300	Ar	4.2	isotopic effect	Ozin and Voet [161]
SO$_2$	1148 (1151.38)	520 716 ν_{intra}(S-S) (517.69)	1347 (1361.76)	Ar			Andrews [13]
SO$_2$	1151.3 1150.0 1146.8 (1151.3)	524.0 522.5 519.5 (518.2)	1342.5 1345.5 1344.0 (1361.5)	Ar Kr Xe	16 17 17		Jodl and Conrad [166]
Se$_3$	313			Ar	15	resonance Raman	Schnöckel et al. [167]
Sn Cl$_2$	353 (352)	(120)	332	Ar	4.2		Huber et al. [168]
SnBr$_2$	244	82 (80)	231	Ar	4.2	polarization measurements, other isotopes and other polymers	Ozin and Voet [161]
SnCl$_2$	353 (352)	(120)	332	Ar	4.2		
Te$_3$	203			Ar	15	resonance Raman	Schnöckel [169]
XeCl$_2$	253	543 ν_{intra}(Cl-Cl)		Xe	4.2		Boal and Ozin [110]
XeCl$_2$	255.0 251.5 248.0			Xe		isotopic effect	Howard and Andrews [170]

Species				Matrix		Notes	Reference
$ZnCl_2$	352			Kr	20	other isotopes and polymers	Givan and Loewenschuss [67]
$ZnBr_2$	222.6						
ZnI_2	163.1						
ZnClBr	273.1 $\nu(Zn-Br)$						
	464 $\nu(Zn-Cl)$						
ZnClI	229.5 $\nu(Zn-I)$						
	460 $\nu(Zn-Cl)$						
ZnBrI	190 $\nu(Zn-I)$						
	378 $\nu(Zn-Br)$						
ZnF_2	595.5			Kr	20	comparison between IR and RS spectra other isotopes and polymers	Givan and Loewenschuss [157]
ZnFCl	412		694				
ZnFBr	303		674				

D. Poly-atomic Molecules

TABLE 5c

Impurity	Vibrational Frequency cm^{-1}	Matrix	T(K)	Remarks	References
Al Cl$_3$	393.5 ($\nu_{sym.}$) (375)	Ar	12	in combination with IR experiments	Beattie et al. [171]
Al$_2$Cl$_6$	342 (ν_2)	Ar	20	in combination with IR experiments	Tranquille and Fouassier [172]
C Cl$_4$	461.9 (ν_1) 220 (ν_2) 789.1 (ν_3) 319.5 (ν_4)	Ar	10-15 isotopic effect		Levin and Harris [173]
C^{35}Cl$_4$ C^{37}Cl$_4$	462.5 (ν_1) 449 (ν_1) (463.6)	Ar	14	isotopic effect in combination with IR experiments	Becker et al. [174]
CH$_4$	2914 (ν_1) 3017 (ν_3)	Kr	13		Cabana et al. [175]
CH$_4$	2928 (ν_1) 2912.5 2900.2 (2916.5)	Ar Kr Xe	16 18 22	in the case of more than one line we propose vibration-rotation lines	Jodl and Regitz [176]
	1531 (ν_2) 1545	Ar	16		
	1526 1545	Kr	18		

	1520 1544	Xe	22		
	(1533.6)				
	3027.2 3045.8 (ν_3) 3058.4	Ar	16		
	3018 3039 3051.3	Kr	18		
	3006 3011 3027.6 3041.2	Xe	22		
	(1306.2) (ν_4)			too weak	
c-C_4F_8	277 (CF_2 bend) 361 (CF_2 bend) 444 (CF_2 bend) 700 ring breathing 1010 ring mode	Ar	15	in combination with IR experiments	Harris et al. [177]
CH_3NH_2	1052 (CN stretch) (1044) 1450 (CH_3 sym. def.) (1430) 2819 (CH_3 sym. stretch) (2820) 3352 (NH_2 sym. stretch) (3360) and others	Ar	4-20	in combination with IR experiments and other isotopes	Purnell et al. [178]

(continued)

TABLE 5c (continued)

Impurity	Vibrational Frequency cm^{-1}	Matrix	T(K)	Remarks	References
D_2O_2	2653.5 (ν_1) (2668) 1021.5 (ν_2) (1028) 871 (ν_3) (867)	Ar	10	in combination with IR experiments	Giguère and Srinivasan [179]
$FeCl_3$	366.0 (ν_1) 68.7 (ν_2) 460.2 (ν_3) 113.8 (ν_4)	Kr	20	isotopic effect	Givan and Loewenschuss [69]
Fe_2Cl_6	426.0 (stretch) (426) 314.5 (ring stretch) (310) 115.0 (ring bend) (163) 108.0 (bend) (127) and others	Kr	20		Givan and Loewenschuss [69]
H_2O_2	3593 (ν_1) (3607) 1385 (ν_2) (1393.5) 869 (ν_3) (863.5)	Ar	10	in combination with IR experiments	Giguère and Srinivasan [179]
$(I_2)_n$	181 ν_{intra}(I-I)	Ar	12	in combination with laser induced fluorescence	Ault and Andrews [180]

Species	Frequencies	Matrix		Notes	Reference
$InCl_3$	352.5 (ν_1) (350) 394 (ν_3) 98.5 (ν_4) (94)	Kr	20	in combination with IR experiments, isotopic effects	Givan and Loewenschuss [70]
$M^+Cl_3^-$	550 ν_{intra}(Cl-Cl)	Ar	15	in combination with IR experiments	Ault and Andrews [181]
Li^+	--- (ν_1 sym.) 410 (ν_3 anti.)				
Na^+	276 (ν_1) 375 (ν_3)			and other alkali halide salt molecules with BrCl and Br_2	
K^+	258 (ν_1) 343 (ν_3)				
Rb^+	253 (ν_1) 335 (ν_3)				
Cs^+	225 (ν_1) 326 (ν_3)				
$M^+ClO_3^-$		Ar	12	in combination with IR experiments, other alkali ions and other dimers	Smyrl and Devlin [182]
K^+	927 (ν_1) 969 (ν_3)				
$M^+HCl_2^-$		Ar	15	in combination with IR experiments, other alkali ions and DCl_2	Ault and Andrews [183]
Rb^+	250 (ν_1) 729 (ν_3)				

(continued)

TABLE 5c (continued)

Impurity	Vibrational Frequency cm^{-1}			Matrix	T$_{(K)}$	Remarks	References
M$^+$F$_3^-$							
Cs$^+$	461 $\nu_{intra.1}$(F$_3^-$)			Ar	15	in combination with IR experiments and other halide molecules	Ault and Andrews [184]
M$^+$NO$_2^-$	1244 $\nu_{intra.1}$(NO$_2^-$)			Ar	16	in combination with IR experiments	Andrews [15]
M$^+$NO$_3^-$	(ν_1)	(ν_2)	(ν_3)	Ar	4	in combination with IR experiments and other dimers	Smith et al. [59]
Na$^+$	1023	825	1283 1484				
K$^+$	1031	830	1291 1462				
Rb$^+$	1033	830	1293 1456				
M$^+$O$_3^-$	$\nu_{intra.1}$(O$_3^-$)			Ar	16	resonance Raman and other oxygen isotopes	Andrews [72]
Li$^+$	1012						
Na$^+$	1011						
K$^+$	1004						
Rb$^+$	1026						
Cs$^+$	1018						
Mg$^+$	1023			Ar	10	resonance Raman in combination with visible absorption	Ault and Andrews [185]
Ca$^+$	1019						

Species				Notes	Reference
N_2O_4	265 (ν_3)	Ne	9	part of the lines, in addition NO^+ and NO_3^-	Jodl and Bolduan [186]
	257 (ν_3)	Xe	15		
Na_2SO_4	1137 (B_2)	Ar	12		Atkins and Gingerich [187]
	1101 (E)				
	1124 (B_2)				
K_2SO_4	1098	Ar	12		
	642 (B_2)				
	610 (E)				
Ni $(N_2)_n$	2088 $\nu Ni(N_2)$	Ar	4	in combination with IR experiments and Pd-atoms or Pt-atoms	Huber et al. [188]
	2187 $\nu Ni(N_2)_2$				
	2246 $\nu Ni(N_2)_4$				
	and others				
Os O_4	963.0 (ν_1)	Ar	12	in combination with IR experiments and oxygenisotopes	Beattie et al. [189]
	332.0 (ν_2)				
	956.0 (ν_3)				
	326.0 (ν_4)				
OsO_3F_2	946.5 (ν_1)	Ar	12		
	619.0 (ν_2)				
	and others				

(continued)

TABLE 5c (continued)

Impurity	Vibrational Frequency cm^{-1}	Matrix	T(K)	Remarks	References
Pb$_4$	111 (A$_1$)	Xe	12		Stranz and Khanna [146]
PrF$_4$	526 (ν_1)	Ar	14		Lesiecki et al. [190]
	458 (ν_3)				
	99 (ν_4)				
SF$_6$	770 (ν_1)	Ar	4		Shirk and Claassen [1]
	640 (ν_2)				
	520 (ν_5)				
S$_2$Br$_2$	280 ν_{intra}(Br–Br)	Ar	17	IR and RS spectra from the self-same matrix	Grzybowski et al. [97]
	359.7 ν_{intra}(S–Br)				
	551.7 ν_{intra} (S–S)				
S$_2$F$_{10}$	245.1 ν_{intra}(S–S)	Ar	8	in combination with IR experiments	Smardzewski et al. [191]
SF$_5$Cl	400.8 ν_{intra}(S–Cl)				
SF$_5$Br	274.3 ν_{intra}(S–Br)				
	and others				
SO$_3$	1065 (ν_1)	Ar	20		Sheng-Yuh Tang and Brown [57]
	(1065)				
	1387 (ν_3)				

Species	Frequency	Matrix		Notes	Reference
	(1380)				
	528 (ν_4)				
	(536)				
Sb_4	241 (ν_1)	Ne	4		Bondybey et al. [115]
	179 (ν_3)				
Tl_2F_2	443.0 ν_{intra}(Tl–F) (473)	Ar	14	in combination with IR experiments, isotopic effect and other polymers	Lesiecki and Nibler [64]
	226.0 (ν_1)				
	93 (ν_2)				
	297.0 (ν_3)				
	263.0 ν_{intra}(Tl–Cl) (282)				
Tl_2Cl_2	131.0 (ν_1)	Ar	14		
	61 (ν_2)				
	190.4 (ν_3)				
UF_5	646 (ν_1)	Ar	10		Jones and Ekberg [101]
UF_6	665.8 (ν_1)	Ar	10		Jones and Ekberg [101]
	(667)				
	530 (ν_2)				

(continued)

TABLE 5c (continued)

Impurity	Vibrational Frequency cm^{-1}	Matrix T (K)	Remarks	References
UF$_6$	(534) 200 (ν_5) (200)			
UF$_6$	667 (ν_1) 533 (ν_2) 197 (ν_5)	Ar 12	in combination with laser induced fluorescence	Grzybowski and Andrews [192]
XeF$_6$	507 (?) (509) 555 (?) (570) 628 (?) (613)	Ar 4	in combination with absorption, IR experiments, on gas and pure solid phase	Claassen et al. [193]
XeO$_3$F$_2$	806.7 (ν_1) 567.4 (ν_2) 892 (ν_5) 316 (ν_6) 190 (ν_7) 361 (ν_8)	Ar 4	in combination with IR experiments	Claassen and Houston [194]

V.   SELECTED EXAMPLES

Our own studies have purposely centered on small molecules which are easily accessible for MI and whose molecular parameters are well known in the literature.   Each subchapter is similarly structured (published knowledge concerning MI-case/physical question/our data and discussion) and only a few species are chosen for presentation and one specific aspect is pointed out.

A.   Hydrogen (H_2 and D_2) - A Model Substance for MIS

For more than twenty-five years hydrogen molecules have been studied extensively in various environments.   For MI-RS investigations this molecule has the advantages that the molecule parameters of the free molecule are completely known, there exist isotopes H_2, HD, D_2 and ortho/para modifications, the vibrational, rotational, and combined excitations are energetically well separated.   Moreover, the structure of the H_2/RGS system is known, sample generation by thin film condensation or crystal growth is possible, Raman activity is characterized.   It is of interest to consider how the various modes (vibration, rotation, libration, vibration-rotation, vibration-translation, lattice modes) are influenced by the matrix.   How do expected multiple trapping sites in various matrices or aggregation during temperature cycles influence the spectra?   Also, how good is the comparison of the theoretical values and experimental data with regard to matrix shift, anharmonic contributions, etc.?   Is there a systematic difference between IR and RS spectral features and why are the discrepancies in experimental data so large between different investigations?   Why are the ortho/para components of the pure vibration hitherto not observed, whereas the o/p rotational lines are well identified?

Recent investigations are briefly mentioned here and extensively discussed elsewhere (Jodl and Bier [127]).   McKellar and Welsh [195] proved the existence of bound states of $(H_2)_2$ complexes and measured the energy levels by IR-absorption.   Kriegler and Welsh [196] and de Remigis et al. [197] studied the induced IR fundamental band of H_2 in Ar at 80K, and of D_2 in Ar at 80K and of H_2 in Kr at 115K.   Besides the pure vibration (zero phonon line) and vibration/rotation they observed always structured phonon sidebands.   Jean Louis and Vu [198] extended near IR measurements at low temperatures (5K and 77K) to high pressure (up to 11 kbar) studies in solid solutions of H_2 in Ar, D_2 and N_2 matrices.   The spectra reveal transitions corresponding to pure vibrations, pure vibration/rotations and to combinations of them with lattice modes. High pressure effects - broadening, frequency shift, increase in intensities - are explainable by anharmonic effects, especially by multiphonon processes. Prochaska and Andrews [58] described the Raman spectra of H_2, D_2 and HD in Ar, Kr, Xe, N_2, O_2 matrices at 12K as a function of impurity concentration and annealing effects. They observe pure rotational, pure vibrational and vibrational-rotational lines and aggregate lines next to them and discuss the data in the light of isotopic effects, ortho/para ratio, interaction with different lattices and internuclear attractions during condensation. Vitko and Coll [199] estimated on the basis of the MI-IRS (Kriegler and Welsh, [196], and de Remigis et al., [197] and MI-RS (Prochaska and Andrews, [58]), data the matrixshift and center of mass frequency (i.e. translational mode) of H_2 and D_2 in Ar, Kr, Xe. The calculations contain the demands mentioned in Sec. I.B. and must not be overestimated; for example, in comparison to experiments a negative sign for the matrixshift and an increase in size of this shift with heavier RG-atoms.

Warren et al. [200] investigated by IR-absorption MI-H_2 and D_2 in Ar, Kr, N_2 and CO. This group extended the former measurements to 13K, but used untypical high concentrations (MIS = 100:10) and added small amounts of H_2O and O_2 to influence the ortho/para ratio. The assignment of the pure rotational region appears to be straightforward, but the coincidence with MI-RS (Prochaska and Andrews, [58]) is far from satisfying. Therefore, they tried to explain this discrepancy, probably errors in frequency calibration, with physical arguments. The stretching region contains two pairs of transitions, the zero phonon and phonon-sideband of the Q branch: vibrational and vibrational-translational modes. A certain confusion arises, if one tries to assign the multiplet structure of the pure vibrational transition and correlate the result to other studies.

We ourselves investigated H_2 in Ne, Ar, Kr, Xe and D_2 in Ar. Experimental parameters for thin film condensation were as follows: copper sample holder; M/S $\geq$ 100:3; deposition rates between 20-300 µmol/h for 1 to 4 hours; deposition temperature 15K; annealing procedure 15K $\rightarrow$ 35K (for 30 min) $\rightarrow$ 15K. Samples with higher concentration (M/S = 100:3 to 100:5) and H_2/Ne mixtures (M/S = 100:10 to 100:50) at T < 10K could either not be maintained on the sample holder or be spectroscopically measured, because the H_2 signal diminished with time. Our Raman equipment is standard and the irradiation geometry similar to the one of Fig. 34a. A typical survey spectrum is presented in Fig. 45; the diagram (Fig. 46) explains the notation and accentuates possible Raman-transitions.

In this section we concentrate only on the question of the splitting of the fundamental band and the difference in its frequency values in the literature. (All the former mentioned interesting aspects will be the content of

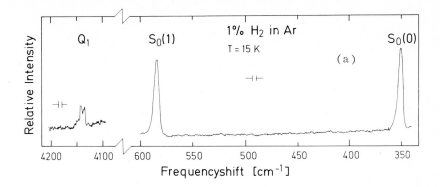

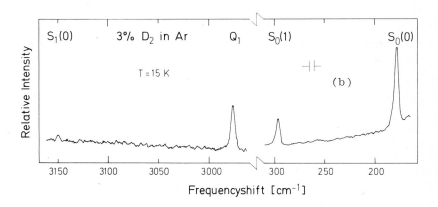

FIG. 45. Raman scattering of MI-H_2 (a) and MI-D_2 (b) at similar conditions.

another publication - Jodl and Bier [127].) The gas data (Stoicheff, [201]) determine an ortho/para splitting ($Q_1(1)$ - $Q_1(0)$) of 5.9 cm^{-1} in the case of H_2 and of 2.1 cm^{-1} in the case of D_2. Kriegler and Welsh [196] and de Remigis et al. [197] could not resolve this zero phonon vibrational line Q_q, which is very weak in IR absorption. Prochaska and Andrews [58] found by MI-RS a triplet of the Q branch (Fig. 48) and associated the low energy line with the unresolved $Q_1(1)$ - $Q_1(0)$ fundamental line of H_2 and the two others with $(H_2)_n$ aggregates; for D_2 they observed a single line only.

Although the authors made concentration dependent measurements (Fig. 48, b-c), one has to keep in mind the general problem of comparing Raman signals (see Sec. III.C. and Fig. 29) and especially the rather poor agreement between insets b and c. If temperature cycles are made (Fig. 48b-d) and aggregates are produced or augmented, the monomer peaks should change reversibly and the area below the triplet structure should remain constant

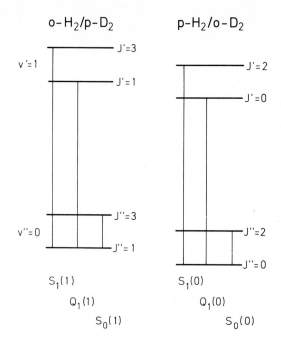

FIG. 46. Vibrational v and rotational J levels of the ortho and para modification of H_2 and D_2: Q and S have the usual meaning of ΔJ = 0 or 2, respectively, while the associate subscript denotes the change in vibrational quantum number; the value in parentheses is the initial J value.

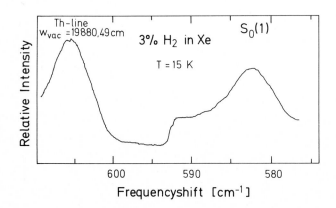

FIG. 47. The Raman spectrum of 3 mol % H_2 in Xe at 15K demonstrates how each Raman line was calibrated. Apparatus conditions: excitation with 4880 Å, 800 mW and spike filter, slit width 120 μm (4 cm⁻¹), signal to noise ratio 2.5.

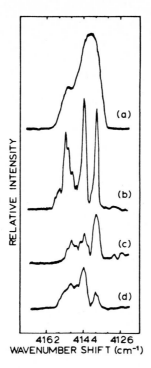

FIG. 48. Raman spectrum of H_2 Q branch in solid argon; spectrum scanned in 10 cm^{-1}/min using 5145 Å excitation (800 mW). Spectrum (a) Ar/H_2 = 50/1, slitwidth = 500 μ/500μ/500μ, scanned on 1.0 x 10^{-9}A range; (b) Ar/H_2 = 50/1, slitwidth = 70 μ/100μ/70μ, 0.1 x 10^{-9}A range; (c) Ar/H_2 = 200/1, slitwidth = 150μ/150μ/150μ, 0.1 x 10^{-9}A range; (d) Ar/H_2 = 50/1, after matrix temperature cycled to 35K, slitwidth = 150μ/150μ/150μ, 0.1 x 10^{-9}A range. Used by permission [58].

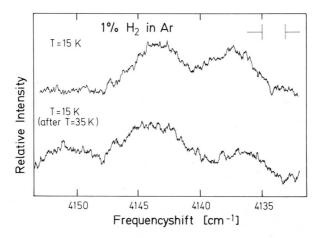

FIG. 49. Raman spectrum of 1 mol % H_2 in Ar: condensed and measured at 15K (upper part), annealed at 35K for 30 min and recooled to 15K (lower part). The low energy doublet is assigned to the Q-branch $(Q_1(1)-Q_1(0))$, the line at about 4150 cm^{-1} to an aggregate of H_2 or to a different trapping site.

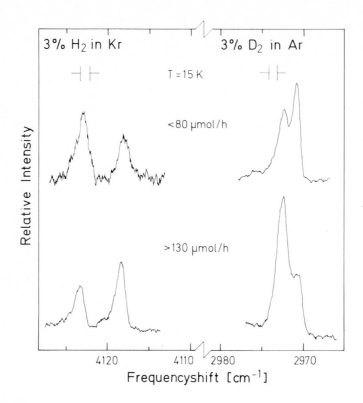

FIG. 50. Splittings of the fundamental band ($Q_1(1)$ - $Q_1(0)$) and intensity ratios as a function of deposition rate are compared for H_2 in Kr and D_2 in Ar.

(uncertainty in intensity ~10%) in contrast to the results in the figure. The structure of the high energy line of the triplet is always small and broad independent of the matrix Ar, Kr, Xe, N_2. In O_2 they observe a doublet, in which only the para modification of H_2 is present. In addition their frequency calibration is based on laser plasma-lines; this may imply an uncertainty of ±3 cm^{-1} to our experience (see Sec. III.C). Warren et al. [200] found four lines in that region of the induced IR fundamental band of 10% H_2 in Ar and tried an assignment on the basis of the RS measurements by Prochaska and Andrews. But they clearly recognized the fact that the IR data of H_2 in Ar and Kr are red shifted by 5 to 6.5 cm^{-1} from the Raman transitions and that the calculated split bewteen $Q_1(1)$ - $Q_1(0)$ of 9 cm^{-1} should be resolved in the Raman spectra.

We can bring into discussion at least three convincing experimentally proven arguments for a correct assignment of the Q-branch components. First, Fig. 49 shows a doublet (6.3 cm^{-1}) of the Q-branch of H_2 in Ar at 15K

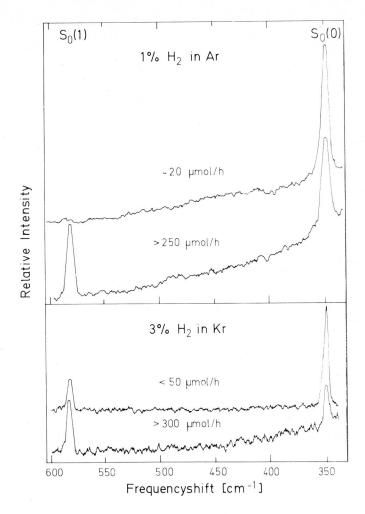

FIG. 51. Change of Raman intensities as a function of deposition rates of the pure rotational lines in H_2/Ar and H_2/Kr.

and an additional peak <u>only</u> after an annealing procedure. The area below the doublet and the triplet is temperature-independent ($\Delta I \sim \pm 10\%$). Second, the same doublet is observed in Kr (see Fig. 50; splitting 5.9 cm^{-1}) and in Xe (see Table 6; splitting 6 cm^{-1}). We were even able to resolve the doublet in D_2 in Ar; this splitting is 2 cm^{-1} and of the order of the gas phase.

Finally the intensity of the doublet is changed systematically if the deposition rate is altered (Fig. 50). This behavior is exactly the same for the pure rotational lines (Fig. 51). For "slow spray on" (few 100 μmol/h) condensation the ortho/para intensity ratio of the $Q_1(1)/Q_1(0)$ Raman lines and of the $S_0(1)/S_0(0)$ Raman lines is the same and of the order of the gas

TABLE 6

Summary of Observed Rotational, Vibrational and Vibration-Translation Transitions of
MI-H₂ and D₂ by Different Authors

Assignments		Matrix	Gas (cm⁻¹)		IR (cm⁻¹)			RS (cm⁻¹)	
			Stoicheff [201]	McKellar & Welsh [195]	Kriegler & Welsh [196] De Remigis et al. [197]c	Jean Louis and Vu [198]d	Warren et al. [200]e	Prochaska & Andrews [58]f	Jodl & Bier [127]g
ROTATION									
$S_0(0)$		Ar	354.381				352.5	358	352.1
		Kr					353	359	351.4
		Xe						356	350.0
$S_0(1)$		Ar	587.055				584.5	591	584.2
		Kr					583	588	583.0
		Xe						587	582.2
VIBRATION									
$Q_1(0)$		Ar	4161.134		(4142)a	(~4150)	4144	4144	4143.2
		Kr			(4132)			(4130)	4124.4
		Xe						(4121)	4116.0
H_2 $Q_1(1)$	±off b	Ar	4155.201	1.6	(4142)	(~4150)	4137	4138	4136.9
		Kr		2.8	(4132)			(4130)	4118.5
		Xe		8.6				(4121)	4110.0
$(H_2)_n$		Ar			4165	~4170	4130	4152	4150
		Ar			4148		4150	4147	4137
		Kr					4140	4138	4130
		Xe							

VIBRATION/ TRANSLATION		Matrix					
$(H_2)_n$		Ar			4244		
		Ar			4242		
$Q_R(0)$		Ar	(4257.5)	(~4270)	4246		
		Kr	(4247)		(4248)		
$Q_R(1)$		Ar	(4257.5)	(~4270)	4248		
		Kr	(4247)		(4248)		
$S_0(0)$	179.056	Ar			176	173	178.9
$D_2\ S_0(1)$	297.521	Ar			296	297	298.2
$Q_1(0)$	2993.548	Ar	(2981)		2978	(2977)	2973.1
$Q_1(1)$	2991.446	Ar	(2981)		2973	(2977)	2971.1

a values in parentheses are not resolved
b transitions of $(H_2)_2$ complexes off H_2-lines at 20K
c at 80K for Ar, 115 for Kr with M/S ≐ 100:1
d at 4K for Ar, with M/S = 100:1
e at 13K for Ar and Kr with M/S = 100:10
f at 12K for Ar, Kr and Xe, with M/S > 100:2
g at 15K for each RGS, with M/S > 100:3

TABLE 7

Fundamental Line of NO in the Electronic Ground State

	$\Delta G_{\frac{1}{2}}(cm^{-1})$			
	Ar	Kr	Xe	References
$B^2\Pi \rightarrow X^2\Pi$	1874	1858	---	Frosch and Robinson [202]
$A^2\Sigma^+ \rightarrow X^2\Pi$	1835 ± 60	1859 ± 30	1854 ± 60	Goodman and Brus [203]
$B^2\Pi \rightarrow X^2\Pi$	1880.3	---	---	Fournier et al. [204]

phase ratio 3:1. For "very slow spray on" (few 10 μmol/h) condensation this ratio shifts in the direction favoring the para modification. As a consequence we definitely assign this doublet to be the ortho/para components $(Q_1(1)-Q_1(0))$ of the pure vibrational mode of the H_2 molecule. Table 6 contains the known IR and RS transitions in which the experimental values of the other groups are reinterpreted according to our assignments.

To close this section we catalogue some remaining problems which need investigating: (1) The variation of the ortho/para ratio as a function of deposition rate using paramagnetic molecules especially O_2 as impurities or as matrix material; (2) H_2 in non spherical matrices like N_2, O_2 because H_2 rotates freely in Ar, Kr, Xe. (3) Mi-H_2 under high pressure to get information about anharmonic contributions to different modes of vibration, rotation, vibration-rotation, vibration-translation. (4) The phonon region of RS of MI-H_2 and (5) H_2 in Ne matrices at T < 8K.

B. Nitric Oxide (NO)-Molecular Constants and Quantitative Annealing Effects

In general, spectroscopic data yield molecular constants, which in turn allow the construction of molecular potentials. In the case of MI-NO recent publications have presented different data with deviations exceeding by far the experimental error (Table 7).

These data were obtained by irradiation with resonance lamps, lasers, or x-ray stimulated emission. If the well known relation for the energy levels of anharmonic oscillators is applied to the measured vibronic series and the term values $\Delta G_{v'' + \frac{1}{2}}$ plotted versus $v'' + \frac{1}{2}$, one can extrapolate to the first vibrational transition $\Delta G_{\frac{1}{2}} = v(1 - 0) = \omega_{Raman}$. The emission lines in the uv or optical region possess an error of a few 10 cm^{-1}. Therefore, it would be very informative to observe the transition $\nu(1 - 0)$ in IR or Raman directly. The second problem, which stimulated us to investigate MI-NO, concerns the experimental difficulty, to trap monomeric NO because of its known high reactivity and therefore its tendency to form polymers. Neither the process $NO \rightarrow N_jO_k$ (i,j,k = 1,2,3) nor the possible structure of N_2O_2 has been sufficiently investigated up to now.

The NO gas phase value of the fundamental mode is accurately known (Johns and Reid [205]), whereas the data concerning the pure liquid or frozen NO, which claim to prove the existence of different polymers, deliver rather inprecise energy values (Smith et al. [206]). The idea of applying the MIS technique to NO, was mainly stimulated by the question of the existence of Rydberg transitions in the uv emission spectra. Frosch and Robinson [202] trapped NO molecules in solid Ar and Kr (M:S $\geq$ 1000:6), the emission was excited by x-rays and spectra were analyzed with regard to energy levels and lifetimes. The role of high lying levels of NO in photochemical reactions and in auroral and airglow spectra was briefly discussed. Goodman and Brus [203] reinvestigated higher Rydberg states in MI-NO in Ne, Ar, Kr, Xe (M:S = 100:1) at 5K; these states would be populated by sequential two photon absorption from frequency doubled pulsed dye lasers. In addition they studied the tendency of NO to dimerize in RGS; condensation of nitric oxide at 4K produces mostly NO, whereas at 15K more N_2O_2 is formed. Fournier et al. [204] observed vibronic series after irradiation of an NO/Ar mixture (M:S = 1000:1) at 6K. The excitation energy was systematically varied below and above the ionization potential of NO to elucidate the fate of Rydberg states in rigid media. Fateley et al. [207] made MI-IRS studies of several oxides of nitrogen in different matrices (M:S = 130:1) at 5K. NO was found as the monomer and as two forms of dimer of cis- and trans-configuration; in addition the NO_2,N_2O_3 and N_2O_5-system was also investigated. Guillory and Hunter [208] executed systematic IR investigations of 1 mol % NO in Ar, N_2, CO by annealing procedures, by isotopic analysis and by photochemical reactions and postulated one stable and two unstable forms of the dimer.

We condensed our samples of NO in Ar, Kr, Xe (M:S = 100:5 to 3) at 15K with a deposition rate of about 20 µmol/h and annealed them later on (15K $\rightarrow$ 35K for 30 min $\rightarrow$ 15K) in a manner described in Sec. III. We did not try to isolate NO in Ne, but we used concentrations up to M:S = 100:50 for N_2O_2 studies. Our Raman equipment is standard and the irradiation geometry is similar to the one of Fig. 34a. We used atomic lines for frequency calibration. Unfortunately the Raman signal is very weak (see Table 4) and the bands are broad (the error in frequency is between 0.3 to 1 cm^{-1}). The

TABLE 8

| Assignment | Matrix | IR (cm^{-1}) | | RS (cm^{-1}) |
		Fateley et al. [207]	Smith et al. [206]	Jodl & Bier [127]
NO vibration	1875.972 gas (Johns & Reid [205])	Ar Kr Xe	1875 --- ---	--- 1872 ---	1873±1 1872.2±0.3 1867±1
cis-N$_2$O$_2$ ν_1 (sym. str.)	1870 liq. (Smith et al., [206])	Ar Kr Xe	1866 --- ---	--- 1865 ---	1864±1 1864.1±0.5 1856±1

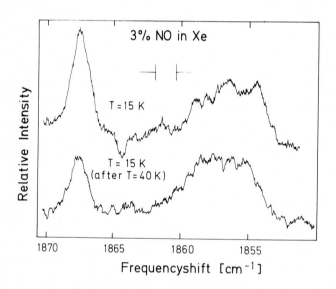

FIG. 52. Raman spectrum of 3 mol % NO in Xe: condensed and measured at 15K (upper part) annealed at 40K for 30 min and recooled to 15K (lower part). The higher energy line is the fundamental vibration of NO, the lower energy band is the symmetric stretching mode of N$_2$O$_2$. Apparatus conditions: excitation with 5145 Å, 800 mW and spike filter, slitwidth 35 μm (1.5 cm^{-1}).

subject is discussed in a wider context: MI-RS of small molecules (Jodl and Bier [127]). Table 8 contains our results and a comparison with MI-IRS data. The improvement is apparent - see Table 7.

We failed in measuring temperature shifts of the fundamental line, either the coefficient $\Delta\omega/\Delta T$ is too small and below our spectral resolution or dimeri-

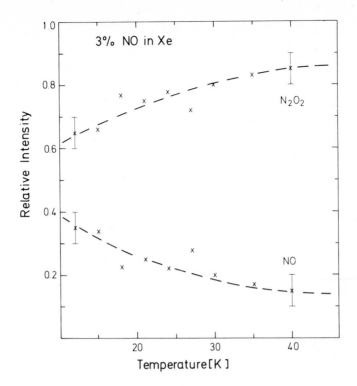

FIG. 53. Relative change of Raman intensities of the NO- line and the N_2O_2-line (from Fig. 52) as a function of temperature. The sum of both intensities is constant to ± 10%.

zation effects obscure it. Figure 52 shows a typical Raman spectrum. The narrower line on the high energy side is attributed to the fundamental mode of the monomer, the broad wing to the symmetric stretching mode of the dimer. After a single temperature cycle the dimer peak grows at the expense of the monomer peak. Figure 53 demonstrates the irreversible conversion up to 40K in steps of 3K.

These experiments demonstrate convincingly that quantitative studies of annealing effects are possible in MI-RS if certain technical and physical conditions like comparison of spectra on the same sample, careful temperature variation, small errors in intensity determination, etc. are fulfilled. Further experiments concerning this complex could be useful: e.g., a larger M:S ratio for better isolation, NO in Ne for completeness, monomeric NO in matrices like N_2, O_2, CO looking for defect - matrix particle complexes, MI-N_2O_2 to solve the question of stable and unstable forms (cis and trans) of this dimer.

C. Carbon Monoxide (CO)-Comparison of Infrared Absorption and Raman Scattering Data

Numerous studies of CO trapped in rare gas matrices have been performed using a variety of spectroscopic methods. This simple heteronuclear diatomic molecule has proven to be interesting to experimentalists and theoreticians alike as a model system for MIS because of the following reasons: First, the exciting and convincing experimental work by Dubost and co-workers [209-212], who developed careful sample preparation of thin films and provided accurate spectroscopic data by high resolution IR-technique. Second, these measurements stimulated matrix shift calculations of a new type (Manz and Mirsky [213]), different theoretical approaches to explain vibrational energy relaxation (Metiu and Korzeniewski [214], Shin [215], Diestler and Ladouceur [216]), various theoretical models for the coupling mechanisms of vibrational, librational, translational and phonon modes (Manz [217], Blaisten-Barojas and Allavena [218]) etc. Third, the more or less accurate molecular constants of CO found by emission spectra of electronic transitions (Fournier et al., [219]), by IR-data (Dubost [90]) and by RS (Jodl and Bier [127]) can now be compared in detail (Table 9).

The reasons why we started MI-RS on CO was that few IR and RS spectra are known for the same MI-species that enable both sets of data to be meaningfully compared. Also, it is known from Dubost [90], that the spectra depend strongly on the sample preparation and sample history. Parameters for thin film condensation (Sec. III.A.) are limited, e.g., M/S ratio (10,000 to 100:1), gas purity ($\leq$ 10 ppm), vacuum conditions in the system, and deposition temperature (T = 4 to 20K). Also, the annealing procedures are restricted by the cryostat. We extended the range of deposition rates from m mol/h (slow spray on) to μmol/h (very slow spray on). This study will be the subject of a forthcoming publication (Jodl and Bier [127]). Here we will discuss only the discrepancies between the IR and RS spectra. However, first we review briefly the recent MI-IRS literature on CO.

Dubost and Abouf-Marguin [209] investigated CO in Ar very carefully and proposed a new assignment of the complex IR-absorption spectrum. His efforts were directed to improve sample preparation, to avoid gaseous impurities, and to use new spectroscopic techniques. He found that the results of MI-CO are very dependent on the preparative conditions and on the history of the condensed thin film. Dubost [90] reviewed the literature dealing with MI-CO up to this date and discussed the different possibilities to interpret the IR spectra. He also extended matrix isolation of CO/Ar to CO in Ne, Kr and Xe. By variation of the relevant parameters (M/S, deposition rate, sample thickness and optical density etc.) over a larger range than

TABLE 9

Fundamental Vibration (cm^{-1}) of CO in Different Rare Gas Matrices

	ω_o vapor	Ne	Ar	Kr	Xe	References
ω_e	2169.8135802					Mantz et al. [220]
$\omega_e x_e$	13.2883076					
ω_o	~2143.23					
$\Delta G_{\frac{1}{2}}$ $a^3\Pi \rightarrow X^1\Sigma^+$ (± 2)			2139.8 (T=5K)	2136.5 (T=5K)	2138.2 (T=5K)	Fournier et al. [219]
IR (± 0.03)		2140.75 2141.17 (T=9K)	2138.56 (T=9K)	2135.68 (T=9K)	2133.24 (T=9K)	Dubost [90]
RS (± 0.2)			2138.4 (T=80K)			Kiefte et al. [129]
RS (± 0.3)		2140.9 (T=K)	2140.1 (T=14K)	2138.2 (T=15K)	2134.8 (T=22K)	Jodl and Bier [127]
matrix			- 4.7			Manz and Mirsky [213]
shift (calc.)		+ 36.3	+ 2.3	- 7.6	- 13.6	Friedmann and Kimel [221]
		- 2.4	- 4.4	- 5.1	- 5.5	Jodl and Bier [127]

hitherto known, by double doping experiments with possible impurity hosts (like CO_2, H_2O, N_2, NH_3, O_2, CH_4) and by an excellent spectroscopic arrangement (resolution 0.03 cm^{-1}; CO gas bulb for reference) he proposed a full and clear assignment of the spectrum. The results contained a comparison of experiment and theory for the matrixshift and the defect induced phonon sidebands as a function of RG host; the possibility of librational motion was also discussed. Dubost and Charneau [210] pioneered a revolutionary technique using CO_2 laser excitation of the vibrational emission in MI-CO to investigate vibrational energy transfer and the relaxation mechanism. Fluorescence spectra and lifetime measurements as a function of

TABLE 10
Measured Vibrational Frequencies (in cm^{-1}) and Their Assignment for CO in Rare Gas Lattices

a) Dubost [90]

	Ne	Ar[a]	Kr	Xe	Assignment
		Matrix			
	2136.4	---	---	---	isolated CO in a large site
	---	2136.7	2134.9	---	CO-N_2 pair isolated librating CO
Strong bands	2140.9	2138.6	2135.7	2133.2	$(\Delta n_{lib} = 0)(\nu_1)$
	---	2140.0	2136.9	---	CO-CO pair
	---	2143.3	2139.6	2137.8	CO-CO_2 pair, CO-NH_3 pair and/or CO-H_2O second neighbor pair
	2152.3	2148.9	2144.3	2141.1	CO_2-H_2O pair (ν_{11})
Weak bands	2156.7 2160.5	2150 ---	2146 ---	2143 ---	isolated librating CO $(\Delta n_{lib} = 1)$
Very weak bands	2113.8 2093.8 2089.8	2111.6 2091.4 2087.4	2108.6 2088.5 2084.5	--- 2086.0 ---	monomeric $^{12}C^{17}O$ monomeric $^{13}C^{16}O$ monomeric $^{12}C^{18}O$
	2172	2195 2204	2179 2185	2163	in-band lattice modes
	2224	2219	2204	2186	localized lattice mode

[a] Frequencies given for Ar matrix differ by 0.2 cm^{-1} from those previously given due to a more precise calibration. Used by permission.

b) Jodl and Bier [127]

Assignment	ν_{vapor}(cm^{-1})	Matrix	ν_{Dubost}(cm^{-1})	$\nu_{this\ work}$(cm^{-1})
Pure vibration	2143.23 (Mantz et al. [220])	Ne	2140.75 2141.17	2140.9
		Ar	2138.56	2140.1
		Kr	2135.68	2138.2
		Xe	2133.24	2134.8
Vibration-libration		Ne	2156.7 2160.5	
		Ar	2150	2149.6
		Kr	2146	2147.0
		Xe	2143	2142.3
Aggregation		Ne	---	---
		Ar	2140.2	---
		Kr	2136.9	---
		Xe	---	2139.6

M/S ratio, of different isotopes and of sample temperature helped to explain vibrational relaxation by phonon assisted resonance energy transfer. Dubost and Charneau [211] reviewed the present philosophy of vibrational energy migration especially in the case of MI-CO and extended laser excited vibrational fluorescence to isotopic ($^{13}C^{16}O$, $^{12}C^{18}O$) enriched MI-$^{12}C^{16}O$. Alternative pathways for vibrational excitation and de-excitation were discussed in the framework of macroscopic rate constants or microscopic probabilities for energy transfer in conjunction with considerations concerning the influence of mechanical and electrical anharmonicities. Dubost et al. [212] explained the broadening and additional fine structure of the fundamental line in IR spectra of CO in Ar by librational modes and by libration/phonon transitions. He compared the experimental line width of the transitions as a function of temperature with the calculated transition probabilities for the different processes.

The discrepancy between IR- and Raman spectra is documented in Fig. 54 and in Tables 9 and 10, and manifests itself in frequency differences of about 1.5 to 2.5 cm^{-1} in spite of sufficient resolution and the use of frequency standards in both techniques. The question is whether this difference arises from different sample preparation or from the excitation of specific IR active and Raman active combination modes.

Both methods of sample preparation are contrasted by comparing the relevant parameters used.

	IR (Dubost [90])	RS (Jodl and Bier [127])
gas purity	< 100 ppm	< 10 ppm
M/S ratio	6000 to 50:1	100:5
deposition Temperature T_d (K)	8(Ne) 20(Ar) 30(Kr) 30(Xe)	9(Ne) 14(Ar) 15(Kr) 22(Xe)
deposition rate (μmol/h)	10^3 to 10^5	~10^2
time for condensation and measurements	~10 h	~10 h
thickness (μm)	10 to 10^3	30 to 50
annealing procedure	$T_d \rightarrow 1/3\ T_{liq} \rightarrow T_d$	$T_d \rightarrow\ \sim 40K(30\ min) \rightarrow T_d$

The amount of time during which the sample holder acts as a cold trap is almost the same (~10h). Therefore, the increase of gaseous impurities by normal leakage is of the order of 10^{-2} of the S/M ratio or absolutely $\leq$ 100 ppm. The initial gas purities are different. In the case of RS we did not observe any of the fundamental lines of impurities X or additional CO-X peaks. In the case of IR absorptions very often CO-X peaks are documented

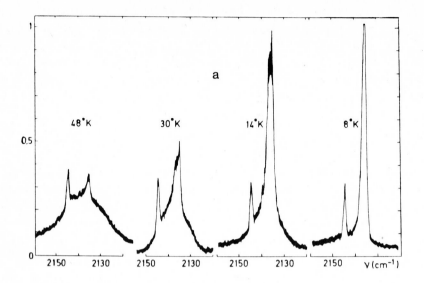

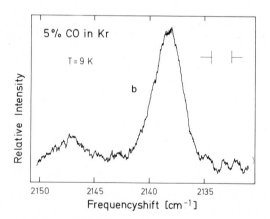

FIG. 54. Comparison of the fundamental line of MI-CO. (a) Temperature dependence of the infrared spectrum of a high optical density Kr-CO
sample with M/R = 1000. (Dubost [90]). (b) Raman spectrum, experimental
conditions: excitation with 4880 Å, 700 mW and spike-filter; slitwidth 60 μm
(2 cm^{-1}) and signal to noise ratio 1.9.

in experiments, in which double doping CO and X is not executed. The
absorption cross-section is much larger than the Raman scattering cross-
section, therefore, we have had to raise the M/S ratio to receive reasonable
signal to noise ratios (3; in IR $\sim 10^3$). For M/S $\sim$ 100:5 some CO molecules

will be found in the first and second shell of the host rare gas lattice while in the case of M/S > 100:1 this separation is larger than the fourth shell diameter. In the case of RS the higher concentration may generate aggregates, but the monomeric peak should be seen at least with low intensity.

The sample temperature is determined by measuring the temperature of the sample holder (CsBr window, or Cu target). Irradiation with a Nernst glower raises this T_{local} by 1 to 2K and with a laser by the same amount; as a consequence it could be that thermal contact is much better in the RS-case. The temperature of the thin film is certainly quite different from that of the sample holder, if light is absorbed or only scattered. Therefore it is to be expected that although the deposition temperatures are the same in both techniques, the difference between T (measured) and T (sample) during measurements is different in IR and RS.

The deposition rate varies by a factor of 10 to 10^3. But if in the IR case the lowest possible deposition rate was chosen (only a factor of 10 larger than in RS) then the position and relative intensities of the CO peaks in the IR spectra and RS spectra are the same. The IR spectra of MI-CO were recorded together with one of a 5 cm long cell filled with gaseous CO at 30 torr pressure. The accuracy then obtained on the measured frequencies is 0.03 cm^{-1}. Wavenumber calibration of standard Raman apparatus (± 1 cm^{-1}) was increased by the use of Thorium lines to be ± 0.3 cm^{-1}, whose light was deflected into the optical path by a moveable mirror. All the Raman values are vacuum corrected.

In spite of these smaller differences in sample preparation the above mentioned discrepancies remain and must have a physical interpretation. Hence, the proposed assignment must be improved (Table 10 a,b).

In RS one can clearly observe the reversible change in frequency shift (Fig. 56) and in the size of the wings symmetric to the fundamental line and the irreversible growth of an additional band between both lines (Fig. 55b), which was therefore due to aggregation.

The published IR-spectra contain lines in the vicinity of the monomeric Q branch which Dubost assigned to wanted (double doping) or unwanted impurity/CO modes (for example, Fig. 57). It is very doubtful to interpret the spectra solely within the framework of vibrational and vibration-librational modes of MI-CO, when the impurity/CO line is larger than the monomeric transition.

We now compare the decisive part of the IR and RS spectral region, illustrated by a schematic diagram and propose different models to overcome this discrepancy. Two similar situations are known to the author in which IR and RS data are presented of the same MI-species and where the spectra reveal a monomeric doublet in IR spectra and a single monomeric line in RS

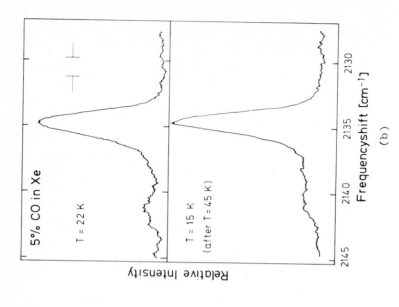

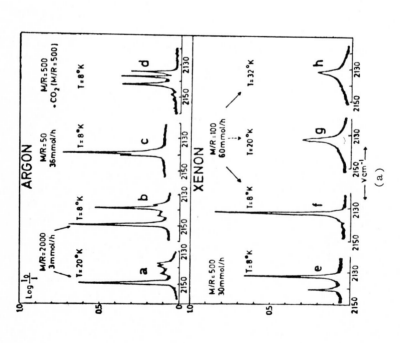

FIG. 55. Comparison of the fundamental line of CO in Xe during annealing. (a) Infrared spectra of CO trapped in solid Ar and Xe for several M/S and deposition rate values. CO$_2$ doping and temperature effects are also shown. Used by permission [90]. (b) Raman spectrum, experimental conditions: excitation with 5145 Å, 800 mW and spike filter; slitwidth 45 μm (1.5 cm^{-1}) and signal to noise ratio 3:1.

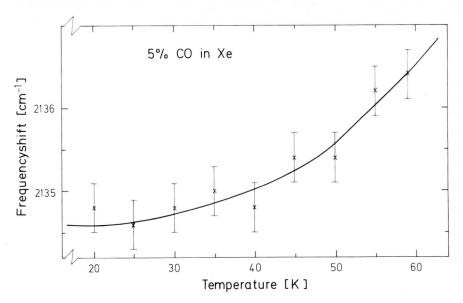

FIG. 56. Frequency shift of the CO fundamental vibration in Xe by Raman scattering as a function of temperature.

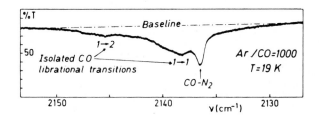

FIG. 57. Spectrum of a Ar/CO = 1000 sample at T = 19K showing the wings of the 0.1 → 1.1 absorption band. The temperature independent line at 2136.6 cm-1 is due to CO-N_2 nearest neighbor pairs. Used by permission [212].

spectra (Loewenschuss and Givan [82], and Barnes and Orville-Thomas [162].

First model: Following the assignment of Dubost [90] (see Table 10a), the intense line in RS (Fig. 54b, 55b and 58) might be due to the CO-CO pair. The broader line on the higher energy side (Fig. 54b, 55b upper part and 58) is then explainable by an isolated librating CO. The reason for this argument is the energetic coincidence of the librational transitions (Table 10b) and the general fact that the Raman samples have had a smaller M/S ratio than the ones used in MI-IRS. There is indeed a Raman counterpart at 2140.1 cm^{-1} for this CO-CO peak 2140.2 cm^{-1} in IR, but only in an Ar matrix and none in Kr and in Xe. On the other hand if the lines due to isolated

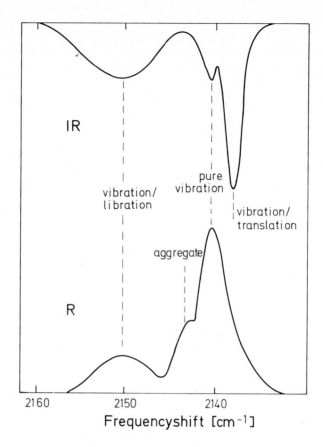

FIG. 58. Schematic diagram to compare part of the IR and RS spectra of MI-CO in the vicinity of the fundamental line.

librating CO are present in each matrix in the case of RS then the monomeric line should at least be detectable. The final arguments against this model are the reversible and irreversible changes observed in RS spectra as a function of temperature; see for example Fig. 55b.

Second model: Another way to look at this question is indicated in Fig. 58. The MI-CO executes a pure vibration which is obviously Raman and infrared active and a vibration-translation mode which is only infrared active. The reason for this coupling of molecule motions might be the fact that the center of mass and the center of interaction is quite different for MI-CO, as already known (Blaisten-Barojas and Allavena [218]; Manz [217]. It is to be expected that the change in dipole moment of CO and therefore the IR absorptivity is larger for vibration-translation than for a pure vibration. The broad structure, identical in both spectroscopic techniques, is due to a vibration-libration (Dubost [90], [211]) and allows no detailed analysis with regard to infrared and Raman active transitions. The separation of vibrational and vibration-translational modes is of the order of 1.5 to 2.5 cm^{-1} (Table 10b) and has a reasonable size, see for comparison MI-CO_2 and CS_2 by Loewenschuss and Givan [82].

Third model: In here the whole feature of IR and RS spectra is altogether explainable in the picture of hindered rotations in conjunction with the molecule vibration. Following the model of Manz [217], in which he calculated in detail librational transitions next to the pure vibration on the basis of spectroscopic data of CO in Ar, one can interpret the broad band to be the R branch of vibration-librations and the single line in RS and the doublet in IR to be the Q branch of vibration-librations. Using general symmetry arguments (for infrared and Raman active allowed transitions), energy spacings (Manz [217]) and a thermal occupation of librational levels in the vibrational ground state according to sample temperature the fundamental line in RS consists of different transitions close to each other, separated only by librational anharmonicities, but not resolvable, whereas the one in IR consists of a well separated pair of different transitions.

This discrepancy in IR and RS spectra of MI-CO must be consolidated by further investigations on CO. One needs IR and RS on MI-CO with the self same matrix. Analogous studies with other heteronuclear and homonuclear diatomic molecules, additional experiments on sample preparation to handle aggregation and to avoid unwanted impurities in the matrix, and measurements of local temperature of matrix film and of sample holder during irradiation in MI-IRS and MI-RS are required. Moreover the study of CO in other matrices like N_2, for instance, might check if the librational model is applicable.

D. Nitrogen (N_2)-Temperature and Pressure Effects

Molecular nitrogen, as matrix or as defect material, has been the subject of many investigations in MIS since the last twenty years. This species has been important with regard to the development of the preparation of matrix material, to the recognition of relevant physical parameters of MI-species and to the conquest of problems in carrying out spectroscopy with MI-films. One would suppose therefore that there exists a clear opinion on the values of molecular constants of MI-N_2; this is not so. Scanning through literature the found values for matrixshift, for instance, scatter by more than a wavenumber (fourth column in Table 11).

For this reason we studied again this common species in order to re-measure the matrix frequency shift with the aid of a frequency standard. A second aim was to determine the influence of anharmonic interaction of the

TABLE 11

Frequency Shifts $\Delta\omega(cm^{-1})$, Halfwidth $\Gamma(cm^{-1})$, Depolarization Ratios ρ (arb. units) of the N_2 Fundamental Vibration in the Gas-liquid and Solid Phase and Isolated in Argon by Raman Scattering. The Matrix $\Delta = \Delta\omega_{solid} - \Delta\omega_{gas}$ is Compared with Theoretical Models. (Jodl and Bolduan [60]).

	T(K)	$\Delta\omega$ (cm^{-1})	Δ	$\Gamma(cm^{-1})$	ρ
N_2-gas	300	2329.66 ± 0.2		<<0.1	0.2
N_2-gas	300	2329.92 ± 0.01		<<0.1	
N_2-liquid	77	2331	+ 1.1	0.066	0.19
$\alpha - N_2$ - solid	14	2329 ± 1.5	- 0.9	1.8	
$\beta - N_2$ - solid	40	2328 ± 1.5		1.6	
$\alpha - N_2$ - solid	18	2326 ± 1	- 3.0		
		2327 ± 1			
$\alpha - N_2$ - solid	5	2327.8 ± 0.8	- 1.4	< 1.5	
		2328.8 ± 0.8			
$\gamma - N_2$ - solid	8	2331 ± 1	+ 1.1		
	20	2330 ± 1		2.0	
	35	2329 ± 1			
N_2 - matrix	12				0.2
N_2 - matrix	5				0.3
N_2 - matrix[a]		2327 ± 1	- 2.9		
1 % N_2 in Ar	55	2323 ± 1	- 6.9	1.0	
5 % N_2 in Ar	77	2328 ± 1	- 1.9	7	
~1 % N_2 in Ar[b]	<10	2330.1 ± 0.4	+ 0.2		
0.3 % N_2 in Ar[d]	80	2326.2 ± 0.2	- 3.7	0.027	
N_2 in Ar[c]	5		- 2.4		
	77		- 3.5		
$\alpha - N_2$ solid	5		- 2.9		
$\alpha - N_2$ solid	5		- 0.8		
$\alpha - N_2$ solid	5		- 1.3 to		
			- 2.0		
1 % N_2 in Ar	8	2326.7 ± 0.3	- 3.2	1.8	0.1
40 % N_2 in Ar	8	2326.8 ± 0.3		1.8	0.1
N_2 - solid	8	2327.6 ± 0.3	- 2.3	2.2	0.1

[a] Infrared spectra of C_2N_2 trapped in N_2

[b] X-ray stimulated fluorescence of the $A^3\Sigma_u^+ - X^1\Sigma_g^+$ transition of N_2 in Ar

[c] Theory

[d] Kiefte et al. [129].

TABLE 12

Frequency-shifts $\Delta\omega(cm^{-1})$ ($\pm$ 0.3 cm^{-1}) of the Matrix-isolated N_2 Fundamental Vibration (1 mol % and 40 mol %) at Different Pressures ($\Delta P = \pm$ 0.2 kbar) and Various Temperatures ($\Delta T = \pm$ 1K).

	5K	$\frac{\Delta(\Delta\omega)}{\Delta P}$ $(\frac{cm^{-1}}{kbar})$	78K	$\frac{\Delta(\Delta\omega)}{\Delta P}$ $(\frac{cm^{-1}}{kbar})$
1 mol % N_2	2327.0 (1 bar) 2330.4 (2.5 kbar)	+ 1.3	2326.0 (1 bar) 2329.4 (3.1 kbar)	+ 1.1
40 mol % N_2	2327.9 (1 bar) 2327.4 (3.5 kbar)	- 0.1	2326.4 (1 bar) 2329.4 (4.5 kbar)	+ 0.7
α - N_2		+ 0.5		
γ - N_2		+ 0.3		

defect with the matrix. The combination of RS at different temperatures and pressures should give information on various contributions to anharmonic phenomena like the volume dependent part or the phonon-phonon interaction. From the experimental point of view this means the variation of temperature at constant pressure and the change of pressure at the lowest temperature. A third topic was the changes of bond length of the molecular defect upon varying the matrix lattice constant by externally applied pressure or by implanting the defect. This and other problems of RS with MI-N_2 are presented in detail elsewhere (Jodl and Boldaun [60]).

For these kind of investigations the technique of MI-RS must be combined with the one of high pressure. An optical cell for temperatures down to 5K and pressures up to 10 kbars, technical advice, and typical results are described in detail by Jodl and Holzapfel [62] (see also Fig. 42 and Sec. III.D). The general ideas and the capability of this technique are extensively discussed in Jodl and Holzapfel [61]. The difficulty of these experiments is demonstrated by the fact that few MI-IRS experiments under high pressure are known (H_2 in Ar by Jean-Louis and Vu [198] or HCl in Ar by Fondère et al. [222]) and none with other spectroscopic techniques.

The expected energy shifts of the fundamental line, such as the matrix shift Δ_M, temperature shift $\Delta(\Delta\omega)/\Delta T$, and pressure shift $\Delta(\Delta\omega)/\Delta P$, are of the order of a few wavenumbers. Hence, special care for wavenumber calibration must be taken (see Sec. III.C). Figure 11 shows the RS of 1 mol % N_2 in Ar at various temperatures and Fig. 13 the one at various pressures and

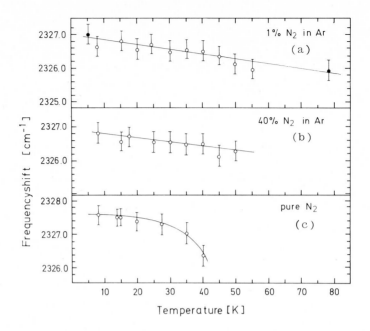

FIG. 59. Frequency shift $\Delta\omega$(cm^{-1}) of the N_2 fundamental vibration for 1 mol % N_2 in Ar (a), for 40 mol % N_2 in Ar (b), and for pure N_2 (c) as a function of temperature. The gas pressure in the MI-technique limit the T-range to about 50K. The full points in (a) are values from the poly-crystalline sample in the pressure cell.

different temperatures. N_2/RG mixture was either sprayed as thin films (m mol/h; 15 K; M/S = 100:1) or condensed as polycrystals in the high pressure cell. Annealing helped to increase the Raman signals enormously (Fig. 32 and Sec. III.A). The Raman equipment was standard and the irradiation geometry similar to that in Fig. 34a for the thin films and Fig. 42b in the case of polycrystals grown in the high pressure cell. Figure 59 demonstrates the influence of the temperature and Table 12 the influence of pressure on the fundamental vibration of MI-N_2: $\Delta(\Delta\omega)/\Delta T$ is about -0.02 (cm^{-1}/K) and $\Delta(\Delta\omega)/\Delta P$ is about +1.3 (cm^{-1}/kbar) for 1 mol % N_2 in Ar.

Raman scattering data from our samples show a pronounced temperature dependence in the frequency shift whereas the linewidth is insensitive to temperature variation.

The phenomenon of the variation of vibrational frequencies with tempera-ture is a manifestation (like thermal expansion and the variation with pressure described in the next section) of the breakdown of the small-oscillation-limit harmonic approximation. The effect is a measure of anharmonicity. In second order perturbation both cubic and quartic anharmonic terms contribute to the frequency shift which is defined by Cowley [223]

$$\Delta\omega_i(T) = \Delta\omega_i(T = 0) - \sum_j B_{ij} \, n_j(T) \tag{5}$$

where $n_j(T)$ is the phonon occupation number of the mode j

$$n_j(T) = [\exp(\hbar\omega_j/kT) - 1]^{-1}. \tag{6}$$

The main ingredients of the complex recipe for B_{ij} are phonon frequencies and anharmonicity coefficients for mode i, mode j and for ij cross terms. An expression for $\Delta\omega(T)$ like the above does reproduce two ubiquitous features of the observed behavior; namely the linear dependence at high T (classical limit) and the bend over to zero slope at low T (zero point motion quantum limit). However, the interesting question here is how many modes j with energy $\hbar\omega_j$ couple to the molecular vibration i and the size and sign of the effective coupling constant B_{ij}. If the Eq. (5) is expanded for high T in the classical limit $\hbar\omega/kT \ll 1$ we get

$$\Delta\omega(T) = \Delta\omega^*(T = 0) - B \cdot \frac{k}{\hbar\bar{\omega}} \cdot T. \tag{7}$$

The experimental data (Fig. 59) are well described by a linear relation between frequency shift and temperature:

$$\Delta\omega(T) = \Delta\omega_{exp}(T = 0) - \frac{\Delta(\Delta\omega)}{\Delta T} \, T. \tag{8}$$

Comparing both equations and taking $\Delta(\Delta\omega)/\Delta T$ between $-0.014 \text{ cm}^{-1}/K$ and $-0.02 \text{ cm}^{-1}/K$ from Fig. 59a and assuming $\bar{\omega}$ to be a specific frequency (here $\bar{\omega} \lesssim \omega_{Debye} = 67 \text{ cm}^{-1}$) we find for the effective coupling constant $B = 1.4$ to 1.9. In this sense all three cases, matrix isolation, mixture and pure solid, show the same behavior for $\Delta\omega(T)$ namely a weak coupling between the defect vibration and lattice phonons describable in the classical limit by an anharmonic interaction. The experimental value $\Delta\omega_{exp}(T = 0)$ (Eq. 8), the mathematical correct value $\Delta\omega(T = 0)$ (Eq. 5) and the one for the classical limit $\Delta\omega^*(T = 0)$ (Eq. 7) are different by definition. But in the case of N_2 in N_2 or Ar the absolute difference between these values is smaller than the wavenumber resolution (0.3 cm^{-1}).

The harmonic approximation treats lattice frequencies as independent of volume and temperature. In the quasiharmonic approximation they are only volume dependent (concept of mode Grüneisen parameters). Whereas both the volume and temperature dependence is considered in anharmonic theories. Equation (5) includes the thermal occupation effect and incorporates the anharmonicity via B_{ij}. Independent measurement of both the pressure and the temperature dependence in $\Delta\omega$ and Γ permits a separation into volume - derivative and temperature-derivative-terms. Temperature is less simple in its effect than is pressure. A change in pressure alters the equilibrium interatomic spacings. A change in temperature, in addition to its effect on the vibrational excursions of the atoms about their equilibrium positions, also alters the interatomic spacings because of thermal expansion. Considering $\Delta\omega$ as $\Delta\omega(T,V)$ and using thermodynamic relations we define (Sapozhnikov [224]):

$$(d\,\frac{(\Delta\omega)}{dT})_P = (\frac{\partial(\Delta\omega)}{\partial T})_V - \frac{\alpha}{\beta}(\frac{\partial(\Delta\omega)}{\partial P})_T. \tag{9}$$

Equation (9) expresses the decomposition of the observed temperature coefficient $\Delta(\Delta\omega)/\Delta T$ of vibrational frequency into the explicit term $(\partial(\Delta\omega)/\partial T)_V$ driven by phonon-phonon interaction and the implicit term $-\frac{\alpha}{\beta}$ $(\frac{\partial(\Delta\omega)}{\partial P})_T$ driven by volume dilatation. The latter is proportional to the observed pressure coefficient $\Delta(\Delta\omega)/\Delta P$. The magnitude $\alpha = \frac{1}{V}(\frac{\partial V}{\partial T})_P$ is the coefficient of the volume thermal expansion and $\beta = -\frac{1}{V}(\frac{\partial V}{\partial P})_T$ is the iso-thermal compressibility. For N_2 and Ar both quantities have been measured experimentally.

Data for the internal mode of $MI\text{-}N_2$ are gathered in Table 12 (pressure shifts) and Table 13 (total and pure temperature shifts). To discuss these data in the light of Eq. (9) it is common (Zallen and Slade [225] to define two additional quantities. The first is

$$(\frac{dT}{dP})_{\Delta\omega} = -(\frac{\partial(\Delta\omega)}{\partial P})(\frac{d(\Delta\omega)}{dT})^{-1}. \tag{10}$$

This quantity is interpretable as the increase in temperature dT which compensates a pressure increase dP in the sense of cancelling the effect on frequency and keeping $\Delta\omega$ constant (fifth column, Table 13). The second is

$$\theta = (\frac{\partial(\Delta\omega)}{\partial T})(\frac{d(\Delta\omega)}{dT})^{-1}, \tag{11}$$

θ denotes the fraction of the total temperature coefficient which is attributable to the explicit contribution $(0 < \theta < 1)$ and if $\theta = 1$ the explicit contribution or phonon-phonon interaction describes the anharmonicity (column 6 in Table 13).

We will compare the different quantities characterizing the N_2 vibration in an Ar-lattice with those in a N_2-lattice and other molecular crystals. The pressure shift $\Delta(\Delta\omega)/\Delta P$ for $MI\text{-}N_2$ is about three times larger than the one for

TABLE 13

Total Temperature Shift and Adapted Pressure Shift of the Fundamental
Vibration for 1 mol % N_2 in Ar and Pure N_2. The Pure Temperature Shift
is Calculated from Relation (9).

	T	$(\frac{\Delta(\Delta\omega)}{\Delta T})_P$	$\frac{\alpha}{\beta_T}(\frac{\Delta(\Delta\omega)}{\Delta P})_T$	$(\frac{\partial(\Delta\omega)}{\partial T})_V$	$\frac{dT}{dP}$	θ
	(K)	(cm^{-1}/K)	(cm^{-1}/K)	(cm^{-1}/K)	(K/kbar)	
1 mol % N_2	5	-0.015	$+8.9 \cdot 10^{-4}$	-0.014	87	~1
	78	-0.015	$+2.9 \cdot 10^{-2}$	$+0.013$		
$\alpha - N_2$	5	-0.01	$+1.0 \cdot 10^{-4}$	-0.01	50	1

pure N_2 (Table 12), i.e., the internal mode as a microscopic parameter is
very sensitive to the volume change of the different lattices caused by
pressure. This behavior is described by macroscopic parameters like the
volume expansion coefficient, α, and the compressibility, β_T. α/β_T for Ar is
three times larger than for N_2. As a consequence, matrix isolation high
pressure experiments provide the method to extract, from readily accessible
spectroscopic data on the defect in the matrix, macroscopic values for the
pure lattice which are normally difficult to be determined otherwise. θ is
about 1 for both systems (Table 13) indicating that phonon-phonon interaction
is the substantial contribution to the anharmonicity via weak coupling to the
lattice. This finding completes the series for different kinds of bonding since
$\theta = 0$ for ionic crystals, $\theta \sim \frac{1}{2}$ for covalent crystals and special modes in
molecular solids, and now $\theta \sim 1$ for molecular solids bonded by weak van der
Waals type forces. The compensation of temperature shift by pressure shift
to keep the internal frequency constant $(dT/dP)_{\Delta\omega}$ (Table 13) mirrors in a
way the fact that the Ar-lattice must be heated to higher temperatures in
order to achieve the same pressure effect as in the N_2 solid. The pure
temperature shift (fourth column, Table 13) in both systems is comparable in
size and sign, but a little bit larger in Ar, as expected.

In summary, if the lattice constant contracts with pressure by 1% per
kbar (Ar matrix) the bond length changes by about 10^{-2}% per kbar (N_2
defect). This result follows by extending the concept of mode Grüneisen
parameter to external and internal modes via the idea of a bond scaling
parameter.

Although nitrogen has been used very often as a matrix material or as a
defect, many interesting questions still remain to be answered, i.e., direct

spectroscopic investigation of whether MI-N_2 rotates or librates in RGS matrices; the effect of increasing the concentration of N_2 in matrices, to investigate the transition from an isolated impurity to a sublattice in the matrix; a determination of relative Raman scattering cross-sections of different species isolated in nitrogen matrices in comparison to the one of the N_2 fundamental line (as a Raman standard like in Table 4); direct spectroscopic measurement of the local temperature of laser irradiated nitrogen matrices by comparing the anti-Stokes to Stokes Raman line intensity of the E_g librational mode at about 30 cm^{-1}; and the change in the crystal field splitting of various defect modes in the spectra by systematically exchanging the RGS matrix (O_h^5 symmetry) by an N_2 matrix (T_n^6 symmetry).

E.   Nitrogen Dioxide (NO_2)-Continuous Resonance Raman Scattering

The absorption spectrum of NO_2 in the visible region has been of interest for decades because of its complexity. This complexity arises from the overlap of two excited electronic states $^2B_1$ and $^2B_2$, whose levels are further disturbed by high lying vibrational levels of the electronic ground state $^2A_1$. These $^2B_1$ and $^2A_1$ states are generated from the Π-ground state via Renner-Teller coupling. There is in addition an $^2A_2$ state in that region which, although it cannot couple to the ground state by electric dipole transitions, may also give rise to perturbations. Therefore the density of states in the visible region is extremely high (0.4/cm^{-1} for vibronic and 100/cm^{-1} for rovibronic levels). The spectral atlas of nitrogen dioxide contains a review concerning the NO_2-problems and a collection of relevant publications till 1977 (Hsu et al. [226]). There have been different kinds of approaches to elucidate the complexity of the NO_2-spectrum. Some groups investigated small details of the whole spectrum, for instance rotational analysis of different vibronic bands. Others measured fine- and hyperfine structures of specific rovibronic levels. But until now no satisfactory analysis of the coarse vibrational structure, especially the origins (T_{00}) of the $^2B_1$ and $^2B_2$ states, was ever done.

Attempts to simplify the spectrum were, for instance, achieved by the use of a supersonic beam of gas-NO_2 mixture obtaining rotational and vibrational temperatures for the NO_2 molecules of 3K and 200K respectively. One hundred-forty vibrational bands were measured in the excitation spectrum of NO_2 and most of them were assigned to the $X^2A_1 \rightarrow {}^2B_2$ transition. A totally different technique of "cooling" molecules is the matrix isolation technique, where molecules are embedded in all rare gas matrices at low temperature (~10K).

Bist et al. [227] analyzed a progression of the ν_2-vibration in the $^2B_1$ state from an absorption spectrum of NO_2 in solid Ne covering the range of 4000-5000 Å. This spectrum is still quite complex and comparable to the low resolution gas spectrum. Resonance Raman Scattering (RRS) of matrix isolated NO_2 (in Ar) was done by Tevault and Andrews [165] but they only observed a few overtones besides the ν_1 and ν_2 fundamental vibrations. Also their wavenumber accuracy was far from good.

The matrix-isolation technique has been used to investigate the vibrational structure of the NO_2-molecule in the ground state as well as in the excited electronic states $^2B_1$ and $^2B_2$. For the ground state resonance Raman scattering seemed to be useful because fluorescence is quenched. The application of this technique to use a physical property of the MI-system has already been mentioned in Sec. II.C and explained with the aid of I_2 in Ar. For investigating the excited states, absorption spectra were used with laser excitation spectra in the range of 5400-6300 Å and transmission spectra (3000-8000 Å). Bolduan [95] and Jodl and Bolduan [78,186], have dealt with this complex extensively. Similar work is in progress to apply MI-RS to SO_2 (Jodl and Conrad [166]), CO_2 (Jodl and Kobel [83]), CS_2 (Jodl and Strelau [158]), etc. in the same manner. The aim of the work on MI-NO_2 was the determination of a nearly complete set of molecular constants (energies of the fundamental modes and their anharmonic corrections), the investigation of the matrix influence on building NO_2-rare gas complexes (by comparing the spectra with regard to matrix shifts, time dependency and line broadening), the controlled matrix isolation of monomeric and dimeric NO_2 and the investigation of the properties of the aggregates, the description of the light scattering mechanism (by the analysis of the overtone and combination progressions with regard to energies, intensities, depolarization ratios and lifetimes).

NO_2 gas (purity 99%; most of the impurity was N_2) was premixed with the matrix gas (1 to 2:100) and sprayed on a cooled sample holder. A highly polished copper mirror, or sometimes a silver coated copper mirror was used to avoid background fluorescence from copper in the red spectral region. Temperature was measured with a gold vs chromel thermocouple with an accuracy of 0.5K and an overall stability of 0.2K. This was necessary to avoid cracking of the sample, which raised background scattering intensity in the Raman spectra enormously. Deposition temperatures of 15K for Ar- and Kr-matrices and 20K for Xe-matrices together with deposition rates of 100 μmol/h gave good results. Sample thickness was about 30-40 μm. Tempering showed no remarkable effects; on the contrary dimerization of NO_2 took place at temperatures above 35K and the brown color, characteristic for NO_2, diminished (see Fig. 33). This fact limits the region for temperature dependent investigations to between 10 and 35K. Resonance Raman Scattering (RRS) was excited by different lines of an Ar^+-laser (4579, 4765, 4880, 4965 and 5145 Å) with usually 100 to 300 mW power. If higher laser power was used, the sample bleached out at the laser spot or even cracked completely, because of local heating due to absorption of the laser light. The Raman equipment was standard and the irradiation geometry similar to the one of Fig. 34a. Wavenumber accuracy of 0.3 to 0.5 cm^{-1} was achieved by using lines of a Thorium hallow cathode lamp, whose light was deflected into the ordinary light path by a moveable mirror. If statements about the change in relative Raman intensities are necessary for physical arguments, enormous care must be taken during the measurements. This is well documented in Fig. 29. One sees that the intensity fluctuates by about a few 100% if the laser focus is scanned over the surface of the MI-thin film. In the case of absorption of 1% NO_2 in rare gas samples (thickness 600-800 μm) were prepared on a sapphire disc, and the transmitted light of a continuous halogen lamp was dispersed by a monochromator and detected photoelectronically.

Figures 23 and 60 show a few samples of RR spectra of NO_2 in solid Ar, Kr and Xe excited with various laser lines. The overall view is nearly the same. On a broad structureless fluorescence with a maximum near 17200 cm^{-1}, independent of laser wavelength and matrix, many discrete lines of 3.5 cm^{-1} width are superimposed, which shift typically with excitation wavelength. These Raman lines can be assigned to overtones and combinations of all three fundamental vibrations of NO_2 in its electronic ground state. Line intensities decrease systematically with increasing overtone progression for symmetric modes. Figure 65 as an example shows the relative intensities of the members of the v_2-progression, which is the most prominent one in all spectra, in various matrices and excited with 4880 Å and 5145 Å. Combinations of v_1 and v_2 with the antisymmetric v_3 mostly show higher intensities, whereas the fundamental v_3 is only barely detectable.

The following results emerge. Firstly, all the Raman lines are shifted about 0.1 - 0.3% in energy, with a blue-shift in Ar and a red-shift in Xe, whereas the Kr matrix has no influence on the energy level positions of NO_2. This observation is also made in the case of absorption. As a consequence the ground and excited electronic states ($^2B_1$ and $^2B_2$) are all shifted in the same direction, to higher energy in Ar, to lower energy in Xe, whereas in Kr there is no measureable shift. Table 14 contains the vibrational constants of the NO_2 ground state in Ar, Kr and Xe at 15K, together with the gas data (Hsu et al. [226]) and matrix data (Tevault and Andrews [165]). The vibrational energies of the fundamentals are slightly changed by the matrix (< 1 - 5 cm^{-1}). Within the rare gas series Ar, Kr and Xe there is an increasing

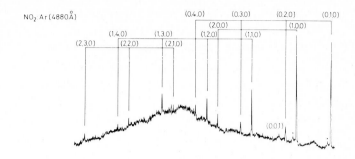

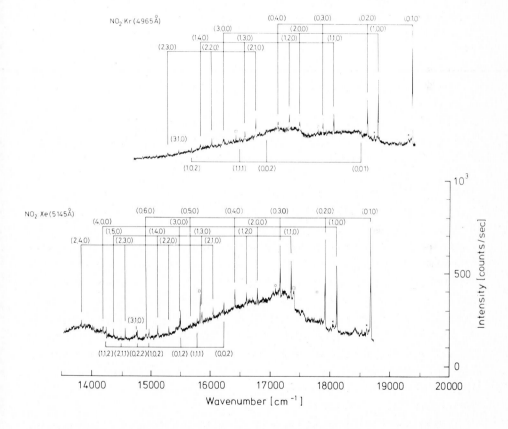

FIG. 60. RR spectrum of 2% NO_2 in Ar, Kr and Xe at 15K excited with different laserlines with maximum power of 300 mW. Spectral resolution is ~7 cm^{-1}, scan speed 1 cm^{-1}/sec. The spectrum of NO_2:Xe was recorded from a sample condensed on a highly polished copper mirror. The raise in background intensity near 14000 cm^{-1} is due to reflections from the copper surface. Notation of special lines: x - ν_1 and ν_2 of N_2O_4; 0 - N_2 vibration - impurity from NO_2 - gas; □ - unidentified lines.

TABLE 14
Fundamental Vibration Energies and Anharmonic Constants
(cm^{-1}) of the Electronic Ground State of NO_2 in Ar, Kr, and Xe
at 15K in Comparison with Gas Data

	gas	gas	Ar	Kr	Xe	Ar
ν_1	1319.2	1320.2	1320.5±0.5	1321.6±0.5	1315.8±0.5	1325
ν_2	750.1	749.6	751.4±0.5	750.3±0.5	748.9±0.5	752
ν_3	---	1618	1616.0±0.5	1615.0±0.5	1614 ±3	---
		gas				
κ_{11}		- 5.47	-6.0±0.5	-7.5±0.5	-6.1±0.5	---
κ_{12}		- 6.43	-6.1±0.5	-6.6±0.5	-6.6±0.5	-9
κ_{22}		- 0.463	-0.6±0.3	-0.7±0.3	-0.8±0.1	-1
κ_{13}		-29.55	-29 ± 3	-31 ± 3	-31 ± 3	-31
κ_{23}		-11.39	---	-10 ± 1	-12 ± 1	---
κ_{33}		-17.06	-25 ± 3	-25 ± 2	-30 ± 3	---

red shift of the energy values with the only exception of ν_1 in Kr matrix.

Secondly, different excitation energies in otherwise identical systems (see for example Fig. 23) do not influence either the discrete line spectrum or the feature of the broad fluorescence band. A first hint for continuous resonance Raman scattering (CRRS). Thirdly, for RS the depolarization ratio ρ is less than 3/4 for Q-branches and for most of the diatomic molecules ρ is usually much less than 0.1. Also in CRRS Q-branches are expected to be polarized, whereas in discrete RRS depolarized bands are observed (Rousseau and Williams [33]). Besides this normal behavior, anomalous depolarization with $\rho > 3/4$ was observed from RR spectra of some porphyrin compounds, a fact, which is ascribed there to interference effects of two excited electronic states in the region of resonance. For NO_2 in Xe values of $\rho = 0.7$ to 1.0 (uncertainty 0.1) were measured after excitation by 5145, 4880 and 4765 Å. The values show no frequency dependence and vary erratically for different vibrational modes. Finally, the influence of temperature and of pressure on the symmetric bending mode of MI-NO_2 is presented in Fig. 61 and 62. The temperature shift is about +0.05 (cm^{-1}/K) and the pressure shift is about +0.7 (cm^{-1}/kbar) for 1 mol % NO_2 in Ar.

The physical question, experimental procedure and data evaluation occurs in a similar manner to that described in detail for MI-N_2 (Sec. V.D).

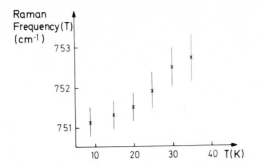

FIG. 61. Frequency shift $\Delta\omega(cm^{-1})$ of the symmetric bending mode of 1 mol % NO_2 in Ar as a function of temperature. Used by permission [96].

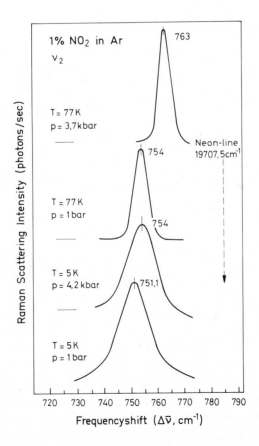

FIG. 62. Part of the Raman spectrum of 1 mol % NO_2 in Ar at various pressures and different temperatures. Used by permission [228].

The calculated differential shifts are of the same size as the ones for MI-N_2. As a consequence there is a weak coupling between defect vibration and lattice modes and the anharmonic interaction is described mainly by phonon-phonon interaction. The sign of the temperature shift is positive, which is remarkable insofar as up to now only negative shifts are known (see for example Fig. 59).

About thirty lines consisting of overtones and combinations allow a calculation of corresponding vibrational constants with rather good accuracy, except the case where v_3 is involved, because here only a few values are available. This calculation follows straight forward from the known formula, neglecting higher order terms:

$$E_{n_1 n_2 n_3} = \sum_{i=1}^{3} \hbar \omega_i (n_i + \tfrac{1}{2}) + \sum_{i,j=1}^{3} \hbar \kappa_{ij} (n_i + \tfrac{1}{2})(n_j + \tfrac{1}{2}) \tag{12}$$

n_1, n_2, n_3 denote the quantum numbers of the v_1, v_2 and v_3 vibrations. From this equation the following relations can be derived immediately:

$$(n_1,n_2,n_3) - (n_1,0,0) - (0,n_2,0) - (0,0,n_3) =$$

$$n_1 n_2 \; \kappa_{12} + n_2 n_3 \; \kappa_{23} + n_1 n_3 \; \kappa_{13} \tag{13a}$$

$$(n_1,0,0) - n_1(1,0,0) = \kappa_{11} \; n_1(n_1 - 1) \tag{13b}$$

$$(0,n_2,0) - n_2(0,1,0) = \kappa_{22} \; n_2(n_2 - 1) \tag{13c}$$

(n_1,n_2,n_3) here denotes the energy values in cm^{-1} of the combination-mode $n_1 v_1 + n_2 v_2 + n_3 v_3$ and $(1,0,0)$ and $(0,1,0)$ are the values of the v_1 and v_2 fundamentals. According to these relations graphical plots can be made, which yield the values of κ_{11}, κ_{22} and κ_{12}, respectively. (See Fig. 63 a-c.)

As can be seen from Fig. 63a the v_1-progression is rather regular in all matrices. Among the v_2-ladder (Fig. 63b) the $2v_2$ values in Ar and Kr deviate distinctly from the fit. In Kr there is even a positive anharmonicity. This effect is beyond accuracy in measurement or calibration and must be due to perturbations by the Ar and Kr-lattices. As Xe does not show this per-turbation the corresponding value in Ne would have been of interest, but we could not produce any NO_2: Ne sample, for reasons which will be explained later.

Rather significant alternations, compared to the gas, are observed for the anharmonic constants κ_{ij} (Table 14). Especially some diagonal elements κ_{ii}

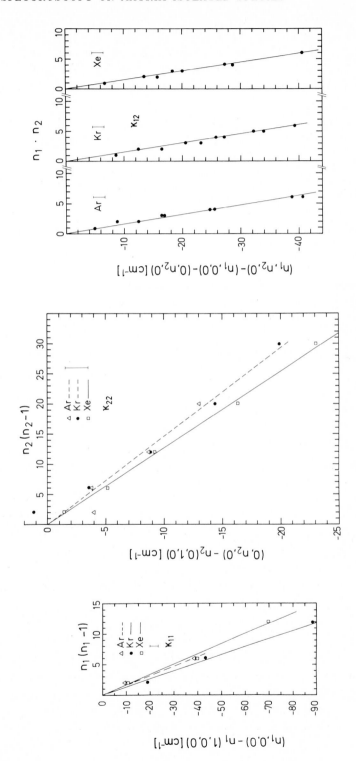

FIG. 63. Calculation of the anharmonic constant κ_{11} of NO_2 in Ar, Kr, Xe (15K) according to equation (13b). Measurement accuracy is marked by a vertical bar representatively; for κ_{22} according to equation (13c) and for κ_{12} to (13a).

increase by about ~50% of their gas phase values, whereas the off-diagonal κ_{ij}, describing combination modes, are changed only about ~1%. Even here the values increase in the direction Ar, Kr and Xe. This shows, that the matrix surrounding causes an increased anharmonic behavior of the NO_2 vibrations with growing tendency among the rare gases Ar, Kr and Xe. This example clearly demonstrates, that simple pictorial arguments do not hold: the bigger the matrix cage the less should be its influence on the impurity. In this sense Xe, with the biggest cage among the rare gas solids, should show the least influence on the NO_2 vibrations, in contrast to our observations.

The reason for us to study in more detail the scattering mechanism was to use the effect of resonance RS, to raise the scattering cross-section, and to present a further example for the question of unique or different models describing resonance fluorescence and resonance Raman scattering. Because NO_2 absorbs in the whole visible region, excitation with Ar^+ laser lines results in a resonant scattering process, where overtone intensities are comparable to the fundamental ones, in contrast to ordinary Raman scattering (see also Sect. II.C). In this sense RS happens, if the excitation energy is far from any real molecular level. If the energy reaches into an excited electronic state, the reemission is called discrete resonance Raman scattering (DRRS), whether or not there is exact resonance. At last, if the excitation energy reaches into a continuum above the dissociation limit of an excited electronic state, the reemission is called continuous resonance Raman scattering (CRRS). All three mentioned types of scattering exhibit different features in their spectra, as far as overtones structure, lifetimes and depolarization ratios are concerned. For example, Rosseau and Williams [33] have done a corresponding calculation for diatomic molecules, applying them to the special case of J_2. Tuning over a resonance, i.e. transition between real states of the defect, may be arranged by various laser frequencies, by different matrices or by variation of sample temperature and external pressure.

Matrix isolated NO_2 seems a rather good example for observing RRS. The absorption spectrum in the visible shows besides a broad continuum many discrete bands of 30-50 cm^{-1} linewidth (Fig. 64); which can be due to inhomogeneous broadening by the crystal, or due to electron-phonon-coupling, resulting in phonon wings. Because of the extremely high density of states in that region ($0.4/cm^{-1}$ - vibronic, $100/cm^{-1}$ - rovibronic) and the above mentioned broadening the spectrum consists of a quasicontinuum, and the reemission takes the features of CRRS, comparable to the case of excitation into a dissociation continuum.

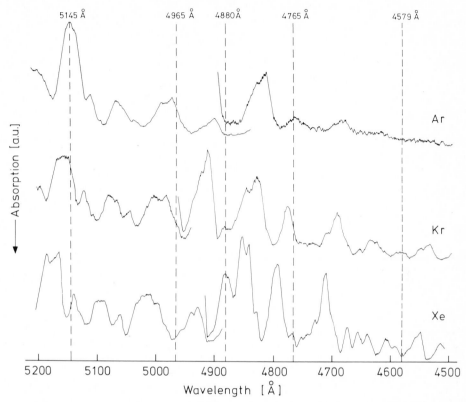

FIG. 64. Absorption of 1% NO_2 in Ar, Kr, and Xe at 15K in the range of 4500-5200 Å. Undispersed light of a halogen lamp was used. Sample thickness was ~700 μm, spectral resolution 5Å, scan speed 0.5 Å/sec.

Excitation with the same energy in different matrices show different intensity distributions among overtones and combinations, which can be explained by environmental effects of the matrix on the NO_2 molecules. The NO_2-potential is slightly changed and shifted differently in Ar, Kr and Xe. Therefore, different levels are excited by the same laser line in all three matrices. This can be seen if the absorption spectra of NO_2 in these solids are compared to each other. Figure 64 shows a part of these spectra in the range of 4500 Å and 5200 Å, together with the wavelength positions of the laser-lines used for excitation. There is an increasing red shift of corresponding absorption profiles beginning from Ar to Kr and Xe. Comparison with the low resolution gas spectrum (Hsu et al. [226]) demonstrates, that energy levels shift as a whole system. There is an absolute blue shift of about 40 cm^{-1} in Ar, nearly no shift in Kr, and a red shift of about 70 cm^{-1} in Xe relative to the gas spectrum in that region. Only excitation with 5145 Å in Ar reaches an absorption minimum and the corresponding RR spectrum

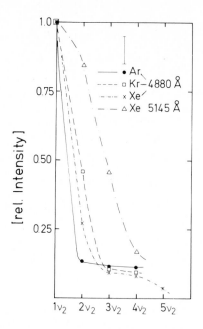

FIG. 65. Relative intensity distribution of the ν_2-progression of NO_2 in Ar, Kr and Xe, excited by 4880 Å. The ν_2-progression excited in Xe by 5145 Å is also shown for comparison. Intensities are corrected for spectral response of photomultiplier and monochromator and normalized on the fundamentals.

shows only few weak overtones, in contrast to NO_2 in Xe.

Many overtones and combinations appear with intensities comparable to the fundamental ones. Common to all our spectra is the decrease in intensity with increasing overtone or combination within a progression. Figure 65 as a typical example shows the relative intensities of the members of the ν_2-progression of NO_2 in different matrices, excited at 4880 Å, together with the corresponding progression in Xe, excited at 5145 Å. The same tendency is also observable in the combined progressions (1, $n\nu_2$, 0) and (2, $n\nu_2$, 0) of Fig. 23 and 60. The intensity of the antisymmetric vibration ν_3 is low and in some spectra not even detectable. Combinations with symmetric vibrations on the other hand are much more intense. This leads to the conclusion, that in the case of matrix isolated NO_2 the scattering process is mainly governed by symmetric vibrations, an observation, also made for other molecules like MnO_4^- and CrO_4^- (Kiefer and Bernstein [229]).

The spectrum of NO_2 deposited with Ne at 9K is very different from those of NO_2 in other rare gases. No NO_2 signal was observed at all (see Fig. 1 in Jodl and Bolduan [186]). It is well known, that NO_2 can form dimers. This species N_2O_4 is still a stable molecule without any excited electronic state in the visible region. Therefore ordinary Raman scattering is

TABLE 15

Molecular Pointgroups, Vibrational Modes and Symmetries of NO_3^-, NO_2 and Its Dimer Configurations Together with Their Schematic Geometric Structures

Species	NO_2	N_2O_4-symm.	N_2O_4-D-type	N_2O_4-D'-type	NO_3^-
Pointgroup	C_{2v}	$D_{2h} = V_h$	C_s	C_s	D_{3h}
Vibrational modes and their symmetry	$\nu_1, \nu_2\ A_1$ $\nu_3\ B_2$	$\nu_1, \nu_2, \nu_3\ A_g$ $\nu_5, \nu_6\ B_{1g}$ $\nu_8\ B_{2g}$	all modes IR and Raman active	all modes IR and Raman active	$\nu_1\ A_1'$ $\nu_3, \nu_4\ E'$
geometrical structure					

seen at excitation with Ar^+-laser lines. Many more lines than expected from the group theory of symmetric N_2O_4 with pointgroup D_{2h} are observed (see Table 16). There is evidence from early IR measurements of matrix isolated and pure solid N_2O_4 that there still exist two antisymmetric $ONONO_2$ configurations of C_s-symmetry besides the symmetric one. These are labelled D and D'-type (see Table 15). Moreover the following reactions are possible (Jones [230]): $O_2NNO_2 \leftrightarrow ONONO_2 \leftrightarrow NO^+ + NO_3^-$ (Nitrosonium Nitrate) resulting finally in two stable molecular ions NO^+ and NO_3^- with their own vibrational modes. The Raman active vibrations of the species NO_2, N_2O_4, $ONONO_2$ and NO_3^- together with their symmetries are listed in Table 15. For symmetric N_2O_4 group theory predicts six Raman active modes with "gerade" symmetry.

The remaining six modes are only IR-active. In the case of D and D'-type all twelve vibrations are IR- as well as Raman active. NO_3^- has three Raman active modes ν_1, ν_3 and ν_4, already well known (Jodl and Bolduan [231]).

It is remarkable at all that molecular ions are produced in van der Waals crystals by separation of the stable molecule N_2O_4. Also, the fact that these species are only observed in Ne matrices is rather surprising at a first glance. In all other rare gas solids the symmetric configuration is mostly favored.

After tempering these samples, the NO_2 signals lose in intensity while

TABLE 16

Measured Frequency Shifts (cm^{-1}) of Vibrations of N_2O_4, $ONONO_2$, NO^+ and NO_3^- in Solid Ne and Xe at 15K. Wavenumber Accuracy is ± 3 cm^{-1}.

	Ne	Xe		
ν_3	265	257	(N-N-stretch)	
ν_6	498	485	(NO_2-rock)	N_2O_4-sym.
ν_8	--	637	(NO_2-wag)	
ν_2	807	815	(NO_2 sym. bend)	
ν_1	1383	1387	(sym. stretch)	
ν_5	--	1718	(NO_2 antisym. stretch)	
	622	626	(O=N-O bend)	$ONONO_2$ D-type
	783	788	(NO_2-bend)	
	1295	1299	(NO_2 sym. stretch)	
	1635	1646	(NO_2 asym. stretch)	
	1806	--	(N=0 stretch)	
	949	953	(N-O stretch)	$ONONO_2$ D'-type
	1873	--	(N=O stretch)	
ν_4	714	--	(in-plane bending)	NO_3^-
ν_1	1042	--	(sym. stretch)	
ν_3	1323	--	(antisym. stretch)	
	2224	--	(N=O vibration)	NO^+

N_2O_4 signals appear. Figure 33, lower half, shows a typical Raman spectrum, observed after annealing of a NO_2:Xe sample at 50K for ca. 30 minutes. Besides the ever present NO_2 vibrations, Raman lines of symmetric and antisymmetric N_2O_4 have appeared. But the NO^+-vibration, the most intense line, is only barely detectable. An explanation of the presence of NO^+ and NO_3^- only in solid Ne could be given by simple geometric arguments. A substitutional vacancy in solid Xe- with fcc-structure has a diameter of 4.33 Å, whereas in Ne it is only 3.15 Å. During condensation with Ne most of the NO_2 dimerizes into the antisymmetric $ONONO_2$ species, which at least occupies two neighboring vacancies in solid Ne. Therefore a separation of $ONONO_2$ into NO^+ and NO_3^- is most probably favored. This effect may be enhanced by the relative softness of the Ne crystal at 9K (melting point 24.6K), our lowest possible temperature, leading to diffusion of NO_2 molecules in the lattice during condensation. On the other hand annealing of NO_2:Xe samples also support diffusion of NO_2 within the Xe-crystal, but the possibilities for NO_2 molecules to "travel" through the solid are restricted by the more or less rigid Xe lattice at these temperatures. A comparison of the geometric dimensions of symmetric N_2O_4 and a substitutional site in solid Xe shows, that the molecule fits into one vacancy with only small lattice distortions, making a separation of $ONONO_2$ into NO^+ and NO_3^- much less probable than in the case of the Ne matrix.

Further work connected with MI-NO_2 and other triatomic molecules will be directed towards establishing a connection between spectra of MI-NO_2 and the highly resolved spectra of gaseous NO_2 (to determine pure electronic transitions), to execute excitation spectra (to receive more information about excited states than by the broad banded absorption spectra) and to investigate charge transfer reactions by matrix isolation of neutral molecules.

VI. SUMMARY AND CONCLUDING REMARKS

By means of well chosen MI-species (H_2, D_2, NO, CO, N_2, NO_2) it has been possible to investigate basic topics such as

(a) the determination of energies of fundamental molecular motions and their matrix shifts, solution of ambiguous assignments in spectra and explanation of differences in IR- and RS-spectra.

(b) the coupling of different defect modes (vibration, rotation, libration, translation) to lattice modes.

(c) changes in Raman frequencies and linewidth as a function of matrix material (Ne, Ar, Kr, Xe), temperature, and pressure, for studies concerning anharmonicity.

(d) use of reproducible annealing procedures and the aimed monomer to polymer conversion and their subsequent investigation.

This has been achieved with the aid of very slow spray on condensation (~10 μmol/h; ~10 μm thickness), frequency standards for wavenumber determination, and the careful determinations of RS intensities along with pressure and temperature dependent measurements of Raman energies. Also effects like Fermi resonance and resonance Raman have been employed to

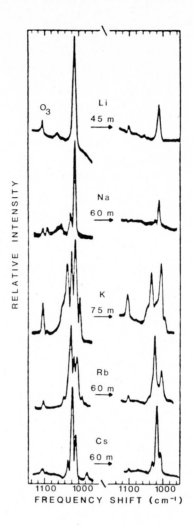

FIG. 66. Raman spectra from 4880 Å irradiation of products of alkali metal atom-ozone matrix reactions, $Ar/O_3 = 100$. All spectra 20 cm⁻¹ min⁻¹ scanning speed, second scan after indicated period of 4880 Å irradiation using same instrumental conditions as first scan. Li, 7 mW of 4880 Å power at sample, 3×10^{-9} range; Na, 180 mW, 0.3×10^{-9} range; K, 13 mW, 1×10^{-9} range; Rb, 16 mW, 3×10^{-9} range; Cs, 66 mW, 0.3×10^{-9} range. Used by permission [72].

enhance and hence determine Raman cross-sections.

At the end of Sec. V.A-E further specific problems for the future are mentioned.

The summary of experimental data (Table 5a,b,c) reflects obviously that only some groups are continuously engaged with MI-RS. The author would like to emphasize their merit for this technique by accentuating the main

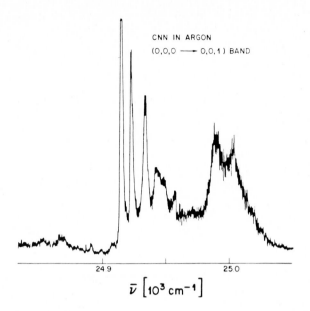

CNN IN ARGON
(0,0,0 ⟶ 0,0,1) BAND

$\bar{\nu}\ \left[10^3\ cm^{-1}\right]$

FIG. 67. High resolution trace of the (0,0,1) band in the excitation spectrum. Used by permission [102].

stimulating experiments.

The group around <u>Andrews</u> applied MI-RS mainly to investigate chemical reactions by photolysis (Fig. 26), to identify and determine new chemical species (Fig. 66) and to make use of resonance RS with halogen molecules and molecular ions.

<u>Barnes</u> et al. used MI-RS and MI-IRS to work on large organic molecules. They studied the effects on the quality of MI-Raman spectra of varying parameters, of kind of sample preparation and discussed the optimum conditions.

<u>Bondybey's</u> group has been mainly concerned with laser induced emission of MI-films. Beyond many interesting experiments he demonstrated the good optical quality of MI-films by resolving the triplet fine structure of MI-radicals (Fig. 67). With the investigation of MI-halogen molecules and their isotopes he provided an important contribution to the understanding of resonance fluorescence, resonance Raman scattering and the relaxation mechanism.

<u>Givan and Loewenschuss</u> improved the technique of MI-RS by the use of frequency standards and of a triple monochromator and combined IR absorption - and RS-investigations on the same MI-species. From the physical point of view they worked mainly on metal halogen complexes and resolved isotopic splitting (Fig. 68) and site splitting in these spectra. They detected Fermi

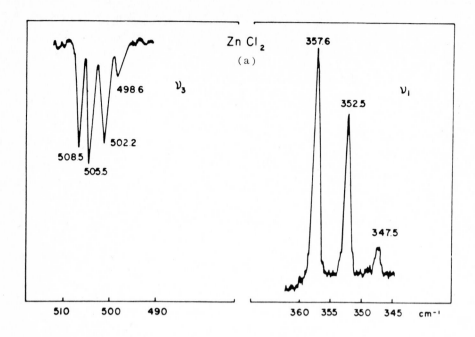

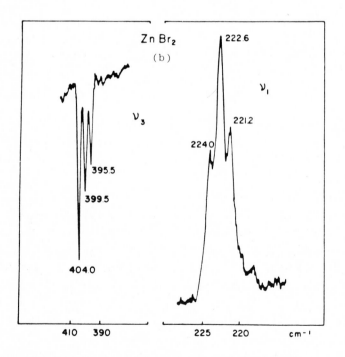

FIG. 68.   (a) Isotopic effects in ν_1 and ν_3 bands of $ZnCl_2$ · ν_1: 5145 Å; 100 cps; slitwidth: 1 cm^{-1}; ν_3: slitwidth: 0.8 cm^{-1}.   (b) Isotopic effects in ν_1 and ν_3 bands of $ZnBr_2$ · ν_1: 5145 Å; 100 cps; slitwidth 1 cm^{-1}; ν_3: slitwidth 0.8 cm^{-1}.   Used by permission [67].

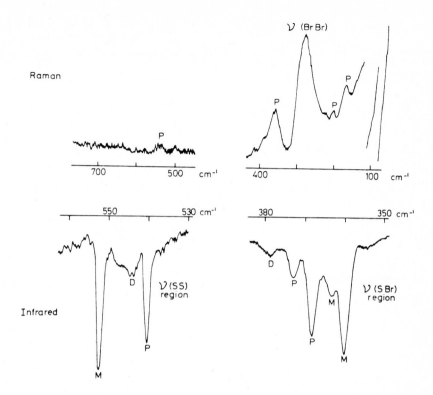

FIG. 69.   Raman and infrared spectra (100-700 cm^{-1}) of S_2Br_2 trapped in Ar (M/S = 1000) at 17K (M = monomer band, D = dimer band, P = polymer band). Used by permission [97].

resonance on gaseous MI-molecules.

Grzybowski developed an experimental system which, for the first time, allows the acquisition of the IR- and Raman spectra from the self same matrix.

Moskovits, Ozin and coworkers have used MI-RS mainly to study chemical questions like the production and determination of new species and of high temperature molecules.   But they also followed physical aspects like measurement of depolarization ratio, Fermi resonance splitting, isotopic splitting and overtone progressions in RS.

Niblers group also used the combination of MI-IRS and MI-RS to investigate isotopic and site splitting effects and dimerization.

Shirk and Claassen executed successfully the first MI-RS experiment on S_2 and SF_6.

Finally, I would like to remind the reader of the questions raised in Sec. I.A., some new trends in MIS outlined in Sec. I.C, and the suggestions for further MI-RS experiments in Sec. V.

Although much success has been achieved in MI-RS by the different research groups, there are still further experiments to be done concerning either more technical or physical aspects. For most experiments it is sufficient to condense just a polycrystalline thin film of defect/matrix mixture. For others, more solid state oriented experiments, it would be better to grow crystals to avoid inhomogeneous line broadening or site splitting. Thus sample generation (thin film condensation or crystal growing), preparation (annealing procedures, controlled diffusion and local temperature during irradiation) and quality (high M/S ratio; sample purity; optical behavior and stray light) remain a key issue. There is still a need to simultaneously execute more than one experimental technique on the same MI-thin film. It would be helpful to find in publications all the necessary details to be capable of comparing results of different groups. The combination of MI-RS with other techniques like ultra high pressure with a diamond anvil cell should yield new and interesting results.

ACKNOWLEDGEMENT

It is a pleasure to thank my teacher E. Lüscher for introducing me to the RGS; M. Klein, A. Loewenschuss and L. Andrews for critical reading of the manuscript; and all members of my group for many helpful conversations and for providing me their results in advance of publication. I am indebted to the Max Planck Institut für Festkörperforschung Stuttgart for cooperation concerning high pressure investigations.

REFERENCES

1. J. S. Shirk and H. H. Claassen, J. Chem. Phys., 54, 3237 (1971).

2. G. A. Ozin, Spex Speaker, 16, 4 (1971).

3. J. W. Nibler and D. A. Coe, J. Chem. Phys., 55, 5133 (1971).

4. H. J. Jodl, J. Wahl and E. Lüscher, Phys. Lett. 38A, 230 (1972).

5. L. Andrews, J. Chem. Phys., 57, 51 (1972).

6. B. Meyer, Low Temperature Spectroscopy, American Elsevier Publ., New York, 1971.

7. H. E. Hallam, Vibrational Spectroscopy of Trapped Species, John Wiley, New York, 1973.

8. M. Moskovits and G. A. Ozin, Cryochemistry, John Wiley, New York, 1976.

9. H. J. Jodl, R. Bruno and E. Lüscher, Impurities and Alloys, TU München (unpublished) 1972.

10. A. J. Downs and S. C. Peake, Molecular Spectroscopy, Vol. 1, (Ed. The Chemical Society), Burlington House, London, 1973.

11. J. W. Nibler, Advances in Raman Spectroscopy, (Ed. J. P. Mathieu), Vol. 1, p. 70, Heyden, London, 1973.

12. G. A. Ozin, Vibratioal Spectroscopy of Trapped Species, (Ed. H. E. Hallam), p. 373, John Wiley, New York, 1973.

13. L. Andrews, Vibrational Spectra and Structure, Vol. 4, (Ed. J. R. Durig), Elsevier, Amsterdam, 1975.

14. M. Moskovits and G. A. Ozin, Vibrational Spectra and Structure, (Ed. J. R. Durig) Vol. 4, Elsevier, New York, 1975.

15. L. Andrews, Cryochemistry, (Ed. M. Moskovits and G. A. Ozin), John Wiley, New York, 1976.

16. L. Andrews, Appl. Spectrosc. Reviews, 11.1, 125 (1976).

17. C. Barraclough, I. R. Beattie, and D. Everett, Vibrational Spectra and Structure, Vol. 5, (Ed. J. R. Durig), Elsevier, New York, 1976.

18. J. K. Burdett, M. Poliakoff, J. J. Turner and H. Dubost, Advances in Infrared and Raman Spectroscopy, Vol. 2, (Ed. R. J. H. Clark and R. E. Hester), Heyden, London, 1976.

19. D. M. Gruen, Cryochemistry, (Ed. M. Moskovits and G. Ozin), John Wiley, New York, 1976.

20. A. J. Barnes and H. E. Hallam, Vibrational Spectroscopy - Modern Trends, (Ed. A. J. Barnes and W. J. Orville-Thomas), Elsevier, Amsterdam, 1977.

20a. International Conference on Matrix Isolation Spectroscopy, West Berlin 21-24, June 1977. In Berichte der Bunsengesell., Phys. Chem. 82 (1978).

21. L. Andrews, Applied Spectrc., 33.3 (1979).

22. B. M. Chadwick, Molecular Spectroscopy, Vol. 6, The Chem. Soc., London, 1979.

23. D. S. King and J. C. Stephenson, Matrix Isolation Raman Spectroscopy, Optics and Lasertechnology, 4, (1980).

24. J. C. Miller and L. Andrews, Appl. Spectrosc. Rev., 16, 1 (1980).

25. A. J. Downs and M. Hawkins, Advances in Infrared and Raman Spectroscopy, Vol. 10, (Ed. R. Clark and R. Hester), Wiley and Heyden, London, 1983.

26. S. Cradock and A. J. Hinchcliffe, Matrix Isolation: A Technique for the Study of Reactive Inorganic Species, Cambridge University Press, Cambridge, 1975.

27.   M. M. Sushchinskii, p. 83 Chapter II, Raman Spectra and Molecular Structure, Israel Program for Science Translations, New York, 1972.

28.   R. Loudon, Adv. Physic, 13, 423 (1964).

29.   R. A. Cowley, Proc. Phys. Soc., 84, 281 (1964).

30.   M. M. Sushchinskii, p. 273 Chapter III, Raman Spectra of Crystals, Israel Program for Science Translations, New York, 1972.

31.   N. R. Werthamer, Phys. Rev., 185, 348 (1969).

32.   N. R. Werthamer, Rare Gas Solids, (Ed. M. L. Klein and J. A. Venables), Vol. I, p, 298, Academic Press, New York, 1976.

33.   R. L. Rousseau and P. F. Williams, J. Chem. Phys., 64, 3519 (1976).

34.   W. G. Fateley, F. R. Dollish, N. T. McDevitt, and F. F. Bentley, Infrared and Raman Selection Rules for Molecular and Lattice Vibrations, Wiley and Sons, New York, 1972.

35.   K. Kobashi, Y. Kataoka, and T. Yamamoto, Can. J. Chem., 54, 2154 (1976).

36.   N. R. Werthamer, R. L. Gray, and T. R. Koehler, Phys. Rev. B, 2, 4199 (1970).

37.   L. M. Klein and T. R. Koehler, Rare Gas Solids, (Ed. M. L. Klein and J. A. Venables), Vol. 1, p. 369, Academic Press, New York, 1976.

38.   P. A. Fleury, J. M. Worlock, and H. L. Carter, Phys. Rev. Lett., 30, 591 (1973).

39.   R. K. Crawford, D. G. Bruns, D. A. Gallagher, and M. V. Klein, Phys. Rev. B, 17, 4871 (1978).

40.   Y. Fujii, N. A. Lurie, R. Pynn and G. Shirane, Phys. Rev. B., 10, 3647 (1974).

41.   L. M. Klein and J. A. Venables, Rare Gas Solids, Vol. I, Academic Press, New York, 1976; Vol. II, 1977.

42.   N. R. Werthamer, R. L. Gray and T. R. Koehler, Phys. Rev. B, 4, 1324 (1971).

43.   R. E. Slusher and C. M. Surko, Phys. Rev. Lett., 27, 1699 (1971).

44.   C. S. Barrett and L. Meyer, J. Chem. Phys., 42, 107 (1965).

45.   C. S. Barrett, L. Meyer and J. Wassermann, J. Chem. Phys., 44, 998 (1966).

46.   L. Meyer, Adv. Chem. Phys. XVI, 343 (1969).

47.   E. Schuberth, M. Creuzburg and W. Müller-Lierheim, Phys. Status. Solidi (B), 76, 301 (1976).

48.   E. Schuberth, Phys. Status. Solidi (B), 84, K91 (1977).

49.   A. S. Barker and A. J. Sievers, Rev. of Mod. Phys., 47, Supp. No. 2, (1975).

50. S. S. Cohen and M. L. Klein, J. Chem. Phys., 61, 3210 (1974).

51. L. C. Brunel, J. C. Bureau, and M. Peyron, Chem. Phys., 28, 387 (1978).

52. P. D. Mannheim, Phys. Rev. (B), 5, 745 (1972).

53. S. Nosé and M. L. Klein, Mol. Phys., 46, 1063 (1982).

54. N. H. Rich, M. J. Clouter, H. Kiefte, and S. F. Ahmad, Can. J. Phys., 60, 1358 (1982).

55. R. E. Benner and R. K. Chang, Fiber Optics, Advances in Research and Development, New York, p. 625, 1979.

56. Tang Sheng-Yuh and C. W. Brown, J. Raman Spectrosc., 2, 209 (1974).

57. Tang Sheng-Yuh and C. W. Brown, J. Raman Spectrosc., 3, 387 (1975).

58. F. T. Prochaska and L. Andrews, J. Chem. Phys., 67, 1139 (1977).

59. D. Smith, D. W. James and J. P. Devlin, J. Chem. Phys., 54, 4437 (1971).

60. H. J. Jodl and F. Bolduan, J. Chem. Phys., 76, 3352 (1982).

61. H. J. Jodl and W. B. Holzapfel, J. Raman Spectrosc., 8, 185 (1979).

62. H. J. Jodl and W. B. Holzapfel, Rev. Sci. Instrum., 50, 340 (1979).

63. R. R. Smardzewski and L. Andrews, J. Chem. Phys., 57, 1327 (1972).

64. M. L. Lesiecki and J. W. Nibler, J. Chem. Phys., 63, 3452 (1975).

65. J. M. Friedman, V. E. Bondybey and D. L. Rousseau, Chem. Phys. Lett., 70, 499 (1980).

66. M. Moskovits and D. P. Dilella, J. Chem. Phys., 72, 2267 (1980).

67. A. Givan and A. Loewenschuss, J. Chem. Phys., 68, 2228 (1978).

68. A. Givan and A. Loewenschuss, J. Chem. Phys., 64, 1967 (1976).

69. A. Givan and A. Loewenschuss, J. Raman Spectrosc., 6, 84 (1977).

70. A. Givan and A. Loewenschuss, J. Mol. Struct., 55, 163 (1979).

71. M. L. Lesiecki and J. W. Nibler, J. Chem. Phys., 64, 871 (1976).

72. L. Andrews, J. Am. Chem. Soc., 95, 4487 (1973).

73. W. F. Howard and L. Andrews, Inorg. Chem. 14, 767 (1975).

74. W. Kiefer, Appl. Spectrosc., 28, 115 (1974).

75. J. Behringer, Mol. Spectr., Vol. 12, (Ed. R. F. Barrow, D. A. Long, and J. J. Miller), The Chem. Soc., London, 1974.

76. J. M. Grzybowski and L. Andrews, J. Raman Spectrosc., 4, 99 (1975).

77. V. E. Bondybey and J. H. English, Chem. Phys. Lett., 60, 69 (1978).

78. H. J. Jodl and F. Bolduan, J. Mol. Spectrosc., 91, 404 (1982).

79. R. A. Teichman, M. Epting and E. R. Nixon, J. Chem. Phys., 68, 336 (1978).

80. M. Moskovits and D. P. Dilella, J. Chem. Phys., 73, 4917 (1980).

81. C. Cossé, M. Fouassier, T. Mejean, M. Tranquille, D. P. Dilella and M. Moskovits, J. Chem. Phys., 73, 6076 (1980).

82. A. Loewenschuss and A. Givan, Spectrosc. Lett., 10, 551 (1977).

83. H. J. Jodl and J. Kobel, Ramanspectr. on Matrixisolated Linear Molecules, Dipolmarbeit, Kaiserslautern, W-Germany, 1982.

84. H. Egger, M. Gsänger, G. Fritsch and E. Lüscher, Zeit. f. Angew. Physik, 26, 5 (1969).

85. W. B. Daniels, G. Shirane, B. C. Frazer, H. Umebayashi and J. A. Lake, Phys. Rev. Lett., 18, 548 (1967).

86. H. Kiefte, M. J. Clouter, N. H. Rich and S. F. Ahmad, Chem. Phys., Lett., 70, 425 (1980).

87. M. M. Rochkind, Science, 160 (1968).

88. R. N. Perutz and J. J. Turner, J. Chem. Soc. Faraday Trans: 2, 69, 452 (1973).

89. A. J. Barnes, J. C. Bignall and C. J. Purnell, J. Raman Spectrosc., 4, 159 (1975).

90. H. Dubost, Chem. Phys., 12, 139 (1976).

91. B. Mann and A. Behrens, Demixing Effects During Matrix Preparation in Int. Conf. on MIS (Berlin 77), 1978.

92. U. Rödler, Sample Production for MIS-Review and Systematic Experiments, FB Physik, University of Kaiserslautern, 1977.

93. M. Moskovits and G. A. Ozin, Appl. Spectrosc., 26, 481 (1972).

94. P. Groner, I. Stolkin and Hs. H. Günthard, J. Phys., E., 6, 122 (1973).

95. F. Bolduan, Thesis, Kaiserslautern, W-Germany, 1981.

96. F. Bolduan, Diplomarbeit, Kaiserslautern, W-Germany, 1978.

97. J. M. Grzybowski, B. R. Carr, B. M. Chadwick, D. G. Cobbald, and D. A. Long, J. Raman Spectrosc., 4, 421 (1976).

98. G. Wehr, K. Böning, J. Kalus, G. Vogl, and H. Wenzl, J. Phys. C, 4, 324 (1971).

99. A. G. Hopkins and C. W. Brown, J. Chem. Phys., 58, 1776 (1973).

100. A. Givan and A. Loewenschuss, J. Chem. Phys., 65, 1851 (1976).

101. L. H. Jones and S. Ekberg, J. Chem. Phys., 67, 2591 (1977).

102. V. E. Bondybey and J. H. English, J. Chem. Phys., 67, 664 (1977).

103. W. Scheuermann and K. Nakamato, Appl. Spectrosc., 32, 251 (1978).

104. M. Hofmann, S. Leutwyler and W. Schulze, Chem. Phys., 40, 145 (1979).

105. P. Felder and Hs. H. Günthard, J. Mol. Struct., 60, 297 (1980).

106. L. J. Allamandola, D. Lucas, and G. C. Pimentel, Rev. Sci. Instrum., 47(7), 913 (1978).

107. M. Poliakoff, N. Breedon, B. Davies, A. McNeish and J. J. Turner, Chem. Phys. Lett., 56, 474 (1978).

108. R. Rosetti and L. E. Brus, Rev. Sci. Instrum., 51, 467 (1980).

109. P. Huber-Wälchli and S. Günthard, J. Phys., E9, 409 (1976).

110. D. H. Boal and G. A. Ozin, Spectr. Lett., 4, 43 (1971).

111. R. C. Hanson and L. H. Jones, J. Chem. Phys., 75, 1102 (1981).

112. W. Schulze, H. U. Becker, R. Minkwitz and K. Manzel, Chem. Phys., Lett., 55, 59 (1978).

113. L. Frommhold, J. Chem. Phys., 61, 2996 (1974).

114. H. P. Godfried and I. F. Silvera, Phys. Rev. Lett., 48, 1337 (1982).

115. V. E. Bondybey, G. P. Schwartz and J. E. Griffiths, J. Mol. Spectrosc., 89, 328 (1981).

116. V. E. Bondybey, G. P. Schwartz, J. E. Griffiths and J. H. English, Chem. Phys. Lett., 76, 30 (1980).

117. F. Ahmed and E. R. Nixon, J. Chem. Phys., 75, 110 (1981).

118. K. Manzel, U. Engelhardt, H. Abe, W. Schulze and F. W. Froben, Chem. Phys. Lett., 77, 514 (1981).

119. B. S. Ault, W. F. Howard and L. Andrews, J. Mol. Spectrosc., 55, 217 (1975).

120. J. M. Friedman, D. L. Rousseau, and V. E. Bondybey, Phys. Rev. Lett., 37, 1610 (1976).

121. C. A. Wight, B. S. Ault, and L. Andrews, J. Mol. Spectrosc., 56, 239 (1975).

122. J. C. Miller and L. Andrews, J. Chem. Phys., 68, 1701 (1978).

123. A. Givan and A. Loewenschuss, Chem. Phys. Lett., 62, 592 (1979).

124. E. S. Prochaska and L. Andrews, J. Chem. Phys., 72, 6782 (1980).

125. V. E. Bondybey and Ch. Fletcher, J. Chem. Phys., 64, 3615 (1976).

126. A. Loewenschuss, J. C. Miller and L. Andrews, J. Mol. Spectrosc., 81, 351 (1980).

127. H. J. Jodl and K. D. Bier, Ramanspectr. on Small Matrix Isolated Molecules, Diplomarbeit, Kaiserslautern, W.-Germany (1981).

128. I. R. Beattie, T. R. Gilson, S. N. Jenny and S. J. Williams, Nature, Vol. 297, 212 (1982).

129. H. Kiefte, M. J. Clouter, N. H. Rich and S. F. Ahmad, Can. J. Phys., 60, 1204 (1982).

130. J. Harris and R. O. Jones, J. Chem. Phys., 70, 830 (1979).

131. W. Schulze and F. W. Froben, in preparation (1983).

132. A. J. Barnes, K. Szczepaniak and W. J. Orville-Thomas, J. Mol. Struct., 59, 39 (1980).

133. A. Givan and A. Loewenschuss, J. Mol. Struct., 78, 299 (1982).

134. W. F. Howard and L. Andrews, J. Raman Spectrosc., 2, 447 (1974).

135. J. C. Miller and L. Andrews, J. Am. Chem. Soc., 100, 2966 (1978).

136. D. P. DiLella, R. H. Lipson, M. Moskovits and K. Taylor, in Proc. VIII Raman Conf., (Ed. J. Lascombe and P. V. Huang), Wiley, London, 561, 1982.

137. J. De Remigis, H. L. Welsh and R. Bruno, Can. J. Phys., 49, 3201 (1971).

138. V. E. Bondybey and J. H. English, J. Chem. Phys., 73, 87 (1980).

139. J. C. Miller and L. Andrews, J. Chem. Phys., 71, 5276 (1979).

140. F. Ahmed and E. R. Nixon, J. Chem. Phys., 71, 3547 (1979).

141. L. Andrews and R. R. Smardzewski, J. Chem. Phys., 58, 2258 (1973).

142. H. J. Jodl and M. Molter, Solid O_2 and N_2 as Matrix Material, Dipolmarbeit, Kaiserslautern, W-Germany, 1983.

143. R. A. Teichman and E. R. Nixon, J. Mol. Spectrosc., 59, 299 (1976).

144. V. E. Bondybey and J. H. English, J. Chem. Phys., 67, 3405 (1977).

145. K. Manzel, W. Schulze, and F. W. Froben, Chem. Phys. Lett., 82, 557 (1981).

146. D. D. Stranz and R. K. Khanna, J. Chem. Phys., 74, 2116 (1981).

147. R. A. Teichman and E. R. Nixon, J. Mol. Spectrosc., 51, 78 (1975).

148. R. E. Barletta, H. H. Claassen and R. L. McBeth, J. Chem. Phys., 55, 5409 (1971).

149. V. E. Bondybey and J. H. English, J. Chem. Phys., 72, 3113 (1980y).

150. V. E. Bondybey and J. H. English, J. Chem. Phys., 72, 6479 (1980).

151. J. C. Miller and L. Andrews, J. Chem. Phys., 69, 936 (1978).

152. F. Ahmed and E. R. Nixon, J. Mol. Spectrosc., 87, 101 (1981).

153. C. E. Hamilton, M. L. Lesiecki, L. J. Allamandola, and J. W. Nibler, J. Chem. Phys., 76, 189 (1982).

154. L. J. Lauchlan, J. M. Brom and H. P. Broida, J. Chem. Phys., 65, 2672 (1976).

155. D. H. Boal and G. A. Ozin, J. Chem. Phys., 55, 3598 (1971).

156. A. Strull, A. Givan and A. Loewenschuss, J. Mol. Spectrosc., 62, 283 (1976).

157. A. Givan and A. Loewenschuss, J. Chem. Phys., 72, 3809 (1980).

158. H. J. Jodl and H. Strelau, Ramanspectr. on Matrixisolated CS_2 and CO_2, Diplomarbeit, Kaiserslautern, W-Germany, 1983.

159. F. K. Chi and L. Andrews, J. Mol. Spectrosc., 52, 82 (1974).

160. D. E. Tevault, J. Chem. Phys., 76, 2859 (1982).

161. G. A. Ozin and A. V. Voet, J. Chem. Phys., 56, 4768 (1972).

162. A. J. Barnes and W. J. Orville-Thomas, J. Mol. Struct., 45, 75 (1978).

163. W. F. Howard and L. Andrews, J. Am. Chem. Soc., 97, 2956 (1975).

164. R. R. Smardezewski and L. Andrews, J. Phys. Chem., 77, 801 (1973).

165. D. E. Tevault and L. Andrews, Spectrochim. Acta 30A, 969 (1974).

166. H. J. Jodl and H. M. Conrad, Ramanspectr. on SO_2 in Ar, Kr and Xe, Diplomarbeit, Kaiserslautern, W-Germany, 1983.

167. H. Schnöckel, H. J. Göcke and Elsper, in Zeitschrift für Anog. Allg. Chemie, (1983).

168. H. Huber, G. A. Ozin and A. V. Voet, J. Mol. Spectrosc., 40, 421 (1971).

169. H. Schnöckel, in Proc. VIII Raman Conf., (Ed. J. Lascombe and P. V. Huang), Wiley, London, 661 (1982).

170. W. F. Howard and L. Andrews, J. Am. Chem. Soc., 96, 7864 (1974).

171. I. R. Beattie, H. E. Blayden and J. S. Ogden, J. Chem. Phys., 64, 909 (1976).

172. M. Tranquille and M. Fouassier, J. Chem. Soc. Faraday Trans., 76, 26 (1980).

173. I. W. Levin and W. C. Harris, J. Chem. Phys., 57, 2715 (1972).

174. H. J. Becker, Hg. Schnöckel and H. Willner, Z. Physik. Chemie, 92, 33 (1974).

175. A. Cabana, A. Anderson and R. Savoie, J. Chem. Phys., 42, 1122 (1965).

176. H. J. Jodl and E. Regitz, Ramanspectr. on Matrixisolated CH_4, Diplomarbeit, Kaiserslautern, W-Germany, 1983.

177. W. C. Harris, D. E. Coe, W. C. Pringle and J. K. Snow, Mol. Spectrosc. 62, 149 (1976).

178. C. J. Purnell, A. J. Barnes, S. Suzuki, D. F. Ball, and W. J. Orville-Thomas, Chem. Phys., 12, 77 (1976).

179. P. A. Giguère and T.K.K. Srinivasan, Chem. Phys. Lett., 33, 479 (1975).

180. B. S. Ault and L. Andrews, J. Mol. Spectrosc., 70, 68 (1978).

181. B. S. Ault and L. Andrews, J. Chem. Phys., 64, 4853 (1976).

182. N. Smyrl and J. P. Devlin, J. Chem. Phys., 60, 2540 (1974).

183. B. S. Ault and L. Andrews, J. Chem. Phys., 63, 2466 (1975).

184. B. S. Ault and L. Andrews, J. Am. Chem. Soc., 98, 1591 (1976).

185. B. S. Ault and L. Andrews, J. Mol. Spectrosc., 65, 437 (1977).

186. H. J. Jodl and F. Bolduan, Chem. Phys. Lett., 85, 283 (1982).

187. R. M. Atkins and K. A. Gingerich, Chem. Phys. Lett., 53, 347 (1978).

188. H. Huber, E. P. Kündig, M. Moskovits, and G. A. Ozin, J. Am. Chem. Soc., 95, 332 (1973).

189. I. R. Beattie, H. E. Blayden, R. A. Crocombe, P. J. Jones, and J. S. Ogden, J. Raman Spectrosc., 4, 313 (1976).

190. M. L. Lesiecki, J. W. Nibler and C. W. Dekock, J. Chem. Phys., 57, 1352 (1972).

191. R. R. Smardzewski, R. E. Noftle, and W. B. Fox, J. Mol. Spectrosc., 62, 449 (1976).

192. J. M. Grzybowski and L. Andrews, J. Chem. Phys., 68, 4540 (1978).

193. H. H. Claassen, G. L. Goodman and H. Kim, J. Chem. Phys., 56, 5042 (1972).

194. H. H. Claassen and J. L. Houston, J. Chem. Phys., 55, 1505 (1971).

195. A. McKellar and H. L. Welsh, Can. J. Phys., 52, 1082 (1974).

196. R. J. Kriegler and H. L. Welsh, Can. J. Phys., 46, 1181 (1968).

197. J. De Remigis and H. L. Welsh, Can. J. Pys., 48, 1622 (1970).

198. M. Jean-Louis and H. Vu, in Phonons Int. Conf. Rennes, France, (ed. M. A. Nisimovici) Flammarion, Paris, p. 268 (1971).

199. J. Vitko and C. F. Coll, J. Chem. Phys., 69, 2590 (1978).

200. J. A. Warren, G. R. Smith and W. A. Guillory, J. Chem. Phys., 72, 4901 (1980).

201. B. P. Stoicheff, Can. J. Phys., 35, 730 (1957).

202. R. P. Frosch and G. W. Robinson, J. Chem. Phys., 41, 367 (1964).

203. J. Goodman and L. E. Brus, J. Chem. Phys., 67, 933 (1977).

204. J. Fournier, J. Deson and C. Vermeil, J. Chem. Phys., 68, 506 (1978).

205. J. W. C. Johns and J. Reid, J. Mol. Spectrosc., 65, 155 (1977).

206. A. L. Smith, W. E. Keller and H. L. Johnston, J. Chem. Phys., 19, 189 (1951).

207. W. G. Fateley, H. A. Bent, and B. Crawford, J. Chem. Phys., 31, 204 (1959).

208. W. A. Guillory and Ch. E. Hunter, J. Chem. Phys., 50, 3516 (1969).

209. H. Dubost and L. Abouaf-Marguin, Chem. Phys. Lett., 17, 269 (1972).

210. H. Dubost and R. Charneau, Chem. Phys., 12, 407 (1976).

211. H. Dubost and R. Charneau, Chem. Phys., 41, 329 (1979).

212. H. Dubost, A. Lecuyer, and R. Charneau, Chem. Phys. Lett., 66, 191 (1979).

213. J. Manz and K. Mirsky, Chem. Phys., 46, 457 (1980).

214. H. Metiu and G. E. Korzeniewski, Chem. Phys. Lett., 58, 473 (1978).

215. H. K. Shin, Chem. Phys. Lett., 60, 155 (1978).

216. D. J. Diestler and H. D. Ladouceur, Chem. Phys. Lett., 70, 287 (1980).

217. J. Manz, J. Am. Chem. Soc., 102, 1801 (1980).

218. E. Blaisten-Barojas and M. Allavena, J. Phys. C, 9, 3121 (1976).

219. J. Fournier, J. Deson and C. Vermeil, Opt. Commun., 16, 110 (1976).

220. A. W. Mantz, J. P. Maillard, W. B. Roh, and K. N. Rao, J. Mol. Spectr., 57, 155 (1975).

221. H. Friedmann and S. Kimel, J. Chem. Phys., 43, 3925 (1965).

222. F. Fondère, J. Obriot, and Ph. Marteau, J. Mol. Struct., 45, 89 (1978).

223. R. A. Cowley, Rep. Prog. Phys., 31, 123 (1968).

224. M. N. Sapozknikov, Phys. Status. Solid. B, 75, 11 (1976).

225. R. Zallen and M. L. Slade, Phys. Rev. B, 18, 5775 (1978).

226. D. K. Hsu, D. L. Monts, and R. N. Zare, Spectral Atlas of Nitrogen Dioxide, Academic Press, New York, 1978.

227. H. D. Bist, J. C. D. Brand, A. R. Hoy, V. T. Jones and R. J. Pirkle, J. Mol. Spectrosc., 66, 411 (1977).

228. H. J. Jodl, High Pressure Science and Technology, p. 769, (Ed. B. Vodar and Ph. Marteau), Pergamon Press, Oxford, 1980.

229. W. Kiefer and H. J. Bernstein, J. Mol. Phys., 23, 835 (1972).

230. K. Jones, <u>The Chemistry of Nitrogen Dioxide</u>, Pergamon Press, Oxford, 1973.

231. H. J. Jodl ad F. Bolduan, <u>Phys. Status. Solidi. B</u>, <u>103</u>, 297 (1981).

Numbers in brackets are reference numbers and indicate that an author's work is referred to although his name is not cited in the text. Underlined numbers give the page on which the complete reference is listed.

Molter, M., 352, <u>422</u>

Mongeau, D., 31 [51], <u>44</u>

Montero, S., 143 [102], <u>158</u>

Montroll, E. W., 49 [5], 57 [5], <u>98</u>

Monts, D. L., 398 [226], 400 [226], 407 [226], <u>425</u>

Moos, H. W., 31 [52], <u>45</u>

Morcillo, J., 123 [75], 264 [121], <u>157</u>, <u>282</u>

Morino, Y., 162 [1], 163, 165 [1], 184 [11, 66], 185 [68], 186 [1, 79], 187 [1, 79, 85], 195 [79], <u>212</u>, <u>214</u>

Morozov, V. P., 164 [17], <u>212</u>

Morrison, L., 273 [143], 274 [143], 275 [143], <u>283</u>

Morse, P. M., 187, <u>215</u>

Mortensen, O. S., 31, <u>45</u>

Morton, T. H., 148 [132], <u>159</u>

Moscowitz, A., 105, 106 [23, 26, 27, 28], 109 [20], 110 [63], 111 [66], 112, 113 [28], 115 [28], 116 [28], 125 [23], 142 [95], 145 [23, 121], 146, 148 [130], 152 [130], 160 [139], <u>155</u>, <u>156</u>, <u>158</u>, <u>159</u>, <u>160</u>

Mosher, H. S., 106, <u>155</u>

Moskaleva, M. A., 241 [66, 67, 68], 242 [67], 245 [66, 68, 77], 259 [109], 264 [120, 122], 265 [122], 267 [120, 122], 269 [125], 270 [135, 136], 273 [145], <u>280</u>, <u>282</u>, <u>283</u>

Moskovits, M., 106 [29], 108, 113 [29], 115 [29], 117 [29], 118, 139, 151 [48, 137], 291, 292, 311, 320, 321 [81], 325, 349, 351 [136], 353 [81], 357, 365 [188], 415, <u>155</u>, <u>156</u>, <u>160</u>, <u>417</u>, <u>419</u>, <u>420</u>, <u>422</u>, <u>424</u>

Mosteller, Jr., L. P., 75, <u>100</u>

Motzfeld, T., 193 [125], <u>217</u>

Moule, D. C., 2 [1], <u>42</u>

Mueller, H., 93, <u>102</u>

Muenter, J. S., 200 [159], 201, <u>218</u>

Mukherjee, R., 15 [14], <u>43</u>

Mukhtarov, E. I., 252, <u>281</u>

Muller-Lierheim, W., 301 [47], 302 [47], <u>418</u>

Munro, I. H., 20 [32], <u>44</u>

Murphy, W. F., 143 [102], <u>158</u>

Murrell, J. N., 38 [75], 41 [88], <u>45</u>, <u>46</u>

Mushlin, R., 3, <u>43</u>

Mutter, R., 172 [45], 179 [45], <u>213</u>

N

Nafie, L. A., 40 [80], 105 [3, 4], 106 [24], 107 [36, 37], 108, 109 [53], 120, 121 [34], 122, 123 [34, 74], 133 [53], 139 [74], 143 [105], 145 [116, 117, 118], 146 [117, 124], 148 [36, 37, 129, 132], 149 [74, 118], 150 [124, 129], 151 [24, 53, 129, 136], 152 [129], 160 [141, 142], <u>46</u>, <u>154</u>, <u>155</u>, <u>156</u>, <u>157</u>, <u>158</u>, <u>159</u>, <u>160</u>

Nakagawa, T., 204 [177], <u>219</u>

Nakamoto, K., 59 [15], 328, <u>99</u>, <u>420</u>

Nakata, M., 162 [3], 182, 183 [63, 64], 184 [64], 187 [63], 194 [63], <u>212</u>, <u>214</u>

Nata, G., 263 [116, 117], 264 [117], 265 [117], 267 [117], <u>282</u>

Nedunghadi, T. M. K., 233 [45], <u>279</u>

Nemes, L., 165 [24], 190 [100, 103, 104, 111], 191 [100], 202, <u>213</u>, <u>216</u>, <u>219</u>

Neto, N., 233 [39], 242 [71], 247 [39], 274 [147], <u>279</u>, <u>280</u>, <u>283</u>

Netzel, T. L., 23 [42], <u>44</u>

Newton, J. H., 123 [76], <u>157</u>

Nibler, J. W., 290, 292, 312, 313 [64], 315 [71], 324, 327 [64], 331 [64], 333, 335, 341, 353 [153], 357, 366 [190], 367, 415, <u>416</u>, <u>417</u>, <u>419</u>, <u>423</u>, <u>424</u>

SUBJECT INDEX

A

B

C